WITHDRAWN

Darwin in Russian Thought

Darwin in Russian Thought

Alexander Vucinich

UNIVERSITY OF CALIFORNIA PRESS
Berkeley · Los Angeles · London

University of California Press
Berkeley and Los Angeles, California

University of California Press, Ltd.
London, England

© 1988 by
The Regents of the University of California

Library of Congress Cataloging-in-Publication Data

Vucinich, Alexander, 1914–
    Darwin in Russian thought / Alexander Vucinich.
        p.   cm.
    Bibliography: p.
    Includes index.
    ISBN 0–520–06283–3 (alk. paper)
    1. Darwin, Charles, 1809–1882—Influence.   2. Philosophy, Russian—19th cen-
tury.   3. Philosophy, Russian—20th century.
I. Title.
B1623.V83 1988
001′.0947—dc19   88–2054
                                                                                   CIP

Printed in the United States of America

1 2 3 4 5 6 7 8 9

To Dorothy

Contents

Preface

This study examines the role of Darwin's theory in the development of Russian thought from the early 1860s to the October Revolution. It describes both the diversity and the unity of Russian responses to the Darwinian revolution in biological thought and in the modern world outlook. In a more general perspective, it attempts to explain the place of Darwinism in the growth of a modern rationalist tradition in imperial Russia. My hope is that it will encourage a continued effort to fill in one of the more critical gaps in the historical study of Russian intellectual culture in general and of Russian science in particular.

The book is organized chronologically and by topic. Chapter 1 deals with the turbulent and exciting decade of the 1860s, the period of the initial reception of the new evolutionary thought. The next two chapters cover the period from 1871—the year of the publication of *The Descent of Man*—to the middle of the 1880s: Chapter 2 treating the spread of Darwin's ideas over an ever-widening realm of science, and Chapter 3 tracing the onrushing waves of criticism directed at the basic principles of Darwin's theory. Chapter 4, covering the short period from 1885 to 1889, analyzes N. Ia. Danilevskii's monumental synthesis of anti-Darwinian arguments and the heated debate it provoked. Chapter 5 stays close to the 1890s, a decade clearly dominated by the full triumph of Darwin's ideas in Russia and by the rise of a new wave of challenges to Darwinian supremacy in evolutionary thought, represented most typically by neovitalism, neo-Darwinism, and neo-Lamarckism.

Chapters 6–9 consider different topics within the same general period, the first seventeen years of the twentieth century. Chapter 6 surveys the guiding ideas of the main defenders of Darwinian orthodoxy. The distinct forms of anti-Darwinism generated by the separate communities of theologians, philosophers, and scientists are the main concern of Chapter 7. Chapter 8 examines the key strategies of deliberate efforts to blend Darwinism with new developments in experimental biology. Chapter 9 takes the Russian celebrations of the one-hundredth anniversary of Darwin's birth and the fiftieth anniversary of the publication of the *Origin of Species* as a convenient vantage point for a panoramic view of the distinctive features of Russian Darwinism.

Chapter 10 digresses from the chronological sequence of the preceding chapters to survey the views of the radical intelligentsia on Darwin's theory and world outlook, from the early 1870s to the October Revolution. This digression makes possible a comparative analysis of the views of the populists, anarchists, Marxists, and Marxist revisionists, the most articulate ideological formations among the intelligentsia.

I am much indebted to the Guggenheim Foundation, the American Council of Learned Societies, and the Russian Research Center at Harvard University for financial grants that helped me bring this study to a conclusion. I am thankful to the Department of the History of Science at Harvard University for a precious opportunity to offer a course on the history of Russian science, to share my thoughts on Russian Darwinism with a delightful group of alert students, and to use the fabulous resources of Widener Library. To Hoover Library I owe gratitude for the richness of documentary resources, the efficiency and kindness of the staff, and the superb working conditions. I am much indebted to the Library of the University of Illinois in Urbana, whose resources in the history of Russian scientific thought are among the richest in the world.

I cannot thank enough Professor Laurence H. Miller for valuable substantive suggestions, fertile bibliographical hints, and cogent insights into the historical flux of Russian thought. To Sheila Levine, editor at the University of California Press, I am grateful for astute criticism and warm encouragement. For valuable editorial suggestions I am much in debt to Mary Stuart.

I am also thankful to my wife, Dorothy, for enthusiastic, dedicated, and tireless help.

Introduction

The 1850s were the age of a true revolution in science. In a rapid succession of brilliant achievements in scientific thought, the foundations were laid for thermodynamics, the theory of electromagnetism, structural chemistry, spectroscopy, cellular pathology, biochemistry, bacteriology, and experimental physiology. The methodology of medical research became an integral part of natural science. Such giants of science as Clausius and W. Thomson, Kirchhoff and Bunsen, Virchow and Pasteur, Helmholtz and Bernard added new power and luster to the Newtonian edifice of scientific ideas and, at the same time, created an enclave of critical thought that represented the first harbingers of post-Newtonianism.

The *Origin of Species,* Charles Darwin's immortal work, appeared at the very end of the decade, in November 1859. It gave biology a new central and integrating idea, fresh domains of research, and original methods of scientific analysis. Although dominated by antitransformism, the pre-Darwinian scientific community produced a mass of empirical data and theoretical insights that became integral parts of Darwinian thought. Charles Lyell, Adolphe de Candolle, J. D. Hooker, Richard Owen, Asa Gray, F. J. Pictet, Karl von Baer, Étienne and Isidore Geoffroy Saint-Hilaire, and scores of other scientists produced rapidly expanding reservoirs of pertinent material that helped make the Darwinian revolution a logical and inevitable step in the history of nineteenth-century biological thought.

Darwin made three general contributions to biology. First, he gave a broader and firmer footing to the genealogical approach to the living

1

world suggested by Lamarck and Goethe and based on the idea that all
species are descendants from one or a very few primordial forms of life.
Second, he created a new theory that recognized only natural causes of
organic evolution, notably the struggle for existence and natural selection. Third, he placed the weight of empirical evidence behind the idea
of organic evolution as an infinite process—a process not limited by
predetermined goals. "The *Origin*," according to Ernst Mayr, "is remarkably free of any teleological language."[1] Darwin's contributions
made biology an inductive science and its laws empirical generalizations. By effecting a clear and irrevocable separation of biology from
metaphysics and theology, he revolutionized both scientific thought and
the world outlook.

Darwin made biology a history—a discipline concerned with random events, the "raw material" of the transformation process. In his
view, only history can reveal the true meaning of life. It alone can reach
the depths of the infinite complexity of vital processes. While making
biology a history, Darwin helped stimulate a lively and persistent modern interest in making history a science, ever in search of harmonies and
regularities in specific universes of inquiry. With its combined interest in
the grand history of human society and in the universal laws of social
structure and dynamics, sociology received its main inspiration from
Darwin's biological ideas. Maksim Kovalevskii was not far off the mark
when he stated in 1910 that every sociologist of his generation was
much indebted to Darwin, his forerunners, and his successors.[2] J. B.
Bury was thinking of the intellectual effects of Darwin's theory of the
origin of man when he wrote: "As one of the objects of biology is to find
the exact stages in the genealogy of man from the lowest organic form, so
the scope of history is to determine the stages in the unique causal series
from the most rudimentary to the present state of human civilization."[3]

Darwinism recognized no rigid academic boundaries. It quietly extended its sway beyond the limits of evolutionary biology and of biology
in general. It emerged as a theoretical orientation firmly rooted in philosophy and history, philology and jurisprudence, ethics and aesthetics,
psychology and sociology, anthropology and political science, comparative anatomy and ethology. The naturalist William Crookes went
even further: he tried to make chemistry an evolutionary discipline.
Boltzmann, an illustrious harbinger of post-Newtonian science, was so
deeply moved by the idea of transformism that he recommended the
creation of a special science concerned with the evolution of matter.
Vladimir Vernadskii worked assiduously to make mineralogy and bio-

chemistry evolutionary disciplines. Einstein's general theory of relativity laid the foundations for an evolutionary view of the universe. Darwin helped elevate individual disciplines to higher positions in the hierarchy of sciences; for example, he transformed paleontology from an "auxiliary" of geology to a full-blown science—"paleobiology"—expected to answer some of the most basic questions of general biology.

The history of Darwin's theory was part of the general transition of science from the age of mechanistic supremacy to the age of quantum and relativity theories. In the beginning, Darwin's ideas helped biologists strengthen their discipline by presenting it as an integral component of Newtonian science, based on mechanical models of the universe, the principle of natural causality as the sine qua non of scientific explanation, and the notion of continuity as an adequate description of the flow of natural processes. As M. J. S. Hodge has stated, Darwin's theorizing stayed close to "the causal, lawful, deterministic Newtonian universe," and it made natural selection akin "to gravitational force in celestial mechanics."[4] At the end of the nineteenth century a typical biologist brought up on Darwin's theory believed that his discipline must reach far beyond the methodological and epistemological limits of Newtonian science to create a basis for a theoretical integration of newly accumulated experimental data. In the twentieth century, Darwin's principle of random variation has assumed the same position in evolutionary biology as Heisenberg's uncertainty principle in quantum mechanics. At no time did Darwinism, or biology in general, veer away from increasing dependence on physics and chemistry. Physics and chemistry, however, found a way out of a total dependence on mechanistic views, models, and metaphors. Jacques Monod had this new development in mind when he said that Darwin's "selective theory of evolution" did not take on its full significance until the advent of molecular biology.[5]

The Darwinian debate reached far beyond the frontiers of science. In philosophy it helped make scientific cognition a topic of primary emphasis. In theology it encouraged a novel and more comprehensive effort to assess the fundamental challenges of Darwinian heterodoxy. In belles lettres the influence of Darwinism helped open the gates for a broader and richer use of naturalist metaphors and also for new forms of social criticism. In the domain of ideology—the articulation of political beliefs and social goals—Darwinian debates echoed some of the most critical intellectual dilemmas of the day.

Darwin's theory exercised a powerful influence on every branch of philosophy from metaphysics and epistemology to logic and moralistic discourses. Whether it evoked praise or negative criticism, it presented philosophers with the most astute challenges in the reinvigorated search for the cosmic proportions of the flow of history. In the words of Josiah Royce, "with the exception of Newton's *Principia*," no single scientific volume had affected philosophy more profoundly than the *Origin of Species*.[6] Charles S. Peirce agreed: Darwin "has stirred science and philosophy as no man since Newton had done."[7] Darwin's theory encouraged the growth of a special branch of philosophy interested primarily in a critical synthesis of modern scientific knowledge.

Darwinism precipitated a significant turn in the evolution of modern theological thought. In the mounting effort to purify the domains of science open to the influence of heretical thought, a typical theological scholar concentrated on Darwinism as a major threat to the moral underpinnings of Christian civilization. In Russia the systematic search for religious answers to the "threat" of Darwinism contributed to the emergence of "theoretical theology" at the end of the nineteenth century, a rather accomplished effort to elucidate the critical areas of contact between scientific and religious thought.

In literature Darwinism became a topic of direct concern or a target of endless allusions. References to the letter or the spirit of Darwinian thought, sometimes of the most subtle nature, became an infallible way of depicting the world outlook and ideological proclivities of the heroes of literary masterpieces. Individual heroes of Dostoevsky's and Tolstoy's literary works provided graphic examples of the myriads of prisms refracting Darwinian science and showing the multiple strands of its impact on current thought and attitudes. More often than not, these heroes were alter egos of their literary creators, giving added scope to Darwinism as an intellectual and social phenomenon. A literary figure took note of Darwinian evolutionism not only by commenting on its scientific principles but also by making use of its metaphors.

On September 22, 1877, Rudolf Virchow, the founder of cellular pathology and an active anti-Darwinist, spoke at the fiftieth national Congress of German Naturalists and Physicians in Munich about the theory of organic evolution and contemporary politics. The speech represented the first major invitation to scientists to discuss the ideological roots and ties of Darwin's biological theory. Virchow warned his listeners that to work for the diffusion of Darwinian ideas was tantamount to working for the cause of "social democracy," which he viewed as a serious threat to the most sacred values of modern culture and social order. He said that Darwin's views had reached every corner of intellectual Europe, giving new strength and encouragement to radical ideologies that allied their political programs with developments in the substance and in the spirit of natural science. Ernst Haeckel, the main target of Virchow's attack, answered the charges by defending the basic principles of Darwin's theory and by outlining the materialistic principles of his personal philosophy, which he identified as monism, an ideology backed up by a combination of scientific and metaphysical propositions. He succeeded only in giving the struggle over organic evolution a sharper ideological focus and a more pronounced political affiliation.

Professional ideologues, whether they accepted or rejected Darwinism, found it necessary to express their views on the avowed merits of the new evolutionary theory as a world outlook, a body of positive knowledge, a scientific theory, and a source of designs for a better social existence. Ideological appraisal—and acceptance or rejection—of Darwinism came also from theologians, literary figures, and scholars working in the humanities. Ideology, as the case of Russia clearly illustrated, pervaded all categories of Darwinian thought.

The relations between science and ideology are not fixed; they are subject to modification caused by changes in either one or the other, or in both. Modifications may also come from changes in the general political climate in the country. As a rule, a Russian scientist of the 1870s was sympathetic to the liberal stirrings in the country; this, however, was not necessarily true during the 1890s and the first decade of the twentieth century, when the sight of a university professor who was both liberal in political convictions and actions and unmistakably critical of Darwin's theory was not uncommon.

At the turn of the century, a typical biologist was neither a pure Darwinist nor a confirmed anti-Darwinist: he occupied a position somewhere between the two extremes. The biologists who considered Darwin one of the true giants of modern science did so for a variety of reasons. Some thought that Darwin's main contribution was in making evolution a central and integrating notion of biology, rather than in presenting a meaningful and acceptable explanation of the mechanisms of the evolutionary process. William Bateson saw Darwin's true victory not in the conceptual triad of natural selection, heredity, and variation but in epochal contributions to the triumph of the evolutionary view in biology.[8] Similar views found a strong expression in philosophical literature. "All modern naturalists of note," wrote the philosopher Josiah Royce, are in a sense "followers of Darwin, not that they all hold his view about natural selection, but that they all teach the doctrine of transformation."[9]

The biologists who criticized Darwin also relied on a variety of arguments. Some critics did not like the mechanistic—or Newtonian—underpinnings of Darwinian science. Others did not like the physico-chemical reductionism with which, they thought, Darwinism was allied. A typical critic thought that the theory of natural selection treated organisms as passive participants in the evolutionary process. He resented the widespread tendency in and outside the scientific community to place natural selection in the same category of the laws of nature with Copernicus's heliocentrism and Newton's gravitation. Most critics argued that Darwin's major claims did not receive adequate empirical support.

National differences in the reception of Darwin's ideas have long been recognized. These differences found a particularly strong expression in the distinctive ways in which individual nations treated the inevitable confrontation of Darwin's theory of selection and Lamarck's emphasis on direct environmental influence on the transformation process and on an inherent tendency within living beings to progress. In the United States, thanks largely to the powerful influence of E. D. Cope's Lamarckian inclinations, the Darwinian emphasis on the evolutionary primacy of natural selection was exceedingly slow in receiving serious recognition. In Great Britain, by contrast, the Lamarckian share of a general evolutionary theory was never of much consequence. In France, at the end of the nineteenth century, a strong group of biologists—led by E. Perrier, A. Giard, and F. Le Dantec—paid homage to Darwin's epochal contributions to evolutionary biology but, at the same time, never failed to regard Lamarck's contributions as more basic and richer in theoretical perspective. Lamarck's theoretical views, they thought, were more firmly lodged in the mainstream science of the nineteenth century. In Russia the situation was just the opposite. Although Lamarck was popular and highly respected, his ideas were usually treated as part of a larger—Darwinian—evolutionary framework.

No nation has shown more diversity and involvement in the treatment of the Darwin-Lamarck confrontation than Germany. In this endeavor one orientation stood out as a uniquely German phenomenon: psychological Lamarckism. Psycho-Lamarckians—as they were often called—were the twentieth-century heirs of the era of *Naturphilosophie*. Their most active leaders were the botanists A. Pauly and R. Francé. They did not hide their high respect for Darwin's achievement, but this did not stop them from building their evolutionary theory on

Lamarck's idea of an inherent tendency of organisms to progress, which helped them restore the metaphysics of teleology as a key explanatory principle of organic transformation and infuse evolutionary thought with the ideas of revived vitalism.

National differences in the reception of Darwinian thought are entwined in the crisscrossing developments in intellectual climate, social dynamics, and politics. To disentangle these forces is an involved task hampered by tortuous detours and dim signals. In Russia, perhaps more clearly than in most other countries, to understand the flow of and the reaction to Darwinian scientific ideas and philosophical elaborations is to understand the main issues and battles on the ideological front and the major dilemmas in the growth of secular wisdom. To study the complexity and dynamics of reactions to Darwin's ideas is to gain pertinent insights into the modernization of what N. A. Berdiaev had called "the Russian idea."

To understand Darwinism as a factor shaping the thought of a modern nation it is necessary to analyze the two major systems of ideas that gave it a distinct intellectual profile. On the one hand, it is necessary to single out and examine the connections of Darwin's evolutionary thought with the general national development in science and with the emergence of a modern scientific community. It is essential to trace the national history of science inasmuch as it responded to the challenges of Darwinian thought. The national scientific communities not only absorb new ideas but also refashion them in such a way as to facilitate their integration into the reigning intellectual traditions. On the other hand, it is necessary to examine the extrascientific reactions to Darwinian thought, particularly those that came from philosophers, religious thinkers, and representatives of secular ideologies. In addition to discussing scientific and nonscientific approaches as distinct categories, it is necessary to take a close look at the multiple forms of their interaction. Religious scholars, for example, may use scientific arguments to discredit Darwin's theory, just as scientists may appeal to religious sentiment to achieve the same goal—and both religious scholars and scientists may be guided by easily discernible ideological precepts.

CHAPTER ONE

Reception

Reforms and the Spirit of Science

After the Crimean War Russia entered a period of national awakening that invited a critical reassessment of dominant values and social bonds and engendered a strong sentiment in favor of sweeping political and social reforms. The reigning ideologies of the time favored natural science as the most reliable source of guideposts for social and cultural progress. A. I. Herzen spoke for the age when he noted that "without natural science there can be no salvation for modern man; without that healthy nourishment, that vigorous elevation of thought based on facts, and that proximity to the realities of life . . . our soul would continue to be a monastery cell, veiled in mysticism spreading darkness over our thoughts."[1]

Another writer saw in the 1860s an accelerated growth of secular interpretations of the universe, both physical and social. This was the time, he said, of the search for "a substitution of anthropology for religion, inductive method for deductive method, materialistic monism for idealistic dualism, empirical aesthetics for abstract aesthetics, and the theory of rational egotism for morality based on supersensory principles."[2] In a reference to university education, a famous embryologist, educated in the 1860s, noted that the students were particularly attracted to scientific articles published in general journals and to translations of popular scientific works. The younger generation showed a particularly strong interest in the new theories of nature and in their close links with positivist and materialist philosophy.[3]

The preoccupation with scientific ideas, and with philosophical views on the cultural preeminence of science, appeared in many forms and came from many sources. The initial outburst came from the rapidly rising popularity of the scientific-philosophical writings of the new breed of German materialists represented by Ludwig Büchner, Karl Vogt, and Jacob Moleschott. The Russian intelligentsia favored the criticism of metaphysical idealism which these studies supplied in great abundance; but they also favored the profuse and laudatory comments on science as a source of models for a radical transformation of society and culture. When Bazarov, the hero of Turgenev's *Fathers and Sons*, suggested that reading Büchner was much more profitable than reciting Pushkin's poetry, he acted as a true spokesman for nihilism, the most radical philosophy of the epoch. Vogt, Büchner, and Moleschott reinforced two guiding ideological principles of the restive intelligentsia: positivist faith in the limitless power of science, and historicism. Their tight interweaving of natural science facts and theories with the precepts of an ideology that opposed the social and cultural status quo found a particularly strong Russian following.

The growth of the popular appeal of science during the 1860s was closely related to the rising popularity of French positivism, as presented by Auguste Comte and his French disciples and as interpreted by John Stuart Mill. Positivism presented science as a new intellectual force that grew on the ruins of religious and metaphysical thought. It produced a separation of sociology from eighteenth-century metaphysics. To give his defense of positivism more weight, a Russian commentator assured his readers in 1865 that the leading minds in western Europe had accepted Comte's designs for a scientific study of society. When G. M. Vyrubov became coeditor of the Paris *Revue de philosophie positive* in 1867 his countrymen went far afield in commenting on the rich promise of the Comtian philosophical legacy. Eager to encourage and to justify this endeavor, A. P. Shchapov noted in 1870 that "French, Italian, English, and German literature is being rapidly enriched by the new books written in the spirit of positive philosophy."[4] P. L. Lavrov informed his readers that positivism was a philosophy built on scientific foundations and that in France "physicians and graduates of the Polytechnical School are in a majority among the followers of positivism."[5] A contemporary scholar described Lavrov's philosophical stance as a blend of Young Hegelianism, Comte's positivism, and Darwin's and Spencer's evolutionism.[6]

Buckle attracted more attention in Russia than anywhere else on the European continent.[7] The intelligentsia favored the *History of Civiliza-*

tion in England (1857–61), Buckle's historical opus, because it treated science as the prime mover of social progress and, above everything else, as a true backbone of democratic institutions. It represented an effort "to interpret the infinite variety of personal and national actions as concrete manifestations of the general and immutable laws of nature."[8] Only science, according to him, can supply knowledge that reaches deep into the mysteries of nature and is essentially objective. The study of society can be objective only if it relies on the models of the physical sciences. Buckle recognized only one history—the history based on the scientific method. He was "particularly popular" because he "erased providentialism from history, denied the freedom of the will, and used statistics to explain social life."[9] The leaders of the intelligentsia relied on Buckle to reinforce their belief in the enormous fecundity of a combination of historical relativism and scientific objectivism. In their eyes, Buckle went one long stride beyond Comte: whereas Comte treated history as a collection of raw material to be used by sociology as a full-blown science, Buckle viewed history as a complete science by itself.

Trying to explain why Buckle's *History* appeared in two Russian translations almost simultaneously and why both translations were ready to come out in second editions, M. A. Antonovich noted in 1864 that "it is possible that Buckle is more popular in our country than in England" and that "to talk about Buckle in Moscow and St. Petersburg has become as common as to talk about the weather."[10] Buckle's ideas reached a wide spectrum of publications, ranging from scholarly papers to feuilletons.

Claude Bernard, whose *Introduction to the Experimental Method in Medicine* was quickly translated into Russian and became a most popular source of arguments in favor of scientific knowledge, appealed particularly to groups avoiding philosophical and ideological extremism. These groups liked Bernard's skillful moves in defending the strengths of scientific thinking without underrating aesthetic expression, philosophical contemplation, and other nonscientific modes of inquiry. Shchapov likened Bernard's type of philosophy to an *excellente gymnastique de l'esprit,* sharpening conceptual thinking and critical judgment and giving the world view a cosmic scope.[11] While philosophy elevates the intellect, science gives it substance that stays close to objective reality. Shchapov surmised that Bernard favored the "impersonal authority" of science over the "personal" orientation of art. The element of impersonality, in his opinion, helps science work for human society as a totality and to propound a democratic ideology.[12] Or, as Bernard put it: "Art is myself;

science is ourselves." The *Origin of Species*, Darwin's classic, stood out as a majestic monument to the Bernardian dictum that real science consists of three phases: "an observation made, a comparison established, and a judgment rendered."[13] N. N. Strakhov's astute discussion of the *Introduction* showed clearly that Bernard appealed not only to persons enamored with the currently fashionable "natural science materialism" but also to those who did not favor an alliance of science and materialism.[14]

In the middle of the eighteenth century, the philosophes, the masters of the French Enlightenment, undertook the ambitious task of writing the *Grand Encyclopedia*, a popular survey and a synthesis of contemporary knowledge about nature and human society and culture. At the very beginning of the 1860s the idea of a grand encyclopedia received strong support in Russia. The organizers of the monumental project had no trouble in recruiting 208 contributors to the vast enterprise and in selecting fourteen editors in charge of particular branches of knowledge.[15] Among the contributors were such eminent scientists as the embryologist and anthropologist Karl von Baer, the mathematician P. L. Chebyshev, the geographer P. P. Semenov, and the crystallographer A. V. Gadolin. The enterprise produced six volumes (five of them covering the letter A) before it collapsed. It was overambitious—conservative estimates called for two hundred volumes for the full set. Serious critics thought that the country was not ready for an intellectual undertaking of such gigantic scope. The spirit of the time favored such an enterprise, particularly when judged by the rapid growth of interest in disciplined and systematic knowledge. The reasons for the collapse of the enterprise must be sought in the disproportionate and asymmetrical distribution of experts in the main branches of organized knowledge, in the persistence of government and church interference, and in the absence of a clear vision of the cultural and social role of such a project.

Russia's Forerunners of Darwin

Russian—particularly Soviet—historians of science have produced voluminous literature on Darwin's precursors in their country. B. E. Raikov's monumental *Russian Biologists-Evolutionists Before Darwin*, in four volumes totaling over fifteen hundred pages, was only the beginning of a determined search for historical information depicting the growth of transformist ideas in the Russia of the pre-Darwinian age. The results of this ambitious undertaking allow us to draw two general conclusions.

First, the idea of evolution received support from a wide spectrum of intellectual endeavor, but mainly from the representatives of the natural sciences: from botanists and zoologists, geologists and paleontologists, embryologists and morphologists. Second, no Russian writer of the pre-Darwinian age had undertaken an elaborate and systematic study in the field. There were no Russian Lamarcks or Chamberses.

M. A. Maksimovich, professor of botany at Moscow University during the 1820s and 1830s, combined an early influence of Oken's *Naturphilosophie* with a strong sense for the concrete offerings of natural history. From Oken he borrowed a cellular outlook on life, an idea of the unity of inorganic and organic nature, a holistic view of living objects, and a strong inclination to stress the evolutionary fluidity of plant and animal species. Nature, he wrote in 1827, represents "an unbroken chain of objects in constant change." Evolution is always gradual and increases the complexity of natural objects. The stability of species is more apparent than real. Species are varieties with deeper roots in the evolutionary process. Maksimovich made no effort either to systematize his evolutionary ideas or to place them within a more elaborate theoretical framework.[16]

G. E. Shchurovskii, another professor at Moscow University, who also passed through a phase of enthusiasm for *Naturphilosophie*, worked first in comparative anatomy and then in geology. In *Animal Organology* he was under the spell of É. Geoffroy Saint-Hilaire's views on the morphological and physiological unity of organisms. Although he did not present a direct discussion of organic evolution, he dwelt extensively in the world of concepts that opened the gates for a scientific approach to the idea of transformism. For example, he was conversant with the evolutionary connotations of the ideas of homology and analogy, the former referring to similarity in structures of corresponding organs, the latter to similarity in corresponding functions.[17] Shchurovskii based his biological arguments on the premise that animal organs are parts of an integrated whole and that each performs not only a specific but also a general function. Each is subject to constant modifications, both on functional and on morphological levels. According to Shchurovskii, a scientific study of modifications should reveal both the unity and the diversity of organic nature.[18] The homology of the corresponding organs of different species and the diversity of animal forms are two sides of the same natural order. The knowledge of this duality reveals the universal laws of nature.

Among the Russian precursors of Darwin three scientists occupied preeminent positions: Karl von Baer, a distinguished member of the St. Petersburg Academy of Sciences, K. F. Rul'e, professor of natural history at Moscow University, and A. N. Beketov, a graduate of Kazan University and subsequently professor of botany at St. Petersburg University.[19]

Karl von Baer represented the culminating point of a strong tradition in pre-Darwinian embryology in Russia built by the scientists of German origin. Caspar Friedrich Wolff, the founder of this tradition and a member of the St. Petersburg Academy of Sciences from 1767 to 1794, is considered the creator of the theory of epigenesis, an essential ingredient of the evolutionary idea in biology. In opposition to the theory of preformation, the theory of epigenesis considered every organism a new formation—a result of unique internal and external influences. Christian Heinrich Pander, a member of the St. Petersburg Academy of Sciences from 1820 to 1827, is in one important respect considered the founder of embryology as a distinct discipline. He was the first to detect and to undertake a study of the three layers of cells that are differentiated in the early stages of embryonic development: the ectoderm, the endoderm, and the mesoderm. He combined his interpretation of embryonic development with a general evolutionary view of organic nature.

Karl von Baer carried the tradition established by Wolff and Pander to new heights. A founder of comparative embryology and a powerful link between the budding embryology, the evolutionary spirit of *Naturphilosophie*, and Cuvier's catastrophism, he contributed enough to the transmutation theory to attract Darwin's attention as one of his distinguished predecessors. Darwin wrote in the *Origin of Species:* "Von Baer, toward whom all zoologists feel so profound a respect, expressed about the year 1859 . . . his conviction, chiefly grounded on the laws of geographical distribution, that forms now perfectly distinct have descended from a single parent-form."[20] Darwin could not but appreciate von Baer's observation that "the embryos of mammalia, of birds, lizards, and snakes, probably also of Cheldonia are in their earliest states exceedingly like one another, both as a whole and in the mode of development of their parts; so much so, in fact, that we can often distinguish the embryos only by their size."[21] Darwin had in mind von Baer's proposition that during their embryonic development animals gradually depart from the more general structural patterns.[22] Only a few months before the publication of the *Origin of Species,* von Baer wrote that many forms which at the present time appear as isolated propagating units had developed from a single species.[23] He added that he was not ready to take a stand on the relative importance of "the development of one species from another" and on spontaneous generation.[24]

L. Ia. Bliakher, a distinguished historian of nineteenth-century biology, claimed that von Baer's pre-Darwinian ideas had gone through two distinct stages of development. During the first stage, covering his early studies in zoology and comparative anatomy, von Baer did not concern himself with organic evolution. He made frequent references to similarities between various taxonomic groups of animals as aids in systematics, but he avoided any suggestion related to the problem of organic evolution. Transformism was clearly outside his sphere of interest. In places he used such terms as "resemblance" and "affinity" interchangeably, but his usage of "affinity" did not imply genealogical relationship or common ancestry.[25]

During the second stage, the period of heavy involvement in embryonic studies and for a while afterward, von Baer thought of the possibility of evolutionary change, occasionally recognizing blood relationship and common ancestry within individual types of animals. He came to the conclusion about the possibility of organic evolution within comparatively narrow limits. He firmly rejected the idea of transformation from one type of animal to another. In a laconic statement, I. I. Mechnikov presented von Baer as a biologist "who opposed the theory of unlimited transmutation of species and of the origin of animals by transformation but who recognized limited variability."[26] Where von Baer allowed for transformation, he based it on the geographical factor, not on embryological dynamics. Where he allowed for evolution as a universal process, he treated it as a gradual triumph of "spirit" over "matter," a metaphysical crutch of *Naturphilosophie*.[27]

S. R. Mikulinskii has made two observations that must be taken into consideration in appraising von Baer's contributions to the pool of knowledge that helped Darwin formulate his general theory of evolution. First, von Baer's division of the animal kingdom into four types was not a copy of Cuvier's division. Whereas Cuvier based his division on comparative-anatomical material, von Baer depended on both comparative-anatomical and embryological material. Cuvier considered "types" as clearly separated and mutually impenetrable entities. Von Baer, by contrast, mentioned the possibility of transitional types, thus suggesting gradual transformation as a universal process in organic nature. Second, von Baer is known in the annals of science as the founder of animal embryology. His embryological work did not lead him to a consideration of transformism. His research, however, did establish the embryological unity of vertebrates, an idea that gave strong, though indirect, support to Darwin's conclusions related to the evolution of animals.[28]

K. F. Rul'e wrote in Russian and was a popular lecturer at Moscow University and at mushrooming public forums. Superb oratorical skills helped his evolutionary ideas reach a considerably wider segment of the educated public in Russia than did those of von Baer, presented mainly in German and strained by a format of tight logical argumentation. Rul'e worked in an unusually wide field, but his most noted and lasting studies dealt with the geology and paleontology of the Moscow basin.[29] Since animals, he said, depended on the environment, it would be paradoxical to assume that changes in the environment did not produce changes in the forms of life. It was not enough, in his view, to establish morphological similarities and differences between presently existing and extinct animals; it was equally important to trace their actual genealogical ties.[30] He thought that the main task of biology—and of paleontology—is to study regularities in the historical development of living forms.[31] In 1881 the zoologist Ia. A. Borzenkov commented on the impression the *Origin of Species* had made on him and his colleagues:

> We read Darwin's book, which reached Moscow (in Bronn's German translation) at the time when the memory of Rul'e was still fresh in our minds. The new book did not repeat exactly what we had learned from Rul'e, but the ideas it presented were so close to those Rul'e had passed on to us that we felt we had been familiar with them for a long time. The only exception was that Darwin's book presented these ideas with more clarity, more scientific rigor, and incomparably richer backing of factual information.[32]

D. N. Anuchin wrote in 1886 that Rul'e advanced evolutionary ideas similar to Darwin's several years before the publication of the *Origin of Species*. Rul'e criticized the reigning—Linnean—idea of "species" as a biological category, talked about the important role of environment in the transformation of organisms, discussed the problems of variation and hereditary transmission of characteristics, and advanced the idea of progress as applied to organic forms. He derived his transformist ideas from two kinds of sources: fossils exhibited in museums and direct observations of organic nature.[33]

In early 1860 the popular journal *Russian Herald* carried an article entitled "Harmony in Nature," written just before Darwin's classic was published. The article, by A. N. Beketov, provided another proof that the idea of organic evolution had attracted Russian naturalists before Darwin had made it the central theme of a new stream of biological thought. Written in a popular vein, the article stated that changes in the environment are direct sources of change in living forms. Obviously impressed with Lamarck's transformist ideas, Beketov argued that the law of the double adaptation of species—the functional adaptation of organs to the organism as an indivisible whole, and the adaptation of the whole to the surrounding environment—is the law that governs the gradual transformation of the living world and determines the survival or extinction of individual species.[34] Implicit in Beketov's argument is the general idea that nature undergoes constant and irreversible change that takes place according to regular and universal laws of nature. It is not important whether Beketov adhered closely to the ideas advanced by Lamarck and É. Geoffroy Saint-Hilaire or anticipated the crucial points of Darwin's theory; what is important is that he showed clearly that Russian scholarship had become ready to accept evolution as the core idea of biology.[35]

Popular Appeal

Educated contemporaries were fully aware of the unalloyed enthusiasm with which Russians received Darwin's works. According to the embryologist A. O. Kovalevskii, who started his distinguished career in the middle of the 1860s, Darwin's theory was afforded a most sympathetic and enthusiastic reception in Russia. The arrival of the new evolutionary thought in Russia coincided with a general national awakening that produced the great reforms of the 1860s. Darwin's book received immediate recognition as a turning point in the history of science.[36] Kovalevskii made this statement in 1882. Writing in the same vein, I. I. Mechnikov, a famous embryologist and evolutionary pathologist, had observed that the evolutionary theory was readily accepted in Russia not only because the general conditions favored it but also because it did not encounter a strong antievolutionary tradition.[37]

Sir Charles Lyell introduced Darwin's theory to the Russian reading public. In January 1860, two months after the publication of the *Origin of Species,* the *Journal of the Ministry of Public Education* published a Russian translation of Lyell's report to the twenty-ninth meeting of the British Association for the Advancement of Science, held in September 1859, in which the eminent geologist referred to the revolutionary significance of Darwin's forthcoming book. It appeared to Lyell that

Darwin had succeeded "by his investigations and reasonings, in throwing a flood of light on many classes of phenomena connected with the affinities, geographical distribution, and geological succession of organic beings, for which no other hypothesis has been able, or has attempted, to account."[38] Lyell gave full support to Darwin's conclusion that the same powers of nature produce both varieties in plants and animals and new species over longer periods of time. In his introductory remarks on Lyell's report, N. N. Strakhov, the Russian translator and commentator, noted that the reason for the dominance of antitransformist views in biology, most categorically expressed by Cuvier, must be sought in the too narrow empiricist orientation of naturalists, who were much more careful in gathering reliable facts than in drawing generalized conclusions. He noted that in the nineteenth century the defenders of the creationist orientation could no longer ignore the mounting empirical evidence in favor of transformism. Strakhov criticized H. G. Bronn, the German zoologist who in 1858 received a prize from the Paris Academy of Sciences for a work that categorically denied the possibility of the origin of species by means of "the transformation of a small number of primeval forms of life."[39]

In 1860 S. S. Kutorga, a professor at St. Petersburg University, presented his students with a general review of the basic ideas contained in the *Origin of Species*. At the end of the same year, N. P. Vagner, professor at Kazan University, read a paper on Cuvier and É. Geoffroy Saint-Hilaire in which he briefly referred to Darwin's contribution to the triumph of the transformist view in biology. It was also in 1860 that a series of translated articles appeared in the *Herald of Natural Sciences*, an organ of the Moscow Society of Naturalists. Among the published articles was Thomas Huxley's report on Darwin's theory read before the members of the Royal Institution on February 10, 1860.[40] All articles selected for publication in the *Herald* were clearly pro-Darwinian. A conservative and aristocratic organization, the Moscow Society of Naturalists stopped publishing articles on Darwin as soon as his theory blossomed into a major controversy of the age. Despite this retreat, the *Herald* deserved credit for having been one of the first Russian journals to offer a detailed description of Darwin's evolutionary views.

In 1861 the journal *Library for Reading* published an essay on the *Origin of Species* that was more systematic and more comprehensive than any previous study on the subject published in Russian. Over two hundred pages in length, the article appeared in two installments. The anonymous writer left no doubt about his wholehearted acceptance of Darwin's evolutionary ideas and about his firm belief in the revolutionary significance of the new theory.[41] Of all theories dealing with the origin of species, Darwin's was superior because, in his view, it offered the simplest, most logical, and most satisfactory explanations. The strength of the article was in the lucidity of its substantive part and in the precision and symmetry of its analysis of the theoretical foundations of the new evolutionary idea. The reader was treated to a bonanza of well-digested information on the leading forerunners of Darwin's theory, particularly on Goethe, Lamarck, Étienne and Isidore Geoffroy Saint-Hilaire, and A. R. Wallace.[42]

The essay was impressive also on two additional counts. First, the author carefully pointed out the ideas Darwin had suggested that required further study and elaboration. Second, he gave his readers a taste of the earliest criticism directed at Darwin's theory. For example, he mentioned Pictet, a Geneva University professor, who thought that variation could not lead to the emergence of new species, and who proposed a kind of limited transformism that satisfied the followers of Cuvier and Agassiz. Pictet, however, went so far as to recognize the role of natural selection in the preservation of heritable variation.[43] The anonymous writer saw the greatest contribution of the *Origin of Species* in its setting the stage for a scientific study of the origin of life—the greatest of all mysteries.

Another journal saw Darwin's evolutionary idea as "an inevitable result of the basic theoretical orientation of modern science" and as the crowning point of a long and careful accumulation of scientific ideas. Darwin's influence, according to this journal, was destined to be felt far beyond the boundaries of biology.[44] The author presented natural selection as an irrefutable generalization derived from a vast reservoir of empirical data.

The first number of the *Russian Herald* for 1863 published "Flowers and Insects," a beautifully written essay, rich in naturalist observations on the relations of plants and animals to each other and to the environment, both geographical and geological. The purpose of the article was clearly to show the marvels of the works of nature as portrayed by the *Origin of Species*. Relying on florid expressions and poetic spontaneity, the author—S. A. Rachinskii, professor of botany at Moscow University—explained every major component of Darwin's theoretical structure in a language that was accessible to the general reading public. He identified the *Origin of Species* as "one of the most brilliant books ever to be written in the natural sciences." The basic contribution of this work, as Rachinskii saw it, was in indicating, first, the genealogical links between organisms of different kinds and, second, the "marvellous adaptation" of these organisms to the inorganic world.[45] It gave a perspicuous description of the work of natural selection, and it explained the divergence of living forms as an index of the work of evolution. It presented the transformation of plants and animals as a result of the evolutionary process characterized by gradual and continuous change. Rachinskii thought that, although Darwin's theory gave an accurate description of the evolutionary dynamics of the living universe, it still had a long way to travel on the road to full empirical corroboration.

In 1864 Rachinskii produced the first Russian translation of the *Origin*. Although not a masterwork of translation, the book sold out so quickly that in 1865 it went through a second printing. By this time Darwin's ideas had reached not only scientists and popularizers but also persons eager to integrate evolutionary thought into ideologically oriented writings. M. A. Antonovich in *Contemporary* greeted Darwin primarily as a master of scientific thought destined to cause drastic changes in the world outlook of the new generation.[46] He viewed the *Origin* as a major victory for the democratic spirit of the scientific method over the authoritarian sway of metaphysical speculation. He left no doubt about his firm belief in the close interdependence of science and democracy. The strengths of the *Origin*, as he saw them, were not only in the emphasis on the natural causation of organic evolution but also in the lucidity of its prose and the power of empirical documentation on which it rested. In Darwin's evolutionary idea and the current triumph of the experimental method in physiology he saw the beginning of a new phase in the growth of biology.

Among the men trained in science, N. N. Strakhov and K. A. Timiriazev became widely known as the first Russian popularizers of Darwin's ideas. In his graduation paper at St. Petersburg University, Strakhov presented three original algebraic theories giving solutions for inequalities of the first degree.[47] This paper, as well as his master's thesis dealing with the wristbones of mammals, appeared in the *Journal of the Ministry of Public Education*. After his effort to obtain a university position had failed, Strakhov became a free-lance writer contributing to several popular journals. In an article published in Dostoevsky's journal *Time* in 1862, he surveyed Darwin's ideas and criticized the efforts of Clémence Royer, the French translator of the *Origin of Species*, to give Darwinian theory a broader sociological meaning. He particularly objected to Royer's claim that Darwin's scientific ideas provided proof that every search for political equality of all human races was predestined to be an utter failure.[48] Strakhov may rightfully be called the first Russian scholar to stand firmly against Social Darwinism, a blend of sociology and ideology which did not have a single supporter among the leading Russian naturalists and social thinkers. Strakhov greeted Darwin's theory as a strong addition to science and a modern world view. Darwin, in his opinion, made two revolutionary contributions to biology and the modern world outlook: he made biology a solid science based on a historical view of nature, and he brought an end to the reign of the metaphysical view of organic nature.[49] Although Darwin's theory did not answer all the intricate questions of biological evolution, it was built on sound foundations.[50]

P. A. Bibikov, an unheralded writer, took it upon himself to refute Strakhov's criticism of Royer's position as a defender of Social Darwinism. His argument was simple and direct: he preferred Royer's faith in "science" to Strakhov's attachment to "sentimental philosophy," which looked for the heights of wisdom outside the domain of scientific knowledge.[51] To him, the social struggle for existence and the resultant social inequality were not only in full accord with Darwin's biological principles but also well-documented historical realities. "The law of the supremacy of the naturally selected and strong over the feeble and degenerate has always been in force."[52] Very few contemporary Russian writers were ready to go along with Bibikov's line of argument.

Despite his deep and enthusiastic involvement in the popularization of Darwin's ideas during the years immediately after the publication of the *Origin of Species,* Strakhov did not represent a typical Russian Darwinist of the 1860s: he did not allow Darwin's natural science "materialism" to displace his own brand of idealistic philosophy.[53] He admitted that Darwin's ideas came to Russia at the time when materialism was a reigning philosophy; but he was equally ready to state that the demise of idealism was a transitory phenomenon and to fight for a revitalization of antimaterialistic philosophy. He endorsed Rudolf Virchow's dictum that "sick idealism" should be transformed into "healthy idealism" rather than be condemned as advocated by the nihilist intelligentsia.[54] A decade later, an idealistic philosophical bias predisposed Strakhov to undertake a vicious attack on *The Descent of Man* as a caricature of empirical science and a flagrant attack on the moral foundations of human society.[55] He became one of the most belligerent Russian anti-Darwinists of the nineteenth century.

Kliment Timiriazev, a pioneer in plant physiology, was the first scholar educated in the atmosphere of the Great Reforms to help spread the basic ideas of the new evolutionary orientation. Like many scientists of the 1860s who came from noble families of rapidly declining fortunes, he combined an insatiable thirst for scientific knowledge with a profound dedication to democratic ideals and to the philosophy of "realism" advanced by the nihilist movement. He enrolled in St. Petersburg University in 1861, a year marked by growing student unrest that led to the frequent closing of the university. Before graduating in 1866 he had written "Garibaldi in Caprera," "The Hunger in Lancashire," and "Darwin's Book: Its Critics and Commentators," all published in the influential journal *Fatherland Notes* (*Otechestvennye zapiski*) in 1863–64. The Darwin article was immediately republished as a small book entitled *Charles Darwin and His Theory*. This study provided several generations of Russian students with a pertinent and vivid exposition of Darwinian evolutionary thought.[56]

Timiriazev discussed three different aspects of Darwin's work on the theory of evolution. First, he gave a detailed discussion of variation among pigeons, which led him to conclude that it was impossible to draw a clear line between a "variety" and a "species" and that variation can best be interpreted as the beginning of a new species. He concluded not only that organic transformation is a universal process but also that the common origin of all species can be assumed. Second, he analyzed the process of change as Darwin presented it. He accepted the struggle for existence and natural selection without notable digression from Darwin's interpretation. He showed a clear inclination to interpret intraspecies competition as the prime mover of evolution.[57] Third, he admitted that paleontology was not an adequate source of information, because it did not throw clear light on transitional forms, indispensable for empirical verification of transformism. While ready to treat the new theory as a hypothesis, he firmly believed in its ultimate triumph. No doubt, Timiriazev, like Darwin, thought that the future of the new theory would depend on the evidence produced by embryology, comparative anatomy, and related disciplines, as much, if not more, than on the data supplied by paleontology. He concluded his long review by citing the concluding paragraph of the *Origin* in which Darwin meditated about the two great laws that governed the planet earth: gravitation and evolution.[58]

Although Timiriazev provided the first link between the interpreters of evolution in the academic community and the nihilist intelligentsia, his book on Darwin contained no statements of direct ideological import. In at least one respect, he stood in direct opposition to D. I. Pisarev, the leader of nihilism. In the famous conflict between Pouchet and Pasteur, centered on spontaneous generation as a source of new forms of life, Timiriazev supported Pasteur's claim that "all life comes from life"—and that no known scientific facts supported spontaneous generation.[59] While science could not answer the question of the emergence of the first forms of life, it had gathered enough strength, thanks to Darwin, to ascertain the unity of organic forms based on common origin and certified by the existence of strong morphological similarities between individual species and by the absence of sharp lines separating different species.

There were indirect links between Timiriazev's essay and nihilist ideology. In the spirit of nihilism and in full agreement with Pisarev's thinking, Timiriazev credited Darwin with presenting the only true picture of nature: a picture that had no room for divine authority and interference with the work of nature. By prefacing his popular book with a statement of Auguste Comte's on selection as a source of harmony between organisms and the changing environment, he was eager to show that the scientific theory of evolution and the philosophical orientation of the nihilist intelligentsia, imbued with the spirit of Comtian philosophy, had much in common.

James A. Rogers has given an apposite description of Timiriazev's Darwinian affiliation:

> Timiriazev's remarkable influence in the propagation of Darwinism came not only from his prestige as a scientist (he was a pioneer in the study of photosynthesis) but also from the widespread popularity which he had won with his open espousal of liberal political views. Darwinism had already been accepted enthusiastically by the radical younger intelligentsia of the 1860s who thought that they saw in the theory of the origin of species the possibility of unifying the development of all organic life under a nonmetaphysical theory which would provide a major support for their materialistic philosophies. Under the influence of Timiriazev's popular writings on Darwinism, this scientific theory soon became a part of the political creed of all those persons who considered themselves progressive in social and political thought.[60]

The *Origin of Species* was Darwin's first book to be translated into Russian. During the 1860s two additional works were translated. A Russian rendition of the *Voyage of the Beagle* appeared in 1865.[61] A. N. Beketov, one of Darwin's most eminent and eloquent precursors, served as the editor of the translation project. In August 1867 Darwin wrote to Lyell that a young Russian, who visited him in Down, "is translating my new book into Russian."[62] The book was *The Variation of Animals and Plants Under Domestication,* and the young man was Vladimir Kovalevskii, who subsequently became a well-known evolutionary paleontologist. At that time *The Variation* was not yet published, and it seems most likely that the translation was made from a set of proofs Darwin had given to Kovalevskii. Thanks to Kovalevskii's rapid work, a Russian translation of the first volume of *The Variation* was published several months prior to the publication of the English original.[63]

The rapid diffusion of evolutionary ideas created favorable conditions for a general discussion of Darwin's contributions. In an article published in February 1864 in the journal *Messenger from Abroad* (*Zagranichnyi vestnik*) J. Schönemann did exactly this; he not only described the Darwinian style in natural history, analyzed Lamarck's theory of evolution, and summed up the main arguments of the *Origin,* but also provided fascinating passages of general comment on the Copernican sweep of the Darwinian evolution in biology. He had no reservations in giving full support to Darwin's conceptualization of natural selection as the prime mover of the evolution of living forms.[64]

The same journal gave its readers an opportunity to become familiar with Darwin's theory of evolution as a method applicable to a wide range of natural and social phenomena. A Russian translation of August Schleicher's "letter" to Ernst Haeckel, published in this journal in 1864, illustrated the applicability of Darwin's idea of the origin of species to the most puzzling question of the origin and evolution of language. He even transposed the Darwinian struggle for existence to the study of the survival and extinction of languages.[65] By his own admission, the editor of the journal was particularly eager to show the applicability of Darwin's theory to the study of the universal attributes of culture.

A few years later, Sir John Lubbock's *Prehistoric Times* (1865) came out in a Russian translation. This work acquired prominence as "the first major study in archeology reconstructing not only human skeletons but the early phases of human society and culture as well." It represented a turning point in making archeology a study of man as both an animal and a creator (and a product) of culture. It offered the first comprehensive survey of empirical data showing the work of evolution on both biological and social levels. Well received in Russia, it created many new openings for the scientific study of evolutionary thought. Petr Lavrov observed in the journal *Fatherland Notes* in 1869 that anthropology was in fact an application of Darwin's theory to the study of the origin and evolution of man. He thought that the idea of the anthropoid origin of man, because of its far-reaching consequences, would soon be recognized as one of the greatest triumphs of the human intellect.[66]

Translation of Western works analyzing the nature of the new ideas played an equally important role in the diffusion of the key principles of Darwin's theory. Lyell's *Antiquity of Man,* T. H. Huxley's *Evidence as to Man's Place in Nature,* Karl Vogt's *Lectures on Man and His Place in the Creation and the History of the Earth,* and many articles on evolution written by leading scientists in and outside biology were translated into Russian in quick succession. While Lyell's work placed paleontological information behind the evolutionary theory, Huxley made challenging excursions into the biology of anthropoids and the major achievements of comparative embryology. Huxley performed a function that Darwin—in the *Origin*—had painstakingly avoided: he made the evolution of man the central theme of his study. He gave scientific backing to the nihilist ideology built on two pillars: the idea of the historical relativity of moral values and social institutions, and the view of the pursuit of natural science as the safest path to a virtuous life and a general betterment of human existence. He gave the evolutionary theory a much broader basis: he carried it from its scientific moorings into social thought. Like Huxley, Karl Vogt made excursions into all the basic sciences providing illustrative material for the grand law of organic evolution; unlike Lyell and Huxley, however, he favored a polygenic theory of the evolution of human races—a view that did not find strong support among the early Russian evolutionists. Translated into Russian in 1865, Vogt's *Lectures* was in many respects an expanded version of Huxley's *Evidence* and was particularly rich in anatomical details.

In addition to these works, all addressing themselves to the general public, there were also translations of technical studies in organic evolution appealing almost exclusively to a narrow circle of specialists. One of these studies was Fritz Müller's *Für Darwin,* the first serious effort to apply Darwin's theory to embryology. In undertaking an empirical study of the crustaceans, Müller was inspired by Darwin's assertion that embryology stands to make a substantial contribution to the transformist view of nature. Darwin thought that embryology could be of particular importance to the study of evolution because it was in an ideal position to throw light on the primeval history of species. He said: "As the embryo often shows us more or less plainly the structure of the less modified and ancient progenitor of the group, we can see why ancient and extinct forms so often resemble in their adult state the embryos of existing species of the same class." "Embryology," he added, "rises greatly in interest, when we look at the embryo as a picture, more or less obscured, of the progenitor, either in its adult or larval state, of all the members of the same great class." [67]

During the early 1860s Russian readers could read in their own language many Western science classics that, while not directly identified with Darwin's ideas, presented the salient theoretical and methodological advances in all branches of modern biology. Included in this list were Matthias Schleiden, *The Plant and Its Life*, Isidore Geoffroy Saint-Hilaire, *General Biology*, and Claude Bernard, *Introduction to the Experimental Method in Medicine*. In 1865 Lyell's *Principles of Geology* was also published in a Russian translation. An earlier edition of this work had inspired young Darwin to dedicate his life to the study of organic evolution.

Countless writers referred to Darwin as a leading scientist but did not undertake an analysis of his contributions. They added significantly to the popularity of the new idea of organic transformism and of the historical orientation in biology. A. P. Shchapov, for example, did not make an effort to scrutinize and diffuse Darwin's ideas; but he made frequent references to Darwin as a giant of scientific thought. He placed him in the exclusive group of reigning scientists made up of Newton, Lavoisier, and Claude Bernard, the masters of "powerful reasoning" and the "real fountain" of great scientific ideas.[68] Elsewhere, he placed Darwin, along with Lavoisier, Laplace, Watt, Cuvier, A. von Humboldt, and Liebig, among "the creators of modern science."[69] Darwin, he contended, represented the type of scientific creativity that must find a functional place in Russian society as the most reliable mechanism for cultural progress. Darwin represented a "highly developed intellectual type" that was still absent in Russia.[70]

Nihilism and Darwinism

Nihilism emerged in an atmosphere that made the liberal reforms of the 1860s not only a possibility but also a reality. It was an ideology that sought no compromise with the social and cultural values built into the autocratic system. It represented a unique combination of materialism, espoused by Büchner and Moleschott, and positivism, a philosophical legacy of Auguste Comte and his followers. Materialism and positivism shared a pure belief in science as the motive force of social and cultural progress. Both viewed Darwinism as a generally successful effort to enhance the power of science in the unceasing war against mysticism, irrationalism, and supernaturalism.

Dmitrii Pisarev has been widely recognized as the most astute and influential architect of Russian nihilism. Materialism guided him in a consistent, but mainly implicit, war on the idealistic metaphysics of conservative writers, and on the most dedicated and bellicose defenders of autocratic values. A firm allegiance to positivism gave support to his views on the close ties between the growth of secular wisdom and social progress, on scientific regularities in the evolution of human society, and on the historical relativity of human knowledge and institutions. In addition to recounting the blessings of science as a major weapon in the war on the ancien régime, he popularized values that Francis Bacon had posited as the necessary cultural base for the advancement of science. He placed particular emphasis on two values closely related to the cultivation of scientific thought, one encouraging critical thought—the challenge to every intellectual authority—and the other stressing social utility as the only reliable indicator of the value of knowledge.

In 1864 Pisarev published "Progress in the World of Animals and Plants," a long essay presenting a fleeting analysis and enthusiastic endorsement of the basic ideas that made the *Origin of Species* a scientific work of epochal significance. The dramatic tone of this essay helped establish Pisarev as one of the most respected and influential popularizers of natural science during the early 1860s. The famous neurophysiologist Ivan Pavlov noted many years later that Pisarev's popular essay helped attract many young Russians to natural science studies in the institutions of higher education. Pavlov himself was one of those students.[71]

No other nineteenth-century work, Pisarev thought, surpassed Darwin's classic as a contribution to the triumph of a world view based on secular wisdom. Nor did any work make a more formidable contribution to the scientific foundations of biology. Pisarev did not hesitate to equate the authority of Darwin's law of evolution with the authority of Newton's law of gravitation. Like the law of gravitation, Darwin's "law" is universal, intolerant of exceptions. Unlike the law of gravitation, however, it must consider individualized external conditions to which organisms or species react in their struggle for existence. In comparison with the law of gravitation, the law of evolution is far more complex and it had not been studied so thoroughly.[72] More than any previous discovery, the law of evolution has shown that the work of nature is not a work of vast complexes of integrated phenomena but one of accumulative effects of "millions of small forces and causes."[73] Only persistent, trained, and minute observation can unveil the work of the universal struggle for existence. The inductive method is the most powerful weapon at the disposal of a naturalist.

The inductive method is one cornerstone of Darwin's theory, as Pisarev saw it. The other is historicism—the treatment of living nature as a continually changing phenomenon. Darwin showed that "not only individuals have their lives, but species and genera too; they gradually come into existence and they too are subject to continual variations according to definite laws."[74] Just as Lyell made the history of the earth the central concern of geology, so Darwin made the history of living forms the central concern of biology.

Pisarev's essay was scarcely more than a detailed and painstaking summary of the *Origin of Species;* it stayed close not only to the basic principles of the evolutionary theory but also to the treasures of Darwin's minute dissections of supporting data drawn from natural history. In addition to the *Origin,* he relied heavily on Lyell's geological work and on Karl Vogt's popular *Zoological Letters.* In the writings of Étienne and Isidore Geoffroy Saint-Hilaire he found both the valuable ideas that contributed to Darwinian thought and erroneous allusions that worked in the opposite direction.

In a way, Pisarev's essay was intended to be a review of Rachinskii's Russian translation of the *Origin of Species.* Pisarev wanted to do something about two major deficiencies in the translation. In the first place, he thought that the translator's Russian was too academic to appeal to a general reader. In the second place, Rachinskii did not write an introduction to the Russian text. "Progress in the World of Animals and Plants" exemplified a kind of introduction Pisarev would have written: it recounts the basic components of Darwin's evolutionary argument in a language comprehensible and appealing to the lay public.

Like Darwin, Pisarev fully rejected the three basic components of Lamarck's theory: the direct influence of the environment on the transformation of living forms, the use and disuse of organs as a propelling force of evolution, and the innate drive for progress. Like Darwin and Lyell, he found Cuvier's catastrophe theory a futile effort to resolve a critical paleontological dilemma.[75] The essay made only scanty references to the contributions of Darwin's precursors. Nor did it make a systematic effort to elaborate the intellectual links between the new theory of organic evolution and contemporary developments in sociology.

Pisarev found Darwin's theory acceptable in its entirety. He treated the enemies of Darwin's way of thinking as his own enemies. His main intent was not to show how Darwin's theory could be made the pillar of a modern ideology, but simply to give a popular account of its basic arguments—and to create a model for popular reviews of developments in science, a type of prose which, he thought, should occupy the first place in the creative work of literary figures. The essay dealt extensively with three major problems: the universality of organic evolution, the mechanisms of evolution, and the geological, geographical, embryological, and comparative-anatomical evidence supporting the evolutionary idea as Darwin saw it. Pisarev's goal was to give a true recapitulation of Darwin's ideas, avoiding criticism and philosophical elaboration. The essay was written in a lively and limpid style, and the illustrative material, despite its magnitude, was closely tied to the theoretical issues at hand. Preoccupied with the details of Darwin's elaboration of evolutionary principles, he chose to ignore the work of Darwin's early supporters and critics. Curious readers learned about Fritz Müller's embryological support and Kölliker's general criticism from other sources.

Pisarev made sure to let his readers know about the immense proportions of the Darwinian revolution:

> A master of vast stores of knowledge, Charles Darwin studied the entire life of nature from so broad a perspective and with so deep a penetration into all its scattered phenomena that he was able to make a discovery that, perhaps, has not been equalled in the history of the natural sciences. His feat was not limited to the discovery of an isolated fact, such as a gland or a vein, or the function of a nerve; he unveiled an entire order of laws, which govern the entire organization and transformation of life on our planet. Making his laws simple and incontrovertible, he built his study on obvious facts, and he made . . . the amateur in the natural sciences wonder why he did not come to these conclusions himself a long time ago. . . .
>
> In nearly all the natural sciences Darwin's ideas have brought about a complete revolution: botany, zoology, anthropology, paleontology, comparative anatomy and physiology, and even experimental psychology, have accepted these discoveries as the guiding principle that promises to unify the numerous observations already made and to open new paths to fruitful discoveries. Darwin's ideas are so far-reaching that at the present time it is impossible to foresee and to enumerate all the consequences of their application to various branches of scientific inquiry. . . .
>
> In [Darwin's] theory the reader will find the rigor of an exact science, the boundless sweep of philosophical generalization, and, finally, the superior and irreplaceable beauty that affects every manifestation of strong and healthy human thought. When the reader has become familiar with Darwin's ideas, even through my feeble and colorless sketch, I shall ask him whether we were right or wrong in rejecting metaphysics, ridiculing our poetry, and

expressing complete scorn for our conventional aesthetics. Darwin, Lyell, and thinkers like them are the philosophers, the poets, the aestheticians of our time.[76]

These statements show clearly that Pisarev was eager not only to give a simple and accurate summary of Darwin's ideas as the most sublime triumph of the modern mind, but also to integrate Darwin's general views into the nihilist world outlook. He argued that science is its own best philosophy and that philosophy is worth its name only insofar as it is anchored in scientific fact and theory. Because he was a great scientist, Darwin was automatically a great philosopher—a masterful synthesizer of the superb achievements of the human intellect. By naming Darwin a great poet, Pisarev merely reaffirmed his strong conviction that the work in science is the most cogent expression of aesthetic quality.

Pisarev viewed Darwin not only as the founder of modern biology and a shining naturalist-philosopher, but also as a most successful expression of the guiding principles of the English intellectual tradition: inductionism, empiricism, skepticism, and utilitarianism. Pisarev's eloquent presentation of Darwin's ideas represented a notable addition to the rapidly swelling literature on English thinkers and philosophical tradition. The educated public was particularly fascinated by Russian translations of the major philosophical studies by or about Francis Bacon, John Stuart Mill, and Herbert Spencer, as well as by the scientific works of Lyell, Huxley, Tyndal, and Lubbock. For several decades, the neurophysiologist Ivan Pavlov could rely on his memory in citing verbatim long passages from Lewes's *Physiology of Common Life,* an assemblage of popular arguments in favor of making physiology the backbone of psychological studies. A typical member of the intelligentsia believed that these works expressed unbounded faith in philosophical realism, empirical science, the idea of the perfectibility of human society, and close interdependence of science and democracy. Kuno Fischer's freshly translated monograph on Francis Bacon made a major contribution to the triumph of "philosophical realism" in contemporary Russian thought. It emphasized the role of skepticism and critical thought in the accumulation of scientific knowledge; it provided strong arguments in favor of a full separation of philosophy and theology.[77]

Advised by D. N. Ovsianiko-Kulikovskii and Vladimir Vernadskii, Emanuel Rádl, the noted German historian of biology, offered a pertinent description of the relations of Pisarev and the nihilists to Darwinism:

In Russia Darwinism became a part of the stream of positivism and materialism which began to flood the country in the middle of the nineteenth century. Before that time intellectual Russia was under Hegel's influence. A
strong reaction against idealism now set in, and natural science was called in
to help in its overthrow. . . . Pisarev held that such abstract subjects as philosophy and psychology merely represent empty scholasticism, and that
natural science must be presented so simply that ten-year-old children and
uneducated peasants can understand it. . . . The works of Darwin, Spencer,
Haeckel, Wallace, and Romanes were translated, and helped to forward the
movement.[78]

Pisarev was the most dynamic and influential nihilist. Sometimes he
was identified as the philosopher of nihilism. His contemporaries knew,
however, that the major contributors to the *Russian Word*, particularly
N. V. Sokolov, V. A. Zaitsev, and G. Blagosvetlov—in addition to
Pisarev—represented a firmly united philosophical and ideological
front backed up by the powerful sway of revolutionary zeal and moral
commitment. All this, however, did not mean that the thinking of nihilists was always true to the norm. There were occasional digressions,
often of sizable proportions.

One such digression came from Zaitsev. In 1864, in a review of de
Quatrefages's book on the unity of mankind,[79] he argued in a vein that
identified him with the sociological-ideological movement subsequently
named Social Darwinism. Contrary to the spirit of nihilism and to
Pisarev's sharply focused philosophy, Zaitsev committed a gross indis-
cretion by making racist comments violating both the substance and the
spirit of science. His Social Darwinist statements contradicted the scientific spirit and the equalitarian sentiment of nihilist philosophy. They
made nihilism appear more reactionary than the most dedicated defenders of autocratic values.

Carrying Darwin's principle of the struggle for existence to what he
considered a logical conclusion, Zaitsev asserted that all efforts to
emancipate the black people from the colonial yoke and to give them an
opportunity to rise on the scale of social and cultural progression were
fated to be total failures. The black people, he wrote, belonged to a
"lower race," which did not have the innate capability to benefit from
the same rights as the members of the white race. In his view, wherever
the white and black races belonged to the same political system, the
black people should be relegated to a subordinate position. Zaitsev did
not repeat his racist arguments. Nor was he particularly apologetic for
having made them in the first place.

Only two popular journals—*Sovremennik* and *Iskra*—took the trouble to refute Zaitsev's reliance on Darwin's theory to justify a racist position.[80] Writing in *Sovremennik*, M. A. Antonovich was particularly annoyed at the blinding rigidity that dominated Zaitsev's effort to apply the evolutionary theory beyond the scope of Darwin's concerns.[81] Nozhin, a young embryologist and social critic, noted that Zaitsev went against Darwin's theory when he failed to view racial differences as transitory characteristics. Zaitsev was the first and the last Russian writer of note to flirt with Social Darwinism on a large scale.[82]

The inappropriateness of biological models in the social sciences attracted the attention of Petr Tkachev, whose general orientation, despite its strong Jacobin and Blanquist elements, was close to nihilist thought. He wrote in the mid-1860s that the idea of "organic progress," highly fruitful in its application to the scientific study of both inorganic and organic nature, becomes "moribund and sterile" when it is transplanted to the scientific study of social phenomena.[83] Tkachev thought that to study "organic progress" meant to rely on organismic analogies (in the Spencerian sense) and organic evolution (in the Darwinian sense). Opposed to the biological foundations of Darwinian sociology, Tkachev made no secret of his unlimited admiration for Darwin's contributions to biology and the scientific world outlook.

Institutional Variation

Russian scientific institutions did not express a uniform attitude toward Darwin's theory: while at one extreme there were institutions that completely ignored organic evolution as a scientific notion, at the other extreme were institutions that not only played a major role in the speedy diffusion of evolutionary ideas but also made Darwin's theory the point of departure in wide areas of scientific research.

With a membership consisting primarily of older scholars whose most productive and creative years belonged to the pre-Darwinian past, the St. Petersburg Academy of Sciences did not produce at this time a single Darwinian scientist of consequence. Foreigners with distinguished records of scholarly achievement but of a decidedly conservative frame of mind dominated the biological sciences in the Academy. As guests of the Russian government, these scholars showed a strong inclination to avoid the great theoretical ideas of modern science which invited interpretations inimical to the sacred values of the autocratic ideology. Karl von Baer became the academic stalwart of anti-Darwinism: in his un-

yielding opposition to the new biological theory he relied on an idealistic interpretation of the ontological foundations of science and on a
teleological view of the processes of nature. The academician G. P.
Helmersen, a geologist of note, made no effort to hide his agreement
with Karl von Baer's anti-Darwinian stance. F. J. Ruprecht, who held
the chair of botany in the Academy and who was well known for his
contributions to plant systematics, opposed "the materialistic orientation of modern natural science" because it had no place for a vital force
in organic life.[84] He did not actively campaign against Darwin, but he
never renounced his vitalistic and teleological views expressed before
the publication of the *Origin*.

The First Congress of Russian Naturalists and Physicians met in St.
Petersburg from December 28, 1867, to January 4, 1868. Selected representatives of the major branches of natural science gave reports on their
current research, adding up to a magnificent display of Russia's involvement in scientific research. In a paper on the sea cow, Johann Friedrich
Brandt, a full member of the Academy since 1833, expressed his views
on the current evolutionary controversy by rejecting both saltatory and
gradual changes as natural mechanisms of the evolution of species.
Every species, he contended, develops from a distinct embryo and follows a predetermined course of development. The boundaries separating species from one another are fixed and permanent. No change can
lead to the emergence of new species.[85] Von Baer's "teleology," rather
than Cuvier's "catastrophism" or Darwin's "gradualism," was in his
opinion the key to understanding the dynamics of living nature.

K. S. Veselovskii, permanent secretary of the St. Petersburg Academy
of Sciences, found it necessary to shed a more favorable light on the perspectives opened by Darwin's work. In this effort he was motivated by
the need to explain the reasons for the election of Darwin as corresponding member of the Academy in 1867. In his annual report on the
activities of the Academy in 1869, he referred directly to the contribution Darwin's theory had made to the widening of the research base of
natural science. He noted the growing interest in the evolution of animal
forms and emphasized the limitless promises of the recent removal of
the boundaries separating botany and zoology from paleontology. "A
comparison of presently existing forms of organic life with fossils buried
in various strata of the earth—and belonging to various geological periods—will pave the way for the ultimate understanding of the general
laws that have governed the transformation of life from its first appearance on earth to its present diversity and profusion."[86] Despite this and

similar pronouncements, all the scientists Veselovskii had mentioned, with the exception of one, worked outside the Academy and were in their twenties. The special committee whose recommendation led to the election of Darwin as a corresponding member of the Academy, a purely honorific title, wrote that none of the works of the noted English scientist contained more errors than the *Origin of Species*.[87]

The learned societies founded before 1860 and typified by the Russian Geographical Society met the new evolutionary ideas with pronounced detachment. These societies acted as closed corporations with a clear aristocratic bias in the election of new members. Some discriminated against younger scholars, as well as against persons who did not speak French or German. They were dedicated to the enrichment of descriptive natural history and had little use for natural philosophy—for scientific theory. The Moscow Society of Naturalists helped start the flow of Darwin's ideas to Russia: in the early 1860s its *Herald of the Natural Sciences* published several translated articles on the new theory. Very quickly, however, the controversial nature of Darwin's theory influenced the Society to withdraw from this kind of activity.

In 1864 a group of Moscow University professors, dissatisfied with, and openly critical of, the caste exclusiveness and scientific conservatism of the Moscow Society of Naturalists, which preferred to use French and German in scientific communications, founded a new scholarly association—the Society of the Friends of Natural Science, Anthropology, and Ethnography—dedicated to attracting private support for scientific research and to organizing systematic surveys of natural resources. The new society manifested a particular determination to keep abreast of most modern theoretical and methodological developments in science. It did not come as a surprise to its members when, in 1864, G. E. Shchurovskii, in his introductory report at the first meeting of the anthropological section of the new society, referred to anthropoid fossils and their significance for the understanding of the evolution of man.[88] The new society became a model for naturalist groups organized in all national universities, including the newly founded Odessa University (1865) and Warsaw University (1869). The *Proceedings of the St. Petersburg Society of Naturalists* became particularly well known by its lively interest in keeping a record of the growth of evolutionary thought in Russia. Most of these societies, however, placed the primary emphasis on fieldwork in various parts of Russia and neighboring Asian countries, surveying natural riches and building an empirical base for a wide spectrum of natural sciences.

During the 1860s the universities became the major centers of scientific research, a position the Academy of Sciences had previously occupied. The university charter of 1863 recognized the rapidly growing popular interest in the natural sciences by allowing for a substantial expansion of the curriculum coverage of these fields. University education and research benefited from the rapid growth of laboratories, museums, and libraries, from the work of naturalist societies affiliated with individual institutions of higher education, from frequent visits of professors to Western universities, and from expanded postgraduate studies of young Russians in the leading German and French universities. While the Academy of Sciences and most older naturalist societies continued to be insulated from the social and ideological fermentation that swept the country after the Crimean War, the universities—to use the phrase of the eminent surgeon N. I. Pirogov—became the true barometers of social pressure generated by new turns in philosophical outlook and intellectual impulse.[89] For all these reasons, it is small wonder that the universities became the centers of evolutionary research. Only a few years after the publication of the *Origin of Species,* Russian universities produced a number of scholars ready to make the notion of evolution the starting point of broadly based research and to lay the groundwork for diverse scientific traditions in Russian Darwiniana.

Moved and inspired by an intellectual atmosphere that emphasized the power and the challenge of science, most Darwinian pioneers belonged to the generation of young people who flooded the natural science departments of the leading universities. The emphasis was on the youth deeply involved in a war against the dominant ideas and habits of the past. I. I. Mechnikov was only eighteen when he submitted his essay on Darwin to Dostoevsky's journal *Time.* At the age of nineteen, K. A. Timiriazev published a long comment on the new theory, which led to the first Russian book on Darwin's contributions to biology. A. O. Kovalevskii launched his distinguished work on the embryology of marine invertebrates and published his first evolutionary paper in the *Memoirs* of the St. Petersburg Academy of Sciences when he was twenty-five years of age.

Russian Pioneers in
Evolutionary Embryology:
A. O. Kovalevskii and I. I. Mechnikov

In the *Origin of Species* Darwin wrote that "various parts in the same individual, which are exactly alike during an early embryonic period, become widely different and serve widely different purposes in the adult state." "So again," he concluded, "it has been shown that generally the embryos of the most distinct species belonging to the same class are closely similar, but become, when fully developed, widely dissimilar." To add weight to his statement, he cited von Baer's professed difficulty in telling apart the embryos of a long series of vertebrates during the early stage of development.[90] Although Darwin did not make extensive use of embryological evidence in favor of evolution, he fully sensed its vital importance for the future development of transformist biology. The *Origin of Species* provided a powerful stimulus for reinvigorated and reformulated embryological research. In Russia the new embryology attracted a group of most promising young scientists, led by Aleksandr Onufrievich Kovalevskii and Il'ia Il'ich Mechnikov. These scientists wanted to answer one of the key evolutionary questions: Does von Baer's description of the vertebrate embryos apply also to the invertebrates? Does embryological evidence support the idea of a fundamental morphological similarity between vertebrates and invertebrates?

A. O. Kovalevskii helped to end the period of exclusive concern with the diffusion and popularization of Darwin's ideas and to open the period of original scientific research in the vast domain of evolutionary thought.[91] A founder of evolutionary comparative embryology, he never doubted the fundamental correctness of Darwin's theory. Darwin approached embryology as a fountain of scientific information confirming the evolutionary point of view; Kovalevskii treated the evolutionary point of view as an interpretive and integrative principle of embryological research. Kovalevskii entered the annals of science as a thorough empiricist, who wrote careful and remarkably precise summaries of his personal research ventures without showing much inclination toward high-level abstractions and complex schemes of logical constructions. He made it abundantly clear, however, that his preoccupation with empirical minutiae did not lead him to lose sight of the challenging world opened by the evolutionary idea of the morphological unity of animal types: he dealt extensively and minutely with homologies and parallellisms in the embryonic growth of animals belonging to different taxonomic groups. Without stating it explicitly, he made a concentrated effort to build the empirical base for a general explanation of embryonic development—to erase the prevalent pre-Darwinian notion of a morphological chasm separating the vertebrates from the invertebrates.

I. I. Mechnikov made a revealing comparison between Kovalevskii and Ernst Haeckel, as opposite types of evolutionary biologists. Preoccupied with a search for universal laws of biogenetic consequence,

Haeckel showed a clear tendency to ignore products of empirical research which did not fit his grand theoretical schemes and to draw conclusions not warranted by available empirical data. He was not essentially a research scholar but an imaginative synthesizer of current ideas and an architect of grand hypotheses—basing much of his theory of gastraea on empirical data Kovalevskii had supplied. Kovalevskii, by contrast, believed that in more delicate areas of evolutionary research—such as embryology—extreme care should be exercised not to overlook a single empirical detail.[92] The empirical basis of his embryological research, however, was sufficiently broad to allow him to draw conclusions of larger theoretical magnitude.[93] By explaining the homologous features of invertebrate and vertebrate embryos he gave evolutionary embryology both a general theoretical orientation and a solid empirical interest.

Kovalevskii's first study—which brought him a magister's degree from St. Petersburg University in 1865—analyzed the growth of the lancelet (*Amphioxus lanceolatus*), a translucent marine animal. Impressed with the morphological simplicity of the lancelet, which at that time was classified as a vertebrate, Kovalevskii considered the possibility that this marine organism might represent a species occupying a transitional position between vertebrates and invertebrates. His hunch brought rich rewards. Painstaking inquiry showed that the embryonic development of the lancelet falls into two clearly distinguished phases. The initial phase follows the pattern of growth common to invertebrates: it is dominated by an even and nearly complete cleavage of the egg and by the emergence of the blastula, a hollow ball filled with a fluid and bounded by a single layer of cells. The second phase produces the embryo, consisting of an external and an internal germ layer, fully corresponding to the primary layers von Baer had described as a vertebrate characteristic. Kovalevskii's research showed that the lancelet should be classified as an invertebrate of the highest order, and that the embryonic growth of vertebrates and invertebrates is basically similar.[94]

Kovalevskii then undertook to study the ascidians, immobile creatures fastened to the sea bottom, at that time classified by many as mollusks and by some as worms. By their external appearance, the ascidians do not show even a remote similarity to vertebrates. A closer study, however, provided Kovalevskii with stunning surprises. It showed that the development of the larvae of this organism displayed features characteristic for lancelets and lower vertebrates.[95] In a later paper, Kovalevskii noted that in their embryonic growth the ascidians were closer to

the vertebrates than any other invertebrate group.[96] His discovery made
a great impression on his contemporaries, more so in the West than in
Russia. In *The Descent of Man* Darwin took serious note of Kovalevskii's interpretation of the embryonic development of ascidians. He
stated:

> M. Kovalevskii has lately observed that the larvae of the Ascidians are
> related to the Vertebrata in their manner of development, in their relative
> position of the nervous system and in possessing a structure closely like the
> *chorda dorsalis* of vertebrate animals; and in this he has been since con-
> firmed by Prof. Kupffer. M. Kovalevskii writes to me from Naples, that he
> has now carried these observations yet further; and should his results be well
> established, the whole will form a discovery of the greatest value. Thus, if we
> may rely on embryology, ever the safest guide in classification, it seems that
> we have at last gained a clue to the source whence the Vertebrata were de-
> rived. We should then be justified in believing that at an extremely remote
> period a group of animals existed, resembling in many respects the larvae of
> our present Ascidians, which diverged into two great branches—the one
> retrograding in development and producing the present class of Ascidians,
> the other rising to the crown and summit of the animal kingdom by giving
> birth to the Vertebrata.[97]

A few years after Darwin commented on Kovalevskii's ascidian
study, Haeckel incorporated the new discovery into popular literature
on the evolutionary idea. He wrote in *The Evolution of Man:*

> Toward the end of the year 1866, among the treatises of the St. Petersburg
> Academy, two works appeared by the Russian zoologist Kovalevskii, who . . .
> had occupied himself in studying the individual evolution of some of the
> lower animals. A fortunate accident had led Kovalevskii to study almost si-
> multaneously the individual evolution of the lowest vertebrate, the *Amphi-
> oxus,* and that of an invertebrate, the direct relationship of which to the *Am-
> phioxus* had not been even guessed, namely the ascidian. Greatly to the
> surprise of Darwin himself, and of all zoologists interested in that important
> subject, there appeared from the very commencement of their individual de-
> velopment, the greatest identity in the structure of the bodies of those two
> wholly different animals—between the lowest vertebrate, the *Amphioxus,* on
> the one hand, and that misshapen lump adhering to the bottom of the sea,
> the sea-squirt, or ascidian, on the other hand. . . . There can be no longer
> any doubt, especially since Kupffer and several other zoologists have con-
> firmed and continued these investigations, that of all classes of inverte-
> brates . . . , the ascidians are most nearly allied to the vertebrates. We cannot
> say the vertebrates are descended from the ascidians, but we may safely as-
> sert that . . . the ascidians are the nearest blood-relations to the primeval
> parent-form of vertebrates.[98]

Kovalevskii returned several times to the study of lancelets and ascidians, but he also expanded his research interests to cover many additional species of invertebrates, represented mainly by the marine microfauna. In his subsequent studies he discovered the existence of mesoderm in annelids and insects, which provided an additional proof for the basic similarity in the embryonic growth of vertebrates and invertebrates. All these studies added essential information in support of embryonic homologies of vertebrates and widely represented invertebrates. The evolutionary basis of his theoretical orientation, the precision and remarkable skill of his research techniques, and the general significance of his findings assured Kovalevskii of a notable place in the mainstream of biological thought during the early Darwinian era. He not only helped strengthen the hold of Darwin's theory on modern scientific thought but also made a noted contribution to the accelerated growth of modern biology in Russia.

In a paper delivered at the Eleventh Congress of Russian Naturalists, held in St. Petersburg in 1903, V. V. Zalenskii, a distinguished member of the St. Petersburg Academy of Sciences, noted that the evolutionary theory provided "the main stimulus for the entire range of Kovalevskii's research." He noted: "Transcending the limits of pure theory, his research manifested a deep and clear awareness of the great importance of the study of animal evolution, the surest path to solving the basic questions of life that have preoccupied the human mind from time immemorial. An evolutionist by general orientation, Kovalevskii contributed more to the theory of the transformation of organic forms than any one of his contemporaries."[99] In 1890 the St. Petersburg Academy of Sciences honored Kovalevskii by electing him to full membership. A particularly important event in the history of the Academy, this election marked a major expression of faith in the rich promise of Darwin's theoretical legacy.

Kovalevskii received wide recognition for extending von Baer's theory of the development of vertebrate embryos to all animals: he demonstrated the presence of germ layers among both vertebrates and invertebrates. He showed that during the early phases of their embryonic development all multicellular animals have common features and that strong differences between various types of animals begin to appear only during the later phases of embryonic growth. This, however, was only part of his contribution to embryology. He not only replaced Cuvier's (and von Baer's) rigid division of the animal kingdom into four types, each with its own separate and fully isolated biological identity, by a unitary system, but also broadened the empirical base of embryology with numerous new concrete facts. He offered, for example, a new empirical method in the study of the early phases in the development of germ layers.[100]

The by-products of Kovalevskii's research were equally impressive: he discovered the presence of several marine species in the Red Sea, previously thought to exist only in the Mediterranean Sea. In a few cases, his information led to a reclassification of individual species. He added a veritable treasure of empirical facts and generalizations to the scientific knowledge on the formation of the body cavity, digestive canal, nervous system, and vascular network—the central problems of comparative embryology.[101]

Despite rough beginnings, Kovalevskii's scientific career was a warm story of success and recognition. He was elected an honorary member of almost all Russian learned societies and universities and of a long series of foreign scientific institutions, including the Paris Academy of Sciences and the Royal Society of London. He received two prizes from the Paris Academy of Sciences. The Russian scientific community recognized the full significance of his scientific contributions and honored him profusely. In addition to the universal grandeur of his scholarly achievement, he made a special contribution to Russian science: by relying on a broad evolutionary framework and microscopic methods, he played a major role in transforming zoology from a narrow description of faunistic facts to a theoretically elevated science, dominated by experimental anatomy and experimental embryology. From a descriptive discipline of local interest, Russian zoology became a theoretical discipline with universal appeal. Thanks to a great extent to Kovalevskii's work, Russian zoology achieved impressive results in two activities: the study of phylogeny (based on the evolutionary principle), and the use of modern instruments of inductive research, surpassing the limits of simple observation. Comparative embryology and comparative anatomy became the scientific mainstay of Russian Darwinism.

Kovalevskii was not the only Russian scientist who helped lay the foundations for evolutionary embryology; the contributions of I. I. Mechnikov belonged to the same category of distinguished achievement.[102] Mechnikov's name is usually associated with the phagocyte theory, built upon a study of intracellular digestion among invertebrates, which helped to explain the origin of multicellular animals and to lay the foundations for evolutionary pathology. His work in the latter field earned him a Nobel Prize, which he shared with the German pathologist Paul Ehrlich.

Mechnikov had much in common with Kovalevskii. Both were models of pure dedication to science, enormous intellectual resourcefulness, and vast reservoirs of energy. But there were also strong differences. Not without some exaggeration, S. Zalkind has pinpointed the differences in their temperaments and styles of work:

> Although Mechnikov and Kovalevskii were personal friends and possessed common scientific interests, it would have been difficult to find two men more unlike in mental make-up, character, and methods of research. Mechnikov was a theoretician attracted to general scientific problems, given to making broad philosophical generalizations, impetuous and quick in his conclusions, sometimes apt to disregard facts but self-confident and persistent in the attainment of his aims. Kovalevskii, on the other hand, was an empiricist, tackling only concrete tasks, avoiding (we may even say, fearing) "all that lofty theoretical stuff," but at the same time extremely precise and thorough in his observations. A modest, mild and yielding man in everyday life, he was firm and indomitable in scientific disputes concerning facts which he knew well and had verified many times.[103]

In 1863, as an eighteen-year-old student at Kharkov University, Mechnikov wrote an essay on the *Origin of Species* in which he scrutinized the pivotal ideas of Darwin's thesis, particularly the derivation of the struggle for existence from Malthus's law of mathematically formulated discrepancy in the growth rates of population and food resources. He also thought that the present existence of many lower organisms disproved Darwin's theory. If Darwin were right, he said, "these beings, the initial steps in the organization of life, would have begun to change a long time ago, giving place to more advanced forms." When pushed against the wall, Mechnikov argued, Darwin did not hesitate to rely on spontaneous generation to account for the emergence of species that could not be accounted for by evolutionary processes. All this, however, did not prevent him from concluding that the theory presented in the *Origin* was destined to have a great future and from considering himself one of its most ardent supporters. Mechnikov submitted his manuscript to F. M. Dostoevsky's journal *Time*, but the journal went out of existence before it could act on the new acquisition. The manuscript waited until 1950 to be published in a volume of Mechnikov's essays.[104]

After intensive study under several leading German biologists and a passing interest in the embryology of insects, Mechnikov made the

Mediterranean marine invertebrates his main research concern. But, even in this activity, he resisted a close adherence to the Darwinian theory: he clung steadfastly to his original idea that Darwin had advanced too many general ideas of a purely hypothetical nature. To Kovalevskii, Darwin's theory served as the incontestable basis of comparative embryology; to Mechnikov, it belonged to the realm of challenging hypotheses requiring careful experimental testing. Mechnikov did not hesitate to criticize the evolutionary conclusions Kovalevskii had reached in his studies of lancelets and ascidians as devoid of a solid empirical basis. He was particularly critical of Kovalevskii's claim to have established the unity of the embryonic growth of invertebrates and vertebrates. Nor did he approve of Kovalevskii's assertion that among lancelets the process of invagination leads to the formation of the digestive tract. He contributed articles to scholarly and popular journals in an obvious effort to discredit Kovalevskii's theses. His conclusions, however, showed that he had left the door open for a more conciliatory attitude toward the idea of the embryonic unity of the animal kingdom.[105] The deeper he became immersed in embryological research, the more closely he became identified with the theoretical foundations of Darwin's legacy. During the 1870s his own empirical research led him to a full acceptance of Kovalevskii's position.[106]

From 1865 to 1869 Mechnikov wrote about thirty papers on the embryonic growth of an unusually large number of animal species, mainly invertebrates. He helped confirm the discovery that all animals have two basic germ layers—ectoderm and endoderm—thus giving added strength to the idea of the evolutionary unity of the animal kingdom.[107] His research concentrated on comparative embryological studies of animals whose morphological affinity had not yet been established. Some of his conclusions were not upheld by subsequent research; yet his evidence in favor of the general relationship of the Echinodermata, the Enteropneusta, and the Chordata had gone unchallenged.

At the beginning of his scholarly life, Mechnikov was not inclined to tie comparative embryology to Darwin's evolutionary theory. As late as 1869 he wrote: "The comparative history of [embryonic] development deals with facts from which it draws direct conclusions without considering the origin of various species." [108] Soon after this pronouncement, Mechnikov, influenced by both Kovalevskii's research and his own, became an evolutionary embryologist in the full meaning of the term. Despite his persistent criticism of certain aspects of Darwin's theory, it would not be an exaggeration to say that after the early 1870s Mechnikov's entire scientific work and all his theories were part of a brilliant search for the deeper meanings of the scientific legacy of the English naturalist.[109]

As he accepted Darwinism as a broad theoretical orientation, Mechnikov saw the basic source of its power in the combination of a historical view of nature, a comparative approach to biological phenomena, and an identification of purposiveness in the organic world with the processes of adaptation as a means of survival. If his work had weaknesses, they stemmed, not from a lack of scholarly dedication and experimental skill, but from the unique features of his temperament: unsettled and excitable, he moved too swiftly—particularly in his embryological work—from one research undertaking to another to do justice to all of them; his embryological research, for example, covered representatives of almost all major groups of invertebrates and some vertebrates.[110]

In 1867 the St. Petersburg Academy of Sciences awarded the first Karl von Baer Prize for outstanding work in biology. A special committee selected Mechnikov and Kovalevskii to be the first recipients of the coveted prize. The committee noted that the work of each scholar showed distinct excellence: Mechnikov was honored for having produced a "complete and integrated" study, Kovalevskii for the "diversity of subjects" covered.[111] The committee noted with approval Kovalevskii's discovery of embryonic links between vertebrates and invertebrates.[112] Leon Bliakher, a modern historian of nineteenth-century embryology, has argued that Mechnikov must also be counted among the discoverers of these links, soon to be recognized as the foundation of evolutionary embryology.[113] Karl von Baer, the leading anti-Darwinian scholar of his age, served as the ranking member of the selection committee. Perhaps because Mechnikov showed signs of reluctance to link embryology with the Darwinian theory of evolution, von Baer praised his scholarship more than Kovalevskii's. Kovalevskii's full and consistent identification with Darwinism was well established from the very beginning.

The scientific work of Kovalevskii and Mechnikov represented the crowning point in the reception and early application of Darwin's ideas by Russian natural scientists. But how were these scientists regarded by the Russian scientific community? Although the St. Petersburg Academy of Sciences published their papers, they had difficulty in finding suitable employment. At the time when St. Petersburg and Moscow provided the most coveted academic positions, Kovalevskii and Mechnikov had no choice but to seek teaching positions in provincial universities, equipped with poor laboratories and libraries and wanting in intellectual stimulus for sustained scientific work. Mechnikov's candidacy for a position in the Medical and Surgical Academy in St. Petersburg proved futile. He served sixteen years on the faculty of the newly founded Odessa University, resigning in 1882 to avoid the grueling pressure of academic intrigue and student unrest. In 1887, before he had reached the peak of his scientific career, he left Russia to join the Pasteur Institute in Paris, where he remained until his death in 1916. A. O. Kovalevskii spent twenty-two years of his academic career in Kazan, Kiev, and Odessa universities fighting the depressing monotony of provincial isolation by extensive correspondence with western European embryologists, frequent scientific trips to the Mediterranean Sea, and cooperative research ventures with eminent foreign scientists.

Whatever the reason for the negative results of their initial search for academic employment in St. Petersburg and Moscow, the two young scholars were primarily responsible for the preeminent role of provincial universities in making Russia one of the early centers of empirical studies in organic evolution. It is most likely that the professional hardships of Kovalevskii and Mechnikov did not result solely—and perhaps not even primarily—from their identification with Darwin's theory. It should be remembered that the academic market—and the growth of employment opportunities—was controlled by the limited purse of the Ministry of Public Education much more than by the efforts of the scientific community to keep up with new developments in individual disciplines.

Among the new breed of evolutionary embryologists N. D. Nozhin occupied a unique position: he began as a searching scientist of great ambition and talent and ended as an ideologist dedicated primarily to emancipating his country from both the decaying feudal law and the burgeoning capitalist relations. After having studied chemistry under Robert Bunsen at Heidelberg University, he moved to Tübingen University, where he studied zoology under H. G. Bronn, the German translator of the *Origin of Species,* but not a Darwinist. In 1863 he went to Italy to conduct research on the embryonic growth of selected species of Mediterranean fauna for the purpose of answering the question of a possible morphological link between vertebrates and invertebrates. It was in Italy that he established close relations with A. O. Kovalevskii.[114] In his spare time he translated Fritz Müller's *Für Darwin* into Russian; this work was generally acclaimed as the first successful effort to combine meticulous embryological research with Darwinian transformism.

The study, according to Mechnikov, marked the first scientific effort to base embryological research on Darwin's theoretical principles.[115] The St. Petersburg Academy published Nozhin's lone scientific paper, a study of coelenterates (primarily medusae).

During his sojourn in Germany, Nozhin participated in several circles of Russian students eager to find ways of bringing modern political and social ideas to their native land. At this time Nozhin began to think of Darwin's theory as a source of ideas for a unitary picture of the evolution of the universe, both natural and social. He viewed Darwin's work as the culminating point of nineteenth-century science, and science as the only source of sound guideposts for purposive action in social development. Like the champions of nihilism, he preached "a visionary faith in science" as the true power of reason and declared that "all scientific knowledge in the hands of its honest servants stands in direct opposition to the existing order" and that "in the world there is only one evil—ignorance—and only one way to salvation—science." [116] He argued that only by knowing the laws of nature could man widen the humanistic base of social existence. However, he rejected the struggle for existence as the moving force of evolution; he called it an aberration, a pathological force exercising a negative influence on both natural and social evolution. He called Darwin a "bourgeois-naturalist" for his emphasis on competition—rather than on cooperation—as the mainspring of biological and social development.[117]

Nozhin supplied populist sociology—to which N. K. Mikhailovskii gave a fully crystallized form—with guiding ideas and logical structure. While Kovalevskii saw in Darwin's theory a fruitful method of scientific analysis, Nozhin saw in it the culminating point in the evolution of the modern scientific world view—a triumphant victory of reason over metaphysical mysticism and religious dogma. Nozhin must be counted among the first Russian scientists to express two thoughts that found strong followers during the subsequent decades: first, not the theory of the struggle for existence but the elimination of supernatural causality in the development of nature was Darwin's major contribution to science; and second, sociology owed a great debt to Darwin's theory—not to the notion of natural selection, but to the unitary developmental scheme and rational models for social analysis.

Among the Russian scientists who supported Darwin's ideas during the 1860s Sergei Usov, professor at Moscow University from 1868 to 1886, occupied a unique position. His translation of Friedrich Rolle's

Darwin's Theory of the Origin of Species, published in 1865, had the unique distinction of having been the first book in the Russian language to offer a comprehensive presentation of Darwin's theory. The author provided a detailed but simple explanation of Darwin's basic principles and made an effort to link the theory of organic evolution with the idea of progress. Usov's doctoral dissertation—*Taxonomic Units and Groups* (1867)—devoted a special chapter to the history of evolutionary theories in biology.[118] The chapter gave a systematic and detailed analysis—the first in the Russian language—of Lamarck's evolutionary theory. In Russia, as in the West, the triumph of Darwin's theory opened the gates for the rediscovery of Lamarck. Usov scrutinized both the similarities and the differences between Lamarck and Darwin. The aim of his analysis was to provide a historical legitimation of Darwin's theory: to show that the new evolutionary idea was built upon the solid foundation of accumulated biological knowledge. Usov played a major role in making Moscow University a true bastion of Russian Darwinism. Although he regarded Darwin's theory as a triumph of modern biological thought, he did not hesitate to point out that some of its grand theoretical conclusions needed stronger empirical support.

Darwin's First Critics

Most contemporaries agreed that the Darwinian evolutionary theory found an enthusiastic reception in Russia and that negative criticism came from isolated quarters that could muster only scattered and feeble support. The triumph of positivism and materialism worked against anti-Darwinian criticism. Eastern Orthodox theology, entangled in spiritualism and ethicism, did not have alert and able spokesmen to fight the new heresy. Once the church recognized the danger, however, the theological journals began to carry anti-Darwinian articles, in most cases translations from Western religious journals. In 1864 the *Creations of the Holy Fathers,* a journal of the Moscow Theological Academy, published a long article, based on a paper carried by the English journal *Athenaeum,* that made no concession to Darwin's theory and pleaded for a crusade against it.[119] A similar translation appeared in *Christian Readings,* a journal of the St. Petersburg Theological Academy. These and similar articles marked not only the beginning of a sustained theological war on Darwinism but also the first step in the rapidly improving quality of church-supported criticism of scientific theories that contradicted scriptural wisdom. In a way, this marked the beginning of the growth of a solid corps of theological writers concerned with the philosophical foundations of modern science.

I. Krasovskii was the first Russian theologian to offer an original, comprehensive, and systematic critique of the *Origin of Species*. He did not overlook the positive side of Darwin's theory, which, he said, made "a significant contribution to the natural sciences" by directing their attention to new areas of inquiry.[120] Darwin showed that many species were actually different varieties of the same species, and he pointed the way to a more efficient "practical application of artificial selection." More than any other theory, Darwin contributed to the understanding of the dynamics of plant and animal modifications "within the limits of existing species." Krasovskii did recognize, however, that the notion of organic evolution was not a radically new development in science but rather the crowning point in the long history of a scientific idea.

Despite the praise, Krasovskii found it necessary to reject all the key postulates of Darwin's theory. He raised many basic questions that were to plague Darwinism for decades to come. Darwin made an unpardonable error, in Krasovskii's view, in limiting his discussion to the "secondary" causes of evolution. The "primary" causes, explained in the holy scriptures, had no place in Darwin's thinking. Darwin's "slight modifications," the source of evolution, eliminated divine interference from living nature. Teleology, as Darwin used it, was only "a play on words"; he recognized only natural causes and firmly rejected the existence of a higher intelligence. By emphasizing the struggle for existence as the motor of evolution, he denied the divine origin of the moral law of human society. Nowhere in his essay did the author try to make the criticism of Darwin's ideas part of a more general criticism of natural science. P. D. Iurkevich, the most eminent and erudite theologian of the 1860s, was too preoccupied with attacks on the materialistic ontology of contemporary experimental physiology to tackle the Darwinian menace.

Criticism of Darwin's theory soon found its way to the so-called thick journals, most with relatively large circulations. In October 1864 the journal *Fatherland Notes* carried, in Russian translation, an article by Albert Kölliker, originally published in the German journal *Zeitschrift für wissenschaftliche Zoologie* earlier in the same year. Kölliker admired the richness of empirical material Darwin had collected, but took serious exception to the basic principles of his theory.[121] The ideas presented in this article quickly became a notable part of the general criticism of Darwinism. During the next three decades, the critics of Darwin's theory showed particular interest in endorsing and elaborating upon the two main criticisms built into Kölliker's opposition to Darwin. First, Kölliker thought that Darwin exaggerated the evolutionary role of the adaptation of organisms to the environment and of natural selection. "The basic idea of my hypothesis," he said, "is as follows: I recognize that all members of the organic world owed their existence to a grand plan of development, which pushes lower forms in the direction of increasing perfection." He admitted, however, that he could not explain how this plan actually worked. Second, Kölliker rejected Darwin's notion of organic transformation as a slow process; instead, he suggested that new species emerged only as a result of leaps in the evolutionary process.[122] He introduced the notion of heterogenesis. Kölliker went out of his way to remind his readers that his general theory of evolution did not assign man an exclusive position in relation to other animals. In his meditation about human origins, he pointed to gorillas, chimpanzees, and orangutans as animal species occupying the positions in the evolutionary scale that were nearest to that of man.

Karl von Baer, an eminent member of the St. Petersburg Academy of Sciences and the greatest embryologist of the pre-Darwinian era, belonged to the group of scientists who prepared the ground for the emergence of Darwin's evolutionary theory. His naturalist historicism had philosophical roots in epigenetic embryology, to which he had made contributions of lasting value. He could not help but recognize that some of his own most formidable contributions had gone into the making of the theory propounded in the *Origin of Species*.[123] It was no surprise that in 1860 von Baer wrote to Huxley: "J'ai énoncé les mêmes idées sur la transformation des types ou origine d'espèces que M. Darwin."[124] Von Baer noted, however, that his ideas on transformism were based on zoogeography. Nor was it a surprise that Darwin greeted von Baer's "approval" of his theory as "magnificent" news.[125] Soon, however, Darwin learned that von Baer had changed his mind and had allied himself with the leaders of the anti-Darwinian movement.

Huxley and Darwin had good reason to expect von Baer's endorsement of the evolutionary idea. They should not have been surprised, however, that he wasted no time in allying himself with the forces of anti-Darwinism. His evolutionary thought underwent constant shifts in interpretation and emphasis. Sometimes it assumed very broad proportions; at other times it was drastically limited in both scope and meaning. In his scientific orientation von Baer did not escape heavy philosophical considerations. In attacking Darwin's theory during the 1860s, he also attacked nihilist materialism and its war on traditional values. To make the situation even more complicated, von Baer wavered in his tireless search for a middle ground between the two extreme views of evolution. At one extreme was the notion of the universal development of nature, embodied in Schelling's *Naturphilosophie,* which cast evolution as a gradual expansion of the power of spirit over matter and favored Aristotelian teleology over Newtonian causality, at least in the interpretation of the organic world. At the other extreme were the new scientific ideas coming from embryology, paleontology, and several other natural sciences, which invited broad causal-mechanistic interpretations. A strong allegiance to the spirit of *Naturphilosophie* prevented von Baer from making these two orientations integral parts of a logically coherent and functional theory; moreover, after he became acquainted with the *Origin of Species,* his speeches and papers showed an increased reliance on the antimechanistic interpretation of evolution.

In his attacks on Darwin's theory, von Baer did not limit himself to the idea of transformism. He bitterly opposed the "materialistic" orientation of modern natural science, which received powerful support from Darwin's theory. In a paper read in 1861, on the occasion of the opening of the Entomological Society in St. Petersburg, von Baer lamented the current popularity of scientific materialism and argued that a science grounded in idealistic ontology would give a much more complete picture of the universe.[126] The basic weakness of materialistic science, he argued, was that it did not—and could not—account for a "spiritual" element in the processes of nature. During the early 1860s, von Baer wavered in his views on organic evolution. For example, in his article "Anthropology," written for the *Encyclopedic Dictionary,* published in 1862, he viewed the brain of anthropoids as occupying an intermediate position between the cerebral cortex of man and that of the higher animals. The article echoed the spirit of transformism, allowing no room for creationist ideas.[127] It made reference to "man and other animals." Since the essay made no allusions to Darwinian transformism, it was most probably written before von Baer had read the *Origin of Species.* In all his subsequent writings concerned with the problems of evolution directly or indirectly, he condemned Darwin's theory on both scientific and moral grounds. In von Baer's view, natural selection, Darwin's principal concept, did not have a basis in empirical data, ignored the role of heterogenesis (as formulated by Kölliker in 1864) in organic transformation, disregarded purposiveness in the processes of living nature, and advanced the "unsupportable thesis" of the anthropoid origin of man. Von Baer found the latter component of Darwin's theory most irritating. Arousing his philosophical and ethical sensitivities, it compelled him to lecture against the new evolutionary idea.

In 1865–67 the journal *Naturalist*, devoted to the popularization of science, carried in numerous installments a Russian translation of von Baer's rambling essay "The Place of Man in Nature," directed mainly against Thomas Huxley's extension of Darwin's evolutionary theory to include man. Too cumbersome to appeal to the general reader, the translation contained many naive and sweeping statements inserted by the translator.[128] In 1868 von Baer repeated the same arguments in a popular lecture at Dorpat University. At this time, however, his anti-Darwinian campaign had only begun to unfold. During the 1870s, particularly after the publication of *The Descent of Man* in 1871, his attack on Darwinism became both more comprehensive and more uncompromising.

CHAPTER TWO

New Horizons

The Descent of Man and
The Expression of the Emotions

At the very beginning of the 1870s Darwin published two major works: *The Descent of Man* and *The Expression of the Emotions in Man and Animals*. These studies helped inaugurate a new phase in the history of evolutionary studies. They carried the transformist orientation to new domains of inquiry, particularly to the study of man as a species and a culture creator. During the 1870s jurisprudence, political science, sociology, linguistics, and anthropology became fully crystallized disciplines riveted to an evolutionary view of nature and society. These studies were also responsible for the emergence of a growing interest in the psychological parameters of organic evolution. The spread of the idea of evolution to previously unexplored domains did not slow down the rapid growth of evolutionary paleontology, embryology, and morphology, the disciplines that provided the first response to the fertilizing influence of the *Origin of Species*. The *Origin* inaugurated Darwinism as a turning point in nineteenth-century biology; *The Descent* and *The Expression of the Emotions* helped make Darwinism a broad system of research designs, theoretical principles, and philosophical outlooks. The *Origin* was strictly a work in biology; *The Descent* and *The Expressions of the Emotions* built a bridge between biology, the social sciences, and the humanities.

The Descent of Man showed that the process of organic evolution, propelled by the struggle for existence and natural selection, applied to man no less than to the rest of the animal kingdom. It gave explicit recognition to the idea of the anthropoid origin of man. This claim surprised no one, for it was clearly hinted at in the great work of 1859 and was elaborated in Thomas Huxley's *Man's Place in Nature* and Vogt's *Lectures on Man*. Nor was it much of a surprise when three Russian translations of *The Descent* appeared within one year after the publication of the English original. Two general ideas represented the essence of *The Descent:* natural selection is not only behind the physical survival of man but also behind the evolution of cultural values; and the differences between animal and human behavior are differences of degree rather than of kind.

The *Expression* helped lay the foundations for a scientific study of the psychological aspect of the evolution of species. The book appeared in a Russian translation only a few months after the publication of the English original. The paleontologist Vladimir Kovalevskii was the translator, and the embryologist Aleksandr Kovalevskii was in charge of editorial tasks. In 1874 Vladimir wrote to Darwin that nearly two thousand copies of the Russian translation were sold.[1] An abridged version of *The Expression* also appeared in a popular edition.

The *Expression* deals much more extensively with selected aspects of human and animal behavior than with general problems of evolutionary biology. The Russian reviewers were generally impressed with Darwin's descriptions and categorizations of animal behavior. The *Journal of the Ministry of Public Education* was unusually profuse in praising the book's content and writing style. The reviewer commended Darwin's impartiality and avoidance of "materialistic trappings." Even the adherents of spiritualism could read the book, he wrote, without the least discomfort. The reviewer thought that psychologists would benefit from the information the book presented on the "physiological" basis of behavior.[2] Indeed, he recommended the book to all readers interested in the scientific foundations of human behavior. The liberal journal *Knowledge* was equally laudatory. It noted that the book was eminently successful on two counts: it offered a "rational explanation" of many expressions of human emotions, and it integrated the study of animal and human behavior into the universal process of organic evolution. In fact, no educated person could afford to ignore it.[3]

N. P. Vagner, professor of zoology and comparative anatomy at St. Petersburg University, called *The Expression* a book with "great strengths and minor flaws."[4] The volume reminded him of Darwin's previous works, which marked "turning points in the history of science." The strength of the book lay much more in its suggestion of new topics for comparative-psychological research than in a presentation of a theoretically and logically integrated system of scientific thought. Insufficient exploration of the physiological underpinnings of mental activities represented the book's major shortcoming. Vagner thought that Darwin would have benefited from a familiarity with current efforts in Germany to advance a "mechanical theory" of behavior, relying on physicochemical analysis of mental activities. He considered I. M. Sechenov a strong representative of this orientation.

This chapter deals with the growth of Darwinian thought in Russia during the period from the publication of *The Descent of Man* in 1871 to the appearance of *Darwinism* by N. Ia. Danilevskii in 1885. It highlights the Russian scientific community's general recognition of the revolutionary sweep of Darwin's scientific ideas and philosophical inferences, the influence of Darwin's theory on the development of individual sciences, and the emergence of a Lamarckian tradition in Russia. It also comments on a sample of writings commemorating Darwin's death in 1882.

Darwin's Contributions: A General View

A typical evolutionist of this era was preoccupied with marshaling empirical data and logical arguments bolstering the foundations of Darwinian principles. Evolutionary biologists were no longer interested in establishing whether Darwin was correct or not; their major aim was to show how evolutionary forces actually worked in the parts of nature covered by individual life sciences. Only isolated scholars worked on refining and amplifying the theoretical principles of Darwin's evolutionary thesis; most were engaged in phylogenetic studies, usually centered on the invertebrates, and were guided more by a general idea of evolution than by Darwin's theoretical specifics. This period did not excel in grand theoretical formulations; its real strengths were in empirical research and derivative theory. It was also a period of carrying Darwin's ideas to the rapidly expanding study of man, particularly of physical anthropology, human geography, ethnography, and evolutionary sociology. A contributor to the *Fatherland Notes* observed in 1872 that Darwin belonged to the small group of eminent scholars, typified by Helmholtz, Tyndall, Huxley, and Virchow, whose contributions to the philosophical thought of their time were as formidable as were their contributions to science.[5]

In the scientific community this was a period of great excitement and prodigious expectations. The biologists worked not only on empirical research imbued with the evolutionary spirit, but also on carving out new domains of living reality that required systematic study. In 1874, A. S. Famintsyn, a young professor at St. Petersburg University, drew a generalized picture of the Darwinian legacy:

> The problems Darwin has investigated are among the most crucial and most interesting problems in biology. Regardless of our specific responses to it, his theory has influenced our views on the surrounding nature and on our position among and relations to other living forms.
>
> Neither his subject of inquiry nor his final conclusions have made Darwin a universally famous person. He was not the first scholar either to raise the question of evolution or to explain the transformation of lower plants and animals into higher plants and animals.
>
> Darwin's main contribution has been in the employment of a rigorous research method, which has enabled him to initiate the study of—if not to give full answers to—some of the most difficult and interesting questions in biology.
>
> Darwin's most notable contributions to science are these:
>
> (1) He has been the first person to adduce definitive proofs in favor of the extensive plasticity of organisms, which, in some cases, has led to blurring the boundaries separating one species from another.
>
> (2) He has explained the factors responsible for this phenomenon. He has brought together an ocean of facts and observations on the transmutability of organisms and on the origin of new species. He has added a massive assortment of new empirical facts supporting the idea of transmutation.
>
> (3) His study of artificial selection has directed the attention of biologists to the role of natural selection in the transmutation of plants and animals, as well as to the role of sexual selection among animals.
>
> (4) He has been the first to study changes in the human species caused by natural or artificial selection.
>
> (5) Finally, in his more recent writings he has explained the true meaning of natural selection: he has shown that natural selection, taken by itself, does not supply sufficient proof for genetic links between lower and higher plants and animals.
>
> Only a combination of love for work, scholarly erudition, and powerful thought can make the labor of a scholar a fruitful contribution to science. Darwin possesses these qualities to the highest degree.[6]

In one respect, however, Famintsyn showed extreme caution: he stood firmly on the position that "at the present time" the question of the origin of man continued to be without a scientifically valid answer. He based his statement on two arguments. First, he thought that there was no sufficient empirical base for viewing the transformation of lower animals to higher animals as an indisputable scientific fact. Second, he was convinced that the material presented by Darwin invited contradictory interpretations. Despite these cautions, he chose not to challenge the idea of the anthropoid origin of man, nor did he question the correctness of Darwin's approach. He wanted to protect the tranquillity of scientific work by avoiding a bitter confrontation with the fast-growing ranks of anti-Darwinists outside the scientific community.

Darwin's theory found an eloquent defender in Ia. A. Borzenkov, professor of comparative anatomy at Moscow University. In 1881 he delivered the annual university lecture at Moscow, selecting for his topic "a historical survey of the main orientations in nineteenth-century biology." The sixty-page essay represented the first Russian effort to analyze the development of modern biological thought. It offered a pertinent analysis of the role of Oken's and Schelling's *Naturphilosophie* in the advancement of two paramount ideas of modern biology: the unity of nature and biological historicism, both of which contributed to the Darwinian triumph of evolutionary thought. Nor did Borzenkov overlook the negative aspects of *Naturphilosophie*, particularly its idealistic base, antiempirical stance, and unwarranted emphasis on introspection.[7]

While recognizing Lamarck as Darwin's most eminent forerunner, Borzenkov opposed the idea of hereditary transmission of characteristics acquired by the use or disuse of organs. The discussion of Darwin's contributions made up the core section of the speech. Carefully and with notable clarity he explained the main principles of Darwin's theory. He left no doubt about his firm belief in the fundamental correctness of the new evolutionary idea. In Darwin's theory he saw a sound basis for work on a universal genealogy of living forms; but he warned against premature efforts to construct genealogical trees. In his view, Haeckel's genealogical tree was a construction with no empirical support.

For tactical reasons, Borzenkov took the position that Darwin's ideas did not represent an established scientific theory, but grew primarily upon a hypothesis. He noted, however, that Darwin's "hypothesis" can be regarded as "completely scientific," a claim for which he advanced three complementary explanations. First, it rested "not on arbitrary propositions but on true facts, some of which can be readily verified by observation and experiment." The verifiable propositions forming the very core of Darwin's theory are the primacy of heredity in the evolutionary process, the inevitable occurrence of variation in transition from one generation of plant and animal forms to another, and the universality of the struggle for existence as the prime mover of the evolutionary process. Second, it meets the standards of the logic of scientific inquiry. Third, it can readily explain all the facts that have been accumulated by zoology and botany: "Neither of the two sciences contains a single fact contradicting this theory." [8]

In 1881 the journal *Mysl'* gave its support to the interpretation of Darwin's theory as a "hypothesis," built on "a solid base of scientific facts and laws." [9] Like Famintsyn and Borzenkov, the author of this article meant to say that Darwin came closer to giving a full scientific explanation of organic evolution than any other scholar or group of scholars. To make his statement more appealing, he also noted that "the struggle for existence" did not refer to "struggle" in the ordinary usage of the word. In Darwin's view, he said, it stood "not for a struggle among men or among animals, but for a struggle between the best and most efficient characteristics and the weak, ungainly, and disadvantageous characteristics." This article, very much like Borzenkov's study, gave a clear picture of the strategy of Darwinian scholars during the period of growing political oppression. This strategy called for two modes of operation. On the one hand, it expressed a somewhat reserved attitude toward the claims that Darwin's theory had received full empirical support. On the other hand, it interpreted the key notions of Darwin's theory in such a way as to make them less discordant with autocratic values. To placate the censorship authorities, Borzenkov categorized Darwin's theory as a "hypothesis" and interpreted the struggle for existence in a way that allowed little room for social conflict and radical ideological involvement.

D. N. Anuchin, known for his effort to make anthropology a modern scientific discipline, put the finishing touches on the edifice of Darwinian accolades. He had no difficulty in identifying the study of organic evolution as the most important contribution to revolutionary changes in the intellectual fabric of contemporary science. Expressing a general view that summed up the reigning sentiment in the Russian scientific community, he wrote:

> In every branch of contemporary science we sense a striving to identify stages in the development of natural forms, to study transitions from one form to another, to search for origins of plants and animals, and to describe the conditions responsible for the diversity of life. This orientation dominates not only the natural sciences but also the humanities—history, philology, psychology, jurisprudence, and political economy. In all these branches of knowledge there is a strong feeling that only a careful study of the causes and stages of evolution can lead to a full understanding of natural phenomena and their mutual relations. [10]

Evolution and Embryology

The pro-Darwinian rhetoric served more to express an attitude than to sum up facts; it achieved more as a statement of faith than as a record of scientific achievement. During the 1870s, however, Darwinian science grew rapidly in Russia. It found a home in many life sciences, where it helped create a penetrating unity of interest and a common research strategy.

With a long and distinguished tradition in Russia, embryology gave Darwin's theory a particularly wholesome welcome and a strong position from which to operate. The fact that Darwin made a direct reference to Aleksandr Kovalevskii's research accomplishments helped bestow a special glamour on embryology and make it particularly appealing to the new generation of Russian scientists. Extensive references to Kovalevskii's and Mechnikov's work in Germany were also an important factor adding to the popularity of embryological research. Russian scholars worked mainly in invertebrate embryology; perhaps they wanted to escape the attention of ideological guardians who were particularly eager to scrutinize the evolutionary studies of vertebrates, more likely to infer, or to state directly, the animal origin of man.

During the 1870s Kovalevskii and Mechnikov continued to dominate embryological research—and to serve as leaders of comparative studies based on evolutionary principles. The most outstanding products of this period were Kovalevskii's study of selected annelids and insects and Mechnikov's work on sponges and coelenterates, both showing that the germ layers of invertebrates and vertebrates are homologous—and, therefore, adding a strong embryological proof for the unity of animal forms.[11] They extended the meaning of "homology," making it not only a morphological (structural) but also an embryological (historical) notion. At this time, only the relatively sparse ranks of biologists welcomed the idea of the embryonic unity of invertebrates and vertebrates.

Kovalevskii's lengthy monograph on the embryology of various annelids and arthropods was easily the most outstanding embryological study to come out of Russia during the 1870s. In addition to establishing the presence of three germ layers in annelids and insects, it showed that among some marine annelids the separation of germ layers comes, not through the process of invagination, as among terrestrial annelids, lancelets, and ascidians, but through a process whereby fast-propagating small cells crowd out larger cells. Kovalevskii was sure to note that the differences in the two processes were differences of degree rather than of kind.[12]

The embryological study of annelids led Kovalevskii to two general conclusions. First, he made it known that the embryonic development of annelids he had described agreed, even to the smallest detail, with the embryonic development of the vertebrates. "In both groups the middle layer appears only later . . . and the basic components of organs agree to an extraordinary degree even down to individual processes."[13] Second, he contended that only by extending the study of homologies beyond the limits of types as these were defined by Cuvier, Agassiz, and von Baer could embryology become a true science. Not the treatment of individual types as natural entities sui generis, but the study of similarity between different types can serve as a key step toward making comparative anatomy and embryology full-fledged sciences. "On the basis of all the evidence I conclude that the organs of the animals of different types are homologous."[14]

Kovalevskii showed little interest in tying embryology to phylogenetic studies aimed at producing a genealogical tree of animal genera and species. He was satisfied merely to note the degree of similarity between various animal forms and to conclude that the greater the similarity between the embryonic development of compared animals, the greater their phylogenetic affinity.[15] The study of the emergence and development of germ layers led Karl von Baer to establish the embryological unity of the vertebrates. The same kind of study led Kovalevskii to postulate the embryological unity of the vertebrates and invertebrates—and to confirm Darwin's idea of a possible common ancestor of all animals. But he did much more than extend von Baer's discovery to the invertebrates: he produced a long series of empirical facts that gave comparative embryology richer substance and more precise methodology.

The study of selected annelids and arthropods gave Kovalevskii an opportunity to add up—and to state explicitly—all the modern arguments in favor of an evolutionary orientation in embryology. Implicitly, his main arguments were directed against von Baer's claim that homologies did not extend beyond the limits of Cuvier's individual types of animals. In a paradoxical twist of history, this study earned him the second Karl von Baer Prize given by the St. Petersburg Academy of Sciences for outstanding studies in biology by Russian scholars. Two years after the publication of the study, von Baer staged a lengthy attack on Kovalevskii's—and Carl Kupffer's—earlier studies on the ascidians which

Darwin regarded as having provided a proof for the biological unity of
the vertebrates and invertebrates.[16] Instead of launching a general attack
on the idea of the unity of the vertebrates and invertebrates, von Baer
concentrated exclusively on pointing out the flaws in Kovalevskii's and
Kupffer's arguments in favor of viewing the ascidians as a link between
the two grand divisions of the animal world. Mindful of von Baer's
scholarly stature and advanced age, Kovalevskii chose not to defend
himself publicly.[17] This did not mean, however, that he was not unhappy
with von Baer's criticism; in a letter to Mechnikov he stated that von
Baer relied on outdated arguments, which showed only that "time has
gone ahead of him."[18]

Kovalevskii encountered criticism from Russian embryologists who
considered themselves evolutionists of an unorthodox variety. While
ready to recognize the vast scope of Darwin's scientific achievement,
they openly voiced criticism of individual propositions of Darwin's evolutionary thought. For years Mechnikov rarely missed the opportunity
to criticize Darwin's method and generalizations. He concentrated his
heaviest attacks on the pro-Darwinian conclusions of Kovalevskii's
studies of ascidians, a field in which he had also worked, drawing quite
different conclusions. According to V. A. Dogel', the ascidian conflict
lasted from 1866 to 1871 and was expressed primarily in a lively correspondence between the two scientists.[19] Kovalevskii not only clung to
his original arguments but also sharpened them by renewed studies of
the same animal. Reluctantly and quietly, Mechnikov finally accepted
Kovalevskii's evolutionary position. Perceptive and tactful, M. S. Ganin
was one of the more prominent critics of Kovalevskii's individual generalizations. For example, he did not always agree with Kovalevskii in
tracing the origin of specific complexes of organs to the same germ
layers. In answering critics, Kovalevskii showed remarkable tact and a
cultivated discipline, which helped create a healthy atmosphere for scientific communication.

Caspar Wolff, Christian Pander, and Karl von Baer, all Germans by
origin and cultural roots, made Russia a major contributor to embryology; Kovalevskii made embryology not only a Darwinian science
but also a Russian science—a preoccupation of Russian scholars. He
did not create a particularly strong school of followers, but, in a sense,
every Russian embryologist of the second half of the nineteenth century
was inspired by his achievement, ingenious research techniques, mode
of combining empirical analysis with theoretical concepts, and wholesome dedication to scientific scholarship. In V. M. Shimkevich's words,

he had few disciples but many followers. Through personal cooperation with individual Western embryologists he helped the Russian scientific community broaden the base of professional contacts with the outside world.[20] In the best spirit of "international cooperation" he conducted joint research projects and carried on an extensive correspondence with A. F. Marion, head of the Laboratory of Marine Zoology at the Marseilles Museum of Natural History.[21]

During the second half of the nineteenth century the embryologists, particularly in Germany, tended to specialize in relatively small and clearly delimited areas of empirical inquiry—an adaptation to the growing complexity and rigor of research procedures. Kovalevskii did not belong to this kind of tradition: his research spanned an unbelievably wide assortment of invertebrates and a rather limited assortment of vertebrates. He moved rapidly from species to species and from problem to problem. His engagement in embryological research lasted until 1887, when he undertook a series of studies of secretory organs among invertebrates.

Only on rare occasions did Kovalevskii present papers that did not deal directly with his research ventures. In 1871, at the Third Congress of Russian Naturalists and Physicians, he spoke on the role of organized national gatherings of scientists in widening the study of Russia's natural resources and human potential, in facilitating the diffusion of scientific knowledge, and in strengthening personal contact among the members of the scientific community.[22] At the Seventh Congress of Russian Naturalists and Physicians, held in Odessa in 1883, he delivered an *éloge* to Darwin, acknowledging the debt of the Russian scientific community to the modern evolutionary idea and its creator.[23]

Kovalevskii's brother Vladimir, an evolutionary paleontologist, exchanged many letters with Darwin. It was primarily through this correspondence that A. O. Kovalevskii maintained close personal contact with Darwin. Kovalevskii was particularly interested in keeping Darwin informed about his continued work on the links between the ascidians and the vertebrates.[24] In 1871, for example, he wrote to Darwin about his research on the common features of ascidian and vertebrate nervous systems.[25] Vladimir informed Darwin about his brother's involvement in the translation of *The Variation* and *The Expression* into Russian. On several occasions Vladimir transmitted to his brother Darwin's requests for clarification of specific biological problems.[26] Darwin praised the originality and great promise of the scientific work of the Kovalevskii brothers.

The embryological work of I. I. Mechnikov equaled that of A. O. Kovalevskii in both magnitude and originality. During the 1870s Mechnikov concentrated on the embryological study of lower marine animals, but he also conducted fruitful research on scorpions. By the middle of the decade his interest began to go in several directions, covering topics not only in embryology but also in physiology, sociology, and the history of science. In search of a general interpretation of the rapid growth of evolutionary biology, he did not hesitate to make pronouncements of a philosophical nature and to challenge the masters of grand theory. At this point in his career, he thought that the idea of evolution needed much empirical buttressing to become a guiding principle in biology, and that Darwin's theory, despite its limitations, offered richer and more promising scientific perspectives than any other evolutionary theory.

Mechnikov devoted much of his embryological research to testing the key assumptions related to the evolutionary theory. An embryological approach to the digestive systems of lower animals attracted much of his attention. His comparative study of echinoderms, medusae, and ctenophores provided information showing that the animals without body cavity and animals with body cavity are products of the same basic evolutionary development. This information, he claimed, provided strong support for the idea of the embryological unity of invertebrates and vertebrates. He rejected Haeckel's gastraea theory on the ground that it assumed the origin of multicellular organisms at a relatively high level of organic evolution. Instead, he proposed a hypothetical animal, which he named *parenchymella,* as a transitional phase between uni- and multicellular organisms. As he envisaged it, this animal evolved a mass of inwardly migrating cells.

The theory of the parenchymatous cell mass was too involved and too daring to make a strong impression on contemporary embryologists. In one respect, however, it was fruitful: it helped Mechnikov initiate one of the most innovative periods in his scientific career. It led him to postulate the widely heralded phagocytic theory, the basis for a strong orientation in comparative pathology.[27] The phagocytes, as he now chose to call migratory cells of mesodermal origin, perform a dual function: they are responsible for intracellular digestion, and they perform a "prophylactic function" by destroying atrophied organs and all kinds of pernicious matter coming from outside, such as infectious germs.[28] This theory led him to postulate the phagocytic theory of inflammation and immunity which he treated as the basis of comparative pathology.[29] In this endeavor he helped widen the Darwinian base of modern biology. In inflammation he found a powerful exhibition of the work of natural selection.

In the early 1880s Mechnikov's scientific interests took another turn. The search for the unity of the animal world, the foundation on which Darwin built his theory of evolution, convinced him that he should extend his research beyond the traditional limits of embryology. Embryology, he said, needed help from other approaches, for it was not fully equipped to study organs in their normal state: it studied organs either in an atrophied state or in the process of formation. He wrote in 1883:

> After many years of studying the genetics of Metazoa, I have become convinced that this problem cannot be solved by adhering strictly to a morphological-embryological approach. . . . Familiarity with physiological history is the essential prerequisite for a genealogical study of organs. For this reason, in my study of the genealogical development of the digestive apparatus I found it necessary to investigate both embryonic growth and physiological functions of the endoderm.[30]

It was fitting—and natural—that in 1884 an embryologist published the first Russian university textbook in zoology and that this book incorporated and propounded the Darwinian idea. The author was N. V. Bobretskii, a young professor at Kiev University and a disciple of A. O. Kovalevskii's. From this time on, zoological textbooks served as the most important vehicles for a systematic dissemination of Darwin's ideas at the university level. At the same time, they became the mirror of an unceasing effort to define the place of Darwinism vis-à-vis the swelling scientific and philosophical output of new biology, particularly of the experimental study of heredity as a factor of evolution. The comprehensive scope of its material and the unity of its theory made Bobretskii's text a winner of the annual prize awarded by Kiev University for the most outstanding publication by the local faculty. An effort to establish phylogenetic links between various groups of animals was the special feature of the book.[31]

V. V. Zalenskii (1847–1919) was the leading representative of a slowly growing group of scientists who carried on the national tradition in evolutionary embryology established by A. O. Kovalevskii and I. I. Mechnikov. During the 1870s he conducted a long series of embryonic and metamorphic studies of annelids, mollusks, and salpas (relatives of ascidians), all inspired by Darwin's theory. His two-volume work on the development of sterlets—small sturgeons living in the Caspian Sea and its rivers—was the first embryological study of this animal.[32] Zalenskii occupied a key position in what became known as the Russian embryological school, which spanned several generations and was united by a firm commitment to Darwin's evolutionary views and a primary interest in homologous features of invertebrate and vertebrate germ-layers and in the development of the mesoderm. Like Kovalevskii—and unlike Mechnikov—Zalenskii relied on evolution as a strategy for empirical research rather than as a topic of theoretical exploration. For a long time a professor at Kazan University, in 1897 he was elected a full member of the St. Petersburg Academy of Sciences.[33] His promotion represented a decisive step in giving Darwinian scholars a preeminent place among the Academy biologists.

Evolutionary Paleontology: Vladimir Onufrievich Kovalevskii

Vladimir Kovalevskii is rightfully counted among the first scientists to carry Darwinian evolutionism to paleontology—and geology in general—a field in which he achieved results comparable to the most signal accomplishments of his older brother, A. O. Kovalevskii, and of I. I. Mechnikov in embryology.[34]

Darwin, who had high praise for the embryological contributions of Aleksandr, thought that Vladimir's work in paleontology was even more significant. Darwin, of course, had in mind the contributions to the empirical base of the evolutionary theory. As a young man, before he chose paleontology as an academic specialty, Vladimir helped Darwin's ideas reach the ever-widening circles of Russia's enlightened public. He made preparations for the publication of a Russian translation of Darwin's *Variation of Animals and Plants*, and he was one of the chief translators of Darwin's *Expression of the Emotions in Man and Animals*. This time, too, the translation was rendered from galley proofs sent by Darwin.[35] Kovalevskii translated some and edited the rest of an entire series of scientific books that, while not necessarily dedicated to the problems of organic evolution, contributed to the advancement of the philosophy of science in which the *Origin of Species* was firmly rooted. Some translations, typified by T. H. Huxley's *Our Knowledge of the Causes of the Phenomena of Organic Nature* (1864), made a significant contribution to the popularization of Darwin's theory. During the same period Kovalevskii published the Russian translations of Charles Lyell's *Antiquity of Man*, Albert Kölliker's *Histology*, Karl Vogt's *Zoological Letters*, and A. E. Brehm's *Illustrated Life of Animals*.[36]

Kovalevskii devoted only four years—from 1870 to 1874—of his eventful and unsettled life to original scientific research. After intensive preparatory work in all the major earth sciences, geological fieldwork in the Permian formations in Thuringia and the cretaceous formations in England and southern France, and after careful study of ungulate fossils deposited in the leading western European museums, he undertook to reconstruct the main line in the evolution of the horse from the late Eocene period to the present time—from palaeotheres and paloplotheres to modern *Equus*. The same problem attracted the attention of T. H. Huxley, who in 1872 pointed out that the rich collections of ungulate fossils made it possible to reconstruct the evolution of the horse through a long period of geological time. While Kovalevskii did not produce a long list of publications, the results of his findings marked a turning point in the development of paleontology from a closed science encumbered by the strong tradition of Georges Cuvier's and Louis Agassiz's antitransmutationism to an open science guided by the evolutionary principle.

The high quality of Kovalevskii's memoirs ensured their publication in the journals of such prestigious learned associations as the St. Petersburg Academy of Sciences and the Royal Society of London. One memoir was published in *Palaeontographica,* edited by Duncker and Zittel. Koelner—in the *Geological History of Mammals*—placed Kovalevskii's studies among the "standard works" corroborating Darwin's evolutionary laws. The French paleontologist Albert Gaudry proclaimed Kovalevskii's essays "one of the basic sources for the study of the evolution of ungulates." [37]

From the very beginning of his involvement in paleontological research, Kovalevskii showed a strong inclination to adhere to an evolutionary approach. In a letter to his brother in 1870 he noted that the purpose of a scientific study of fossils should be "to describe the course of the history of nature, to study causes of the transformation of species, and to unveil the path this transformation has followed." [38] This was at the time when a large number of established paleontologists continued to resent and resist the rise of Darwin's evolutionary ideas.

Kovalevskii concentrated on modifications in the skeletal structure of horses as a result of morphological adaptations of various organs to changing environmental conditions. He not only studied fossil genera of ungulates but also succeeded in arranging them in a linear series according to their geological age. The series showed progressive increases in size, reduction of the digits to a single toe, modifications in skull proportions, and gradual enlargement and functional differentiation of teeth.[39] In the opinion of a modern expert, Kovalevskii "made out an essentially correct story of the mechanical evolution of the horse's foot and dentition," even though he studied the fossils representing genera ancestral to *Equus*.[40] Particularly in the evolution of the feet of the ungulates—characterized by increasing morphological simplification and reduction—Kovalevskii adduced formidable evidence supporting natural selection as a mechanism of variation leading to the emergence of new species.

Kovalevskii found out that changes in environmental conditions produced two types of skeletal modifications in feet: "adaptive" and "inadaptive." "Adaptive" modifications represented more drastic deviations from established forms and occurred much less frequently than "inadaptive" modifications. He viewed evolution as a process of many phases and many lines: evolution means the survival of animal forms with adaptive morphological features and the extinction of forms with inadaptive features. There is no direct and one-sided relationship between morphological specialization and progress; in some cases specialization and progress are synonymous; in other cases a trend opposite to specialization is equal to progress. Among the ungulates the evolutionary simplification of the skeleton was a sign of progress, that is, of better adaptation to the rigors of the environment.[41] Kovalevskii was also firmly convinced that without close scrutiny of aberrant forms there could be no meaningful understanding of the full scope of the evolutionary process.

The American paleontologist Henry Fairchild Osborn gave the following appraisal of the quality, historical meaning, and methodological bent of Kovalevskii's work:

> The remarkable memoirs of Vladimir Onufrievich Kovalevskii (1842–1883) . . . are monuments of exact observation of the details of evolutionary change in the skull, teeth, and feet, and of the appreciation of Darwinism. In the most important of these memoirs, entitled *Versuch einer natürlichen Classification der Fossilen Hufthiere* (1875), we find a model union of detailed inductive study with theory and working hypothesis. These works swept aside the dry traditional fossil lore which had been accumulating in France and Germany. They breathed the new spirit of recognition of the struggle for existence, of adaptation and descent.[42]

During the early phase of evolutionary studies, Kovalevskii had had more personal contact with Darwin than any other Russian scientist.[43] He made several visits to Darwin's home in Down, and the two exchanged occasional letters. In a letter that has been preserved, Darwin expressed high hopes for the promising scientific contributions of the Kovalevskii brothers and informed Vladimir that his paleontological ideas had found influential supporters in England. He was pleased to hear from Vladimir that he had decided to dedicate his forthcoming monograph on *Anthracotherium* to him.[44]

Kovalevskii published eight monographs on the paleontology of ungulates, one of them over three hundred pages in length. Most essays consisted of two parts: an introductory section stating the general views on the importance, complexity, and challenges of undertaken studies, and a descriptive section, staying close to the empirical material. The purpose of introductory statements was to help the reader learn about the general problems of paleontology. Kovalevskii relied on introductory remarks to spell out the evolutionary bent of his own orientation. In dedicating his monograph on *Anthracotherium* to Darwin, he stated that new paleontology should be devoted to a theoretical and empirical elaboration of the basic principles enunciated by the founder of modern evolutionism.[45] Under Darwin's influence, Kovalevskii's paleontological work displayed three distinct features: a consistent search for evolutionary significance of the study of fossils; a strong emphasis on the biological side of the paleontological equation; and a carefully cultivated concern with the paramount role of the environment in the evolution of living forms.

A. A. Borisiak gave a true summary of Kovalevskii's paleontological work when he stated: "Instead of the usual characterizations of species, the descriptive part of Kovalevskii's monographs concentrates on a comparative study of the given genus: in his works every bone is treated as part of a living organism in motion and in the process of feeding. Every form or component part of the organism is related to its function in the process of adaptation to the surrounding environment."[46]

Daniel P. Todes has offered a general appraisal of Kovalevskii's allegiance to Darwin's scientific thought:

> For Kovalevskii, Darwinism was a synthesis, breaking down artificial boundaries between species, time periods, and disciplines. It allowed paleontologists to meaningfully classify and interpret fossil forms and had precipitated "a revolution in the evolution of mammals [in which] the excellent works of Gaudry, Rütimeyer, and Huxley laid the foundations of a new era in the science of fossils." But the revolutionary consequences of Darwinism were, in Kovalevskii's view, still more profound. Natural selection did not operate on fossils, but on living organisms in a specific environment. The paleontologist, therefore, was required to employ the theoretical perspective

of Darwinism and the techniques of the zoologist, comparative anatomist, and geologist to ascertain the demands of environments and their effect on the evolution of life forms. Kovalevskii could never have agreed with Gaudry that paleontologists should leave the study of process to life scientists, for he believed that, with the publication of the *Origin of Species,* paleontology itself had become that branch of the life sciences charged with explaining the significance of fossils.[47]

Kovalevskii did not fare well in academic employment; indeed, for a long time all his efforts to secure a regular teaching position came to naught. It is not difficult to identify the main reasons for his misfortune. The authorities did not forget that he published an entire series of translations of Western scientific and semiscientific works with "materialistic" leanings. His undergraduate training was in law rather than in natural science. The Ministry of Public Education neither sponsored nor supervised his study abroad. He had failed his examination for the degree of magister of science at Odessa University; he was the victim of personal revenge by an examiner—a geology professor whose scholarly ethics he had seriously challenged a few years earlier. He echoed the nihilist motto that science should work not only toward understanding nature but also toward ushering in a better society, free of feudal vestiges. He was a paleontologist at a time when the university curriculum did not have a clearly defined slot for paleontology. Like his brother and Mechnikov, he encountered formidable opposition from men occupying influential positions in the scientific establishment, who had achieved scholarly eminence by their work in nontransformist biology and paleontology of the pre-Darwinian era. Upon his return to Russia at the end of the summer of 1874, he complained that "no one understands my work and cannot even read it. . . . In general, I find the life here unbearable and I bitterly resent my decision to return from abroad, since I see no prospect of employment here."[48] Mechnikov, who during the 1870s wrote extensively about the history of the evolutionary idea, ignored Kovalevskii's work altogether. Timiriazev, the great advocate of Darwinism, was equally quiet.

In 1876 Kovalevskii received recognition and praise from an unexpected source: Karl von Baer. In the second volume of his *Addresses,* von Baer discussed, one by one, the salient points of Kovalevskii's major studies. He stated directly that if the study of transmutation was to gain solid empirical grounding it would do well by following Kovalevskii's manner of "careful, assiduous, and cautious" research. This, however, did not mean that he considered Kovalevskii's evolutionary interpretations unchallengeable.[49]

In 1880 Kovalevskii was appointed a lecturer at Moscow University, for which he received a warm congratulatory note from Darwin.[50] The much belated recognition did little to mend his broken spirit. In 1883 he committed suicide. D. N. Anuchin, a much respected geographer and ethnographer, wrote the only obituary for Kovalevskii in the Russian language. Published in 1883, this was also the first discussion in Russian of Kovalevskii's scientific achievements and of the reception of his ideas in the West.[51] Slow in receiving recognition from the scientific community, Kovalevskii did not exercise a strong influence on the national interest in paleontological studies. In 1881 S. N. Nikitin wrote the first Russian survey of post-Darwin developments in paleontology; he did not refer either to Kovalevskii or to any other Russian paleontologist, making it clear that he was convinced that the scientific study of fossils did not belong among the more advanced, and more attractive, national engagements in science. In 1886 Nikitin wrote favorably about Kovalevskii's scientific work, but his reference was limited to "geological" rather than to "paleontological" matters. He referred specifically to Kovalevskii's original thoughts on the Western, Mediterranean, and Russian Jurassic provinces as independent geological creations.[52]

In 1882 Nikitin was appointed senior research associate of the newly founded Geological Committee. His work covered many areas of geological research; in paleontology he was particularly known for his studies of ammonites of the group *Amalth funiferus,* which he approached from clear but unelaborated Darwinian positions. In recognition of his work, the St. Petersburg Academy of Sciences made him a corresponding member in 1902. To Nikitin goes the credit for the first Russian survey of the work and theoretical and methodological ideas advanced by Neumayr, Zittel, Laube, Hyatt, Waagen, and other western European and American leaders in evolutionary paleontology. Nor did the ideas of such known antievolutionists in paleontology as Barnard, Davidson, Heer, and Fuchs escape his attention. He also supplied a detailed analysis of Darwin's treatment of paleontological material. His general orientation rested on two premises: the incompleteness of the fossil record makes it impossible to reconstruct the entire genealogical tree of living forms; and, despite all the handicaps, the fossil record has produced surprisingly rich results and has become the most successful "method" for the study of organic evolution.[53] Nikitin himself provided interesting thoughts on "transitional forms" and on the role of fossils in geological periodization.

Darwinian Theory in Anthropology, Physiology, and Comparative Anatomy

Darwin wrote that embryology, paleontology, and comparative anatomy are the three sciences that would pass the decisive judgment on the scientific underpinnings of his theory. He also showed—particularly in *The Descent of Man*—that the evolutionary theory would provide useful tools for a scientific study of man as a biological and a cultural phenomenon. Russian scientists and popular writers were quick to grasp the paramount role of Darwinian thought in the development of modern physical anthropology and ethnography. A contributor to the journal *Russian Thought* wrote in 1884 in reference to new developments in anthropology, ethnology, psychology, history, economics, and jurisprudence:

> For some of these sciences the emergence of the evolutionary theory coincided with the beginning of their renaissance, or, in any case, of their rapid growth. Indeed, this development had not been evoked only by Darwin's theory but also by many other favorable conditions; it must be admitted, however, that without the idea of gradual change the analysis of many phenomena of human life would not have gone so far and to such depths, and would not have produced such interesting results.[54]

A. P. Bogdanov, who worked in several areas of biology, deserved the main credit for carrying the idea of evolution to the expanding field of anthropology, a discipline that, from the beginning, supplied fertile ideas of both a scientific and an ideological nature.[55] Anthropology, he said, made evolution a social force of large magnitude.[56] He saw Darwinism not only as one of the main tools for scientific study of the species *homo sapiens* but also as a fountain of significant and impressive guideposts for the organization of anthropological museums, the most effective displays of the biological indicators of human evolution. The Anthropological Museum of the Academy of Sciences in St. Petersburg and the Anthropological Museum of Moscow University, both relying on the evolutionary idea in organizing displays of specimens, contributed to the spread of Darwinian ideas far beyond the scientific community.

In 1876 the authorities of Moscow University, swayed by Bogdanov's arguments and enthusiasm, made anthropology part of the regular curriculum. Bogdanov's student D. N. Anuchin, the first Russian professor of anthropology, went far beyond his teacher in giving the new discipline a firm footing in Russian scholarship and in linking it with archeology and ethnography.[57] A consistent and thorough Darwinist, he wrote the first Russian study of the anthropoid origin of man. This work led him to a study of human races, consistently defending the thesis of the common origin of mankind. Infusing theory into ethnography, he treated evolution as a vital scientific link between "natural-historical" and "humanistic" approaches to the growth of human society.

In a general survey of current developments in anthropology and ethnography, published in *Russian Thought* in 1884, Anuchin noted that modern research had corroborated Darwin's claim that the human species was a product of gradual change over a long period of time. He noted that not only Darwin but also Huxley, Haeckel, Vogt, and Wallace had established man's biological relation to primates. He admitted, however, that there was no direct evidence in favor of this claim: no existing primate can be considered a direct ancestor of man; and paleontology had not yet discovered fossils representing an immediate link between man and ape.[58] This did not deter him from relying on such indirect evidence as isolated cases of anatomical and mental "retrogressions" to the "stages" of human development that featured pronounced anthropoid characteristics. Nor did he hesitate to make use of Lombroso's claim that in some instances the "anthropoid" features of various "criminal types" represented a recurrence of man's earlier physical features. He expressed a generally optimistic view about the future of paleontological discoveries as a possible source of anthropological information confirming Darwin's theory of the descent of man. Anuchin echoed the popular opinion that, despite their key role in the accumulation of modifications leading to the origin of species, the struggle for existence and natural selection could not explain the full sweep of the evolutionary process.

Dominated by an evolutionary perspective, anthropology did not have smooth sailing. The university charter of 1884 eliminated it from the curriculum, a penalty for its close association with Darwin's heresy. The same charter established geography as part of the university curriculum. As professor of geography at Moscow University, Anuchin played a major role in strengthening the ethnographic and anthropological parameters of the much-expanded geographical exploration in Central Asia, carried out under the auspices of the Russian Geographical Society and various naturalist associations. As a physical an-

thropologist—whether he worked in paleoanthropology or in racial anthropology—Anuchin maintained an evolutionary position with great
clarity and persistence. As an ethnographer he made a deliberate effort
to avoid the grand schemes of cultural and social evolutionism as typically advanced by Herbert Spencer and Lewis H. Morgan. He preferred
the concreteness of the historical method. In general, however, he is remembered more as a popular champion of the evolutionary view in anthropology and ethnography than as an original contributor to the
body of scientific theory.

Despite the compounded difficulties in advancing the sciences of man,
the annual report of the St. Petersburg Academy of Sciences in 1878 was
generally optimistic. It noted that during the last ten years, anthropology had developed "more swiftly than any other science."[59] The report
stated that the rapid accumulation of exact knowledge on the presence
of human fossils in the relatively old strata of the earth provided a particularly strong impetus for the development of anthropology. The
study of human fossils, in turn, encouraged an "anthropological and
ethnographical study" of present-day "tribes and national groups." The
function of anthropology, as the Academy report saw it, was not only to
describe human fossils and the present-day "remnants" of primordial
man but also to discover the main stages in the universal evolution of
man as a species and the creator of culture. The optimistic tone of the
report reflected the general, rather than specifically Russian, situation in
anthropology and related disciplines. It was also a bright look into the
future rather than a realistic summation of past achievement.

Russian scientists did not show a particularly strong interest in exploring the physiological parameters of organic evolution. This did not
mean, however, that Ivan Sechenov, the founder of a strong national
tradition in physiology, did not have a healthy respect for Darwin and
his biological ideas. A translator of *The Descent of Man* into Russian,
Sechenov readily acknowledged Darwin's major contribution to the
study of the relationship between man and animal and to the linking of
"variation" to the "struggle for existence." He claimed that natural selection, "one of the most fruitful and brilliant hypotheses," was supported
by information from everyday experience. He even noted that a great
majority of physiologists had accepted Darwin's theory of the origin of
species as a framework for the study of plant and animal behavior.[60]

Actually, Sechenov's enthusiasm for Darwin's theory was more philosophical than scientific. While expressing much respect for Spencer and
his effort to transpose Darwin's biological principles of evolution to psy-

chology, the idea of evolution remained generally outside his research vision.[61] He referred to Darwin much more in his popular and polemical papers in which he defined the philosophical premises of his world view than in his essays related to empirical research in physiology. Inspired by "evolution" as a philosophical idea much more than as a scientific subject, Sechenov leaned on Spencer more extensively than on Darwin.[62] There were two reasons for Sechenov's hesitation to make physiology a Darwinian discipline. In the first place, he was far more concerned with introducing the methods of physics and chemistry in physiology than with designing independent biological methods. At this time, to emphasize physicochemical analysis of physiological processes meant to find very little use for Darwinian evolutionary ideas.[63] In the second place, the wedding of physiology to evolution in Western scientific scholarship was still in a much underdeveloped state. Kliment Timiriazev was close to the crux of the matter when he stated a decade or so later that it was much easier to tie the vast quantities of information assembled by morphology in the nineteenth century to a universal evolutionary approach to living nature than to make physiology an evolutionary discipline.[64]

The combination of philosophical commitment to evolutionary philosophy and scientific involvement in the study of the physicochemical underpinnings of physiological phenomena made Sechenov a particularly inviting target for all kinds of crusades against the materialistic leaning of the natural sciences of the Darwinian era. Conservative scientists, idealistic metaphysicists, and theology scholars interested in critical relations between scientific theories and religious thought viewed Darwin's evolutionary notions and Sechenov's psychology as two classical expressions of "natural science materialism," incompatible with the sacred values of Russian culture.

There was another—and indeed stronger—tradition in Russian physiology: the confluence of experimental developments that gave birth to the I. P. Pavlov school, which at this time was in a stage of infancy. Pavlov's experimental research had two distinguishing features: a heavy reliance on surgical methods and a strong emphasis on the unique attributes of the biological nature of physiological processes, transcending the competence of physicochemical analysis. In his research, however, Pavlov dealt much more with the mechanisms of neurophysiological activity than with the dynamics of evolutionary processes. He did not work directly either on extending the evolutionary principles to physiology or on applying the tools of physiology to the ever puzzling problem of the evolution of species. In spelling out his philosophy of science, however, he made no statements that could be interpreted either as antievolutionary or as anti-Darwinian.

No Russian scientist of the 1870s and early 1880s had contributed more to making Darwinism the mainstay of comparative anatomy than Ia. A. Borzenkov. This discipline, he thought, benefited enormously from Darwin's successful effort to advance an approach that concentrated on genealogical links between species rather than on specific characteristics of individual taxonomic groups. It replaced the old method that emphasized the discrete nature of morphological traits of larger taxa by a method that featured the morphological unity and common origin of various groups. Borzenkov thought that the new approach made three contributions to biological theory.

First, Darwin predicted an early existence of "synthetic" forms of life, sharing morphological characteristics that have been preserved until the present time and are widely scattered among larger taxonomic groups. The wide distribution of certain characteristics at the present time has historical roots in limited and concentrated distribution of the same characteristics during the early course of organic evolution. Borzenkov claimed that paleontologists had produced ample evidence corroborating Darwin's claim of an early existence of synthetic groups. The struggle for existence may have wiped out individual groups of plants and animals, but it did not eradicate the characteristics that went into making synthetic groups. Common characteristics were preserved by groups that were successful in their struggle for existence. Through the process of divergence the synthetic characteristics became morphological components of an ever widening array of species and larger groups. Divergence produced not only new idiosyncratic characteristics but also a wider distribution of common characteristics.

Second, Darwin's theory reemphasized the evolutionary significance of the study of homologies É. Geoffroy Saint-Hilaire had started at the end of his distinguished scientific career. To study homologies, as Borzenkov saw them, meant to recognize that the organs of all animals or plants belonging to related taxonomic groups are built and arranged in a similar manner. Darwin deserved the main credit for creating an empirical base for the scientific study of the evolutionary significance of homologies. He asked the same questions related to homologies that troubled Saint-Hilaire, but he demanded stronger safeguards in carrying out comparative analysis. Saint-Hilaire's conclusions are logical constructions arrived at with the help of deductive reasoning. Darwinian homologies, by contrast, are generalizations distilled from rich empirical accounts of the evolutionary process. Only by depending on Darwin's method could A. O. Kovalevskii—according to Borzenkov— conduct the empirical study of the evolution of the lancelets and ascidians which led him to conclude that "the Vertebrata and Tunicata are similar to such a degree that the unity of their organizational plan must be taken for granted."[65]

Third, Darwin's theory explained the meaning of so-called rudimentary organs, that is, organs that are underdeveloped and without importance for the survival of organisms. Inherited from primeval ancestors, these organs were more developed and more functional in the past. The changed conditions of life were responsible for their transformation into nonfunctional relics of the past. In Borzenkov's view, they have become a sort of documentary material shedding important light on the history of living forms. Vestigial structures, so profusely displayed in the living world, make sense only when viewed in the light of organic evolution. Borzenkov did not make it clear, however, whether his notion of evolution explained rudimentary organs or his view of rudimentary organs explained the course and the meaning of evolution.

Borzenkov concluded his discussion with an optimistic note on the place of Darwin in contemporary biology:

> Darwin's theory made a profound impression on the scholarly world. It was almost impossible to meet a scholar of his time who had no opinion on his theory. Some scholars, typified by R. Wagner, Wigand and a number of other German biologists, but particularly L. Agassiz in America, took an antagonistic position toward him. This antagonism, however, did little damage to Darwinism. Much more damage was done by persons who became too enthralled with its ideas . . . but who adopted only some of its views and remained completely alien to the spirit of careful inductive research that produced the theory of the origin of species. These persons took a directly opposite path, the path of the defunct Naturphilosophie. Among them, Ernst Haeckel occupied the leading position.[66] In his effort to popularize the new evolutionary theory, Haeckel, "one of the most talented zoologists of our age," actually constructed an original system of ideas, steeped in metaphysics, that had little in common with Darwin's principles. Darwin handled the question of the origin of life with extreme caution; Haeckel, by contrast, answered the question in a few metaphysical suppositions, devoid of any trace of empirical support.[67]

Contemporary historians did not overlook Darwin's work as a fertile source of guides for a reinterpretation of the national past. In 1876 the *Journal of the Ministry of Public Education* published a lengthy review of a new study of the pre-Mongolian period of national history, which emphasized "struggle" as the main motor of social and political evolution. Adorned with a Darwinian epigraph, the book—written by Zatyrkevich—aimed at no less than a total historical reconstruction of the Russian past.[68] It evoked negative criticism not because of its Darwinian bias but because of major flaws in its use—and misuse—of documentary sources.

Evolution in a Historical and Sociological Light

I. I. Mechnikov was the most productive and influential member of the scientific community interested in the historical depth and social implications of Darwin's theory. He was particularly successful in tracing the growth of the evolutionary idea from the scattered hints of eighteenth-century naturalists and philosophers to the first generation of post-Darwinian specialists in various branches of biology.[69] Starting with a historical commentary on the evolutionary thinking of Charles Bonnet and Georges Buffon, expressed in general pronouncements that lacked scientific precision and empirical grounding, he ended with a review of the evolutionary ideas built into the embryological work of Fritz Müller, the general schemes of Ernst Haeckel's comparative morphology, and Karl von Nägeli's theoretical adumbrations.

In tracing the modern growth of evolutionary thought Mechnikov adhered to two general principles. First, he believed that the true clues to an understanding of modern evolutionary thought must be sought not only in the rudimentary transformist thought of the eighteenth century—typified by the work of Buffon, La Mettrie, and Diderot—but also in the systems of nontransformist thought. He treated Charles Bonnet, the author of *Contemplation de la nature,* as a noteworthy precursor of Darwin, even though he placed the entire living world into a universal "chain of being" that gave every plant and animal species a fixed, rather than a transitional, place in the unbroken line of organic complexity. Bonnet translated the idea of gradation into a principle of continuity built into the laws of Newtonian mechanics and into differential calculus. Mechnikov noted a close connection between Bonnet's "chain of being" and Darwin's rejection of "leaps" in the evolutionary process.

Second, Mechnikov argued that the history of the evolutionary idea was not tied to a specific philosophical orientation in science. The roots of evolution, he reasoned, can be traced to both materialistic and idealistic thought. For example, in the process of tracing the scattered and faint lines of transformism, he did not overlook Schelling and Oken, the architects of the romantic *Naturphilosophie,* who, despite their deep commitment to idealistic metaphysics, articulated a grand evolutionary view of living nature as part of cosmic unity.

Mechnikov must be recognized as the first Russian biologist to work on the historical legitimation of Darwin's grand synthesis. He made a broadly based and systematic effort to show that the ideas Darwin built into his theory grew out of a gradual maturation of modern scientific thought. This effort added new ammunition to the war on anti-Darwinism and gained popular support for the challenging—and un-orthodox—ideas of the new biology. Mechnikov's essay on the past and the contemporary history of evolutionary thought contained one serious omission: it paid little attention to the notable contributions of Russian scholars to the triumph of Darwinian thought—particularly those of A. O. Kovalevskii in comparative embryology and V. O. Kovalevskii in evolutionary paleontology.

At this time Mechnikov was deeply involved in the sociology of Darwinism. He was particularly attracted to the idea that overpopulation, expressed in the Malthusian formula, does not explain all the complexities of the struggle for existence. In human society, he noted, the struggle for existence is particularly complicated and diverse. Here it operates within two distinct systems of causal factors: one responsible for natural inequality, the other for cultural inequality. Mechnikov dealt with both the biology and the sociology of inequality.

In contemporary economics Mechnikov recognized two major attitudes toward inequality. The articulators of the Manchester school, as he saw them, shared the vision of a future society that would eliminate the cultural sources of inequality and would recognize only natural inequality. The second school, represented by Adolph Wagner, took the opposite position. Wagner advocated cultural equality as a means for overcoming natural inequality; led by humanistic motives, he went so far as to emphasize the moral obligation of human society to extend special favors to victims of natural inequality. Although Mechnikov did not show particular enthusiasm for either of the two orientations, his thinking was closer to that of the Manchester school. Every serious student of human society, he said, must recognize three kinds of struggle for existence: natural, cultural, and mixed.

Mechnikov showed particular eagerness to point out the unrelatedness of the struggle for existence to ethical norms. Whereas the former, in his opinion, has a firmly established empirical basis and can be shown by inductive reasoning, the latter come from deductive thinking and do not have a concrete substratum. To talk about the struggle for existence and about ethical precepts is to talk about two completely different sets of sociologically significant phenomena. He criticized the views of Albert Schäffle, a German sociologist who tried to base morality and law on the struggle for existence and natural selection.[70] Mechnikov succeeded in adducing compelling arguments that anticipated statements, like Thomas Huxley's, that the mechanisms of "biological progress" are in certain respects directly opposite to the inner fabric of "social progress."[71] Preoccupied with the criticism of various theories dealing with the elusive region where sociology and biology meet, he did not give his own views enough depth and precision to make a serious impression on contemporary readers. Mechnikov's occasional excursions into the unwieldy domain of social science produced literary pieces that in most cases suffered from imprecise organization and vague statements.

Mechnikov never doubted the reality of organic evolution and never challenged Darwin's major role in the triumph of the transformist point of view and in the irrevocable separation of biology from vitalistic metaphysics and teleological involvement. He accepted natural selection as a factor of evolution, but not without notable reservations. In his opinion there are two serious limiting factors related to the work of natural selection.[72] First, among the lowest animals and human beings—the "two poles of the animal kingdom"—natural selection plays an insignificant role in producing morphological changes that in due time give rise to new species. Second, morphological characteristics that separate one species from another are not the main wheel of natural selection. Mechnikov preferred to think that such "physiological characteristics" as fecundity, viability, and endurance were much more important, but he made no effort to explore this problem. Contrary to the prevalent opinion, Mechnikov refused to identify "evolution" with "progress": at a given point in history, individual groups or subgroups of living forms may be termed "progressive," "regressive," or "conservative." The first type shows clear signs of morphological or physiological improvement; the second type displays signs of degeneration; and the third type is in a static state. Spencer erred in identifying "evolution" and "progress": he achieved this simply by considering the differentiation and integration of organs the only reliable index of evolutionary change. Static types, Mechnikov contended, play a sizable role in organic evolution, but cannot be related to natural selection.[73]

Mechnikov's unsettled and uncertain attitude toward the struggle for existence came at a time when this concept encountered serious challenge not only from populist ideologues but also from influential representatives of the scientific community. In 1879 K. P. Kessler, zoologist and former rector of St. Petersburg University, read a paper entitled "The Law of Mutual Aid" before the members of the St. Petersburg Society of Naturalists. He presented his case in simple and straightforward terms: the struggle for existence, as Darwin saw it, placed too much emphasis on competition and ignored cooperation, which in fact provides the real key to the understanding of the dynamics of the living world. He viewed the struggle for existence as a primary factor in interspecific relations; he made mutual aid, operating on the intraspecific level, a weapon of the struggle for existence operating on the interspecific level.[74] Organic evolution, according to him, is propelled by interspecific struggle for survival and intraspecific mutual aid.

All organisms, according to Kessler, must satisfy two basic needs: the need for nutrition and the need for procreation. The need for nutrition leads to the struggle for survival—to competition between groups for limited food supplies. The need for procreation leads to intragroup cooperation and mutual aid. The evolution of the organic world depends much more on the unity of individuals belonging to the same species than on the struggle between them.[75]

Kessler's ideas found strong, but not prevalent, support in the scientific community. They concurred with the sociological postulates of Russian populism, built on the principle of cooperation as the pivot of social life. Kessler's view became a basis for P. A. Kropotkin's grand theory of social evolution, popular during the early decades of the twentieth century. In a way, Mechnikov tried to reason out the important role of the struggle for existence in nature and society without disregarding the role of cooperation. Kessler and Kropotkin, by contrast, were so taken by the idea of cooperation that they looked at the struggle for existence as unnatural and contrary to the ideals of human society.

Mechnikov's critical and ambitious survey of the historical growth of evolutionary thought and of the major pitfalls of Darwin's theory produced two notable results. First, it gave Russian readers a fair and symmetrical analysis of the continuity and accumulating effects of the growth of evolutionary thought from a vague philosophical notion to a precise scientific theory. It helped create a documented historical base for the legitimation of Darwin's theory. Second, it showed that, despite

their unquestionable strengths, Darwin's principles—particularly the
struggle for existence—were much in need of critical reassessment and
modification.

Ia. A. Borzenkov's critical survey of the main currents in nineteenth-
century biological thought produced similar effects.[76] It gave a more detailed picture of the contributions of Russian scientists to both pre-
Darwinian and Darwinian thought. While ebullient in expressing a
strong pro-Darwinian sentiment, it was not symmetrical in presenting
the arguments of various pre-Darwinian orientations. For example, it
allocated more space to the metaphysical arguments of the leaders of
Naturphilosophie than to Lamarck's scientific arguments. Borzenkov
presented his ideas in a long report to an assembly of the professors and
students of Moscow University. On April 12, 1881, he expressed a strong
belief in the future of biology as a discipline dominated by Darwin's
ideas.

From Photosynthesis to Evolution:
K. A. Timiriazev

Next to Mechnikov, K. A. Timiriazev showed the most interest in the
general makeup of Darwin's evolutionary theory and in its philosophical foundations. He brought a unique theoretical background in empirical research to the Darwinian debate. During the 1870s he was
deeply involved in the study of photosynthesis. Experimental work
in various western European laboratories and at Moscow University
earned him a magister's degree in 1871 and a doctorate in 1875. It
also earned him teaching positions at the Petrovsk Agricultural Academy and Moscow University. Numerous research papers which he presented at various international and national congresses of scientists, the
Russian Physical and Chemical Society, and the St. Petersburg Society of
Naturalists dealt without exception with questions related to photosynthesis. Once established as a respected member of the scientific community and a popular teacher, Timiriazev began to expand his activities
in various directions, always within the domain of academic expectations. From that time on he devoted most of his enormous reserves of
energy to three kinds of activity: the defense of science as a base of the
modern world outlook, the struggle for academic autonomy, and the
championing of Darwinism. During the last forty years of his life he
returned with predictable regularity to the task of making Darwinian

historicism part of the Newtonian mechanistic picture of the universe, and of waging a relentless war against the major developments in biology that, in his opinion, went against the Darwinian grain.

From 1864, the year in which he published a series of articles entitled "Darwin's Book, Its Critics and Commentators," until 1878, Timiriazev made no published contribution to the rapidly expanding world of Darwinian thought. In 1878 he published *The Life of the Plant*, a series of ten lectures delivered in Moscow's Polytechnical Museum in 1875–76, and "Charles Darwin as a Model Scientist," delivered as a public lecture at Moscow University. While the first publication dealt with Darwin peripherally, the second made him a direct and central topic of discussion.

In *The Life of the Plant*, Timiriazev explained the cellular makeup of plant organs and emphasized the chemistry of cellular activity as the prime mover of vital processes. In describing the "harmony" and the "perfection" of plant life, he relied on evolution as the central strategy in the biological study of nature. Darwin, he wrote, deserved the credit not only for placing biology on evolutionary foundations but also for showing how evolution worked specifically in the plant universe. He also suggested that the study of "variation, heredity, and rapidly increasing reproduction" was the central concern of the historical orientation in biology. This recognition, in turn, led him to the formulation of the principle of natural selection—a logical sequel to the struggle for existence—as the chief mechanism of organic evolution. In brief, Darwin established evolution not only as the central and integrating theme of biology but also as the universal law of the living world.[77] Timiriazev was particularly eager to show the unity of the Newtonian mechanistic orientation and Darwinian historicism.

Speaking in the largest lecture hall of Moscow University in 1878, Timiriazev offered a warm and appealing analysis of Darwin's work as a synthesis of the essential attributes of exemplary devotion to scientific scholarship. This address reaffirmed Timiriazev's position as the country's most enthusiastic and most eloquent supporter of Darwin's scientific legacy. It concentrated on two themes.

The first theme elaborated Timiriazev's conviction that Darwin's theory rested on granite foundations and that it needed only minor refinements and improvements to reach the stage of perfection. In defending the scientific validity of natural selection, Timiriazev relied heavily on the authority and prestige of the German naturalist Emil du Bois-Reymond:

As an empirical rule, natural selection will not be overturned in the near future. True, we cannot yet attribute to it the infallibility of the physico-mathematical laws that govern the material world. The result of an entire chain of deductions, it is a rational conclusion that occupies a middle position between a guiding rule and a natural law, perhaps closer to the latter, and is mandatory for scientists.

In appraising a theory, we must base our judgment on information available at the present time. The future is beyond our clear vision. Perhaps, in time there will emerge a new theory that will advance and transform the theory of selection, but to dwell on that question is to invite useless guessing. What is certain, however, is that the break with the past is complete. While the scientific conflict still goes on, the outcome is already clear: the history of science knows few examples of such a decisive and brilliant victory.[78]

Returning to du Bois-Reymond, Timiriazev informed his listeners that the conflict between the old and the new biology had ceased to be acute because the new generation had made up its mind to abandon "the theory of the independent creation of species advanced by Cuvier and Agassiz," and to take a firm position on the side of Darwin's theory. In all this, Timiriazev represented the views of a strong segment of Russian biologists who made no secret of their full concurrence with the basic principles of the new evolutionary theory. He told his audience:

Sensing the impotence of their criticism, the enemies of Darwinism quickly changed their war strategy: now they claim that Darwin's theory is only a *hypothesis*. This assertion is particularly bizarre when it comes from persons known for their acceptance of the arbitrary and groundless hypothesis of the independent origin of individual species. Let us concede that Darwin's theory is only a hypothesis, but what a difference between the two hypotheses! On the one side is the evolutionary hypothesis that agrees with the facts of many different branches of knowledge and has the support of countless examples of predicted developments. In no need of arbitrary assumptions, this hypothesis rests on the general laws of nature. It not only satisfies man's idealistic striving to explain natural phenomena, but also, according to Asa Gray's judicious comment, has been highly productive in creating fresh approaches and in opening new areas to scientific study. On the other side is the hypothesis of the independent creation of species, founded on a totally arbitrary and unproved assumption that not only does not explain anything but hampers the search for a true explanation by spreading darkness and ignorance.

Darwin's theory is not a hypothesis in the sense of simple guesswork; it is a necessary and logical conclusion from which there can be no deviation. To deny variation, heredity, and the rising rate of reproduction is to deny the geological record that covers countless centuries after the emergence of life on our planet. In the meantime, the theory of *natural selection* continues to be an undeniable logical conclusion. We cannot tell how far it reaches, but we can demonstrate its real existence in nature.[79]

The second theme dealt with the three stages in the development of Darwin's engagement in advancing evolutionary theory.[80] Here again Timiriazev portrayed Darwin as a model scientist. During the first stage Darwin formulated the basic principles of a consistent and unified evolutionary theory. The second stage was dominated by a broadly conceived inquiry into the topics suggested by his general theoretical position. The purpose of this inquiry was to eliminate inconsistencies and to carry the theory to new research areas. In all this, Darwin avoided unnecessary speculation. The third stage concentrated on "a factual verification of achieved results."[81] Most of Darwin's profuse writings after the *Origin of Species* dealt with "very complex questions" and with the light their answers had thrown on the working of natural selection. Nor did he hesitate to ask and answer the most puzzling question: the origin of man and his physical, mental, and moral characteristics.[82] Like the two preceding stages, this stage produced the kind of information Timiriazev needed to portray Darwin as an example of the "ideal scholar."[83]

The evolutionary studies of Timiriazev and Mechnikov served as issuing points for two strong currents of Russian Darwinism that became fully crystallized during the first three decades of the twentieth century. One tradition, with Timiriazev at the helm, fought the new theoretical developments in biology that appeared to be antithetical to Darwinian principles. The other tradition, dominated by the towering figure of Mechnikov, recognized both the preeminent place of Darwinian biological views in modern life sciences and the need to modify these views in the light of rapidly emerging new theoretical insights. From the very beginning of their scientific activity, Timiriazev and Mechnikov agreed in viewing Darwin's theory as a gigantic victory for modern science and a primary contribution to the intensified growth of rationalist tradition in Russia.

That Timiriazev quickly became the most consistent, determined, and influential Russian defender of Darwinism was somewhat of a paradox. He was an expert in plant life at a time when the study of zoological phenomena served as the unquestionable stronghold of Darwinian views and studies. Most leading "botanists" worked in various branches of plant physiology and in the experimental study of the chemistry and physics of plant life with very little concern for the problems of evolution. Those "botanists" who studied the role of the environment in the morphological changes of plant forms tended to be on the side of Lamarck, even when they professed an affinity with Darwin's ideas.

Ivan Fedorovich Shmal'gauzen, professor at Kiev University and an expert on the flora of the St. Petersburg region and southwestern Russia, played a rather unusual role in the community of botanists. Fully aware of the stream of ideas Darwin contributed to biology, he went out of his way to make laudatory statements about Gregor Mendel's experimental and mathematical work on plant hybridization—"one of the paths to the origin of new species."[84] This happened at the time—1874—when Mendel's ideas had not yet attracted the attention of the guardians of authoritative thought in biology. Shmal'gauzen credited Darwin—and Nägeli—with giving the idea of the evolutionary role of hybridization a firm grounding. He made no effort to compare Mendel's emphasis on the stability of heredity with Darwin's emphasis on the instability of heredity.

The Lamarckian Line

To be a Darwinian scholar did not require one to make direct references to Darwin. Most biologists made no—or very rare—references to the creator of the theory of natural selection. This was the period of mounting political oppression and of deep conservatism in the high offices of national education. To talk or write about Darwin was to preach heresy. To avoid intimidation by authorities, the evolutionists found it helpful to avoid categorical and blanket endorsement of every idea Darwin put forth and to use a more subtle idiom in expressing their favorable interpretations of general evolutionary views.

Darwinism also competed with a Lamarckian strain in biological theory that emphasized the direct influence of the changing environment on the transformation of plants and animals and made the inheritance of acquired characteristics the main mechanism of evolution. The Lamarckian substratum of the Russian evolutionary theory served as a springboard for the early resistance of a strong segment of Russian biologists to August Weismann's theory, which had no room either for the role of environment in genetic modifications or for the inheritance of acquired characteristics.

A. N. Beketov, known for his work in plant morphology and anatomy, as well as in the popularization of modern scientific ideas, earned an enviable reputation by his effort to give prominence to the Lamarckian residues of Darwin's theory.[85] A professor of botany at St. Petersburg University (after having first taught at Kharkov University) and an honorary member of the St. Petersburg Academy of Sciences, he exercised a strong influence on forging a community of Russian biologists and on the flow of modern biological ideas in and outside the academic world. Shortly before the appearance of the *Origin of Species*, Beketov wrote "Harmony in Nature," which, as he stated in his "Autobiography," emphasized the direct influence of the surrounding environment on the evolutionary role of the inheritance of acquired characteristics. Subsequently, he worked on redefining—rather than on negating—Darwin's theory.[86]

Despite his strong Lamarckian bias, Beketov was ready to admit that "the great contribution of Darwin and Wallace has been in showing that the struggle for existence and heredity are the leading factors in natural selection: they determine the adaptation of organisms to the surrounding environment and the main course of the evolutionary process."[87] Beketov treated the struggle for existence as a mechanism of equilibrium—or harmony—in the interaction between organisms and their environment. He viewed competition as a special ramification of cooperation, a result of the constant pull of natural forces toward a state of equilibrium. Translated into sociological terms, equilibrium becomes "harmony," and "harmony," translated into psychological terms, becomes "sympathy." Guided by a loose interpretation of the principles of Newtonian mechanics—and its conceptualization of equilibrium— Beketov built a world made up of Newtonian metaphors and poetic vision, a philosophical view which he chose not to elaborate.

Beketov may be considered the founder of a strong tradition in Russian biology which gave Darwin's theory a strong Lamarckian base. He did not hesitate to make the struggle for existence the keystone of the evolutionary theory; nor did he hesitate to superimpose three Lamarckian ideas upon it: the inexorable evolutionary pull of an "inner impulse for perfection";[88] the use and disuse of organs as adaptive instruments of organic transformation; and the hereditary transmission of acquired characteristics.

Beketov thought that Malthus's law of discrepancy in the ratios of the growth of population and food resources referred only to "quantitative changes" in nature; only the "inner striving" of organisms for "progressive change" can explain the "qualitative" aspect of evolution. He thought that Alphonse de Candolle, in his *Histoire des sciences et savants depuis deux siècles* (1873), exaggerated the power of science to enter the magic world of natural selection.[89] Science was not yet in a position to tackle the relationship of intellectual development to physical evolution, de Candolle's basic concern. Beketov also opposed the claim of the uniformitarians that the same causes of organic evolution have operated in all geological eras.

In a paper presented before the members of the St. Petersburg Society of Naturalists in 1882 Beketov asserted that "almost all leading contemporary biologists are followers of Darwin."[90] In his opinion, Darwin was the primary contributor to the elimination of "crude teleological explanations" from modern biology.[91] He argued, however, that Darwin made the struggle for existence too broad and imprecise and suggested that it be limited exclusively to relations between organisms and the "general physical environment." Under the spell of Kessler's recent pronouncements, he acknowledged the important role of mutual aid in the life of plants and animals.[92] Loyal to the Lamarckian legacy, Beketov thought that the future work of biologists would concentrate on achieving a full union of evolutionary theory and the general laws of physics.[93]

While Beketov supported Lamarckism on a theoretical level, V. I. Shmankevich supported it on experimental and practical levels and received wide publicity for it both in Russia and abroad.[94] In the 1870s Shmankevich, a science teacher in an Odessa secondary school, put forth resounding claims that by changing the salinity of water he was able to observe the transformation of the brine shrimp (*Artemia salina*), found in local lagoons, into two different species: whereas increased salinity led to a transformation of *Artemia salina* into a species similar to *Artemia Mühlhausenii,* reduced salinity led to the transformation of *Artemia salina* to animals very similar to various species of the genus *Brachipus.* Shmankevich found the latter transformation particularly exciting because, according to at least some zoologists, *Artemia* and *Brachipus* are different genera.[95] In all his studies, he combined a phase of observation with a phase of experimental research. In his experiments he depended on two variables: salinity and temperature of the lagoon water. Specific combinations of salinity and temperature led either to progressive or to regressive transformation.

In 1875 the *Zeitschrift für wissenschaftliche Zoologie* made Shmankevich's research findings readily accessible to the Western scientific community. During the next two decades, Shmankevich was one of the most cited Russian biologists in western Europe. Even the friendliest supporters mixed guarded praise with strong reservations; the opponents were firm and irreconcilable. In 1893 William Bateson presented the details of Shmankevich's claims, but he was inclined to think that *Artemia salina* and *Artemia Mühlhausenii* were actually the same species.[96] Especially in Russia and Germany the role of the direct influence of the environment continued to be recognized as a challenge of large proportions. At the Eleventh Congress of Russian Naturalists and Physicians, held in St. Petersburg in 1902, a speaker informed the audience that Shmankevich's experiments and findings continued to be referred to favorably in most university textbooks in zoology.[97]

Shmankevich noted that it was too early to give a precise description of the relation of his conclusions to Darwin's theory of natural selection. He observed in 1875 that, unlike his (Lamarckian) mode of explanation, Darwin's theory depended exclusively on indirect evidence.[98] This did not prevent him from expressing high respect for Darwin's contributions and an optimistic view about its future triumph. At no time did he consider Darwin's and Lamarck's theories mutually exclusive. In his view, only the future generations of biologists, armed with the methods of physics and chemistry, would be in a position to shed sufficient light on the process of the evolution of organic forms and on the general validity of the theory of natural selection. In one respect, Shmankevich was very much like Beketov: both praised Darwin without abandoning their primary loyalty to Lamarck. Both contributed to the strength of Lamarckism as a strong undercurrent of the Darwinian tradition.

The 1870s witnessed the rise of yet another form of extreme Lamarckism, presented as an orientation consonant with the spirit and the substance of Darwin's scientific legacy. It was at this time that V. I. Michurin, inspired by the atmosphere created by the diffusion of Darwinian ideas, had barely begun his long-term activity of inducing heritable characteristics in fruit trees by changing the environmental conditions under which they grew. He was guided by the idea of the possibility of adding a new dimension to the Lamarckian theory: the artificial inducement of predetermined and accelerated transformation of characters. His method was the crossing of geographically distant plants; his aim was to produce varieties best adapted to specific environments. Not recognized by the scientific community, mainly because he operated in the realm of folk science, Michurin became part of a popular movement devoted to improving domestic plants in Russia and to extending their cultivation to new areas. With exemplary devotion, Michurin worked on developing new varieties of fruit trees in central Russia. Keeping Lamarckism alive, Michurinism became part of a general cultural setting that encouraged a union of Lamarck and Darwin and that stood in the way of a faster diffusion of the theoretical ideas of modern genetics. To its chief articulators, Russian Lamarckism of this period represented a modification rather than a negation of Darwinism.

Douglas R. Weiner has suggested that the roots of Michurinism go back to the pre-Darwinian period—to the acclimatization movement of the mid-nineteenth century. In this movement he sees a distinct Russian adoption of the views of Lamarck and É. and I. Geoffroy Saint-Hilaire on the "direct environmental induction of hereditary adaptations." Weiner traces the roots of Lysenkoism to the strong influence of the "French school" on the early acclimatization movement in Russia.[99]

Darwin did not stand in the way of the current effort to merge his theory with Lamarck's ideas. In fact, he encouraged such an effort. In his preface to the second edition of *The Descent of Man* he made his position eminently clear:

> I may take this opportunity of remarking that my critics frequently assume that I attribute all changes of corporeal structure and mental power exclusively to the natural selection of such variations as are often called spontaneous; whereas, even in the first edition of the *Origin of Species,* I distinctly stated that great weight must be attributed to the inherited effects of use and disuse, with respect both to the body and mind. I also attributed some amount of modification to the direct and prolonged action of changed conditions of life.[100]

In recounting the landmarks of his scientific achievement, a surprisingly large number of Russian scientists and ideologues credited Darwin with saving Lamarck from obscurity and with making his evolutionary idea an organic part of the mainstream of modern biology.

Darwin's Death: Commemorative Remarks

Charles Darwin died on April 19, 1882. A few months later, M. A. Menzbir, a young scholar associated with Moscow University and a future member of the Academy of Sciences of the U.S.S.R., commented on the passing of the eminent naturalist:

> [With the death of] Charles Darwin, the biologists have lost their Newton. His death did not end the life of the ideas that he presented to mankind. Generations of students have been raised on his ideas, despite all the attacks on them. . . . A quick survey of the achievements of science during the last twenty-three years shows that there is no branch of knowledge that has escaped Darwin's influence. The evolutionary theory has embraced all the sciences dealing with man and has been extended to the studies of art. It has

become so popular that it is impossible to find an educated person who does not make use of the expression "the struggle for existence." Briefly, this theory has changed our world view, has opened countless new paths for human thought, and has given science—to use Wallace's expression—a new life and new powers.

The broad sweep of his work, particularly of the *Origin of Species*, made Darwin a target of massive attacks. The ideas he presented in this book were not exactly new: [for example], earlier writers had referred to the genealogical unity of organisms. What Darwin gave was a new explanation of this unity. In a relatively small volume he collected such a mass of masterfully arranged facts that his conclusions stood out clearly before our eyes. Earlier, it was customary to talk about the limitless variation of organisms only in jest; Darwin was the first to introduce serious talk about this matter. Earlier, most questions relating to the life of animals were not recognized as pertinent questions and were ignored; he showed that ignoring these questions did not mean proving that they were not significant. Briefly, Darwin gave much attention to questions of this kind and he raised others that could no longer be ignored.[101]

In August 1883 the Seventh Congress of Russian Naturalists and Physicians met in Odessa. A. O. Kovalevskii, Russia's leading evolutionary embryologist, was given the task of presenting an *éloge* to the great man. He said:

> Darwin . . . is our great common teacher and a man whose name and ideas are at home in every branch of science. Everywhere, there is a striving to explain all phenomena of individual and social life in accord with the laws which he formulated. Darwin died before he had completed his work; his death interrupted a steady stream of publications, representing the result of fifty years of unremitting scientific activity. Darwin's theory was received in Russia with particular sympathy. While in western Europe it encountered many established traditions and a strong opposition, in our country, awakening after the Crimean War, it lost no time in acquiring the rights of citizenship in the scientific as well as in the social world. Admiration for Darwin has continued until the present time.[102]

The scientific community was not alone in commemorating Darwin's death. In a thoughtful article published in the popular journal *Fatherland Notes*, N. K. Mikhailovskii, the most influential ideologue of populism, provided a brief analysis of the main attributes of Darwin's scientific stature. Darwin, according to Mikhailovskii, was blessed with the two attributes indispensable to leadership in scientific achievement. His unusual gift of "observation" enabled him to uncover astounding minutiae of the infinite complexity of nature, much of which other naturalists had overlooked, and to gain insights into the marvels of the living world surrounding him. His equally unusual gift of "imagination" gave him a fruitful vision of the forces and mechanisms that had shaped nature in the past and had given the keys for dependable predictions of future developments.[103] By uniting the gifts of "observation" and "imagination" into a "creative force," Darwin became a great scientist who relied on the most precious assets of a great poet.

In his earlier writings Mikhailovskii questioned the universal validity of natural selection as the prime mover of evolution. In the commemorative article, however, he took a more conciliatory position. It was incorrect to assume, he contended, that the theory that emphasized the struggle for existence was "a dark and pessimistic theory." On the contrary, he added, Darwin's theory was thoroughly optimistic, for it envisaged the struggle for survival as a path to more advanced forms of life. Moreover, the struggle for survival figured, not as an inviolable axiom, but as a general idea inviting different and often conflicting interpretations. Darwin himself belonged to the group that preferred a more cautious interpretation of the role and magnitude of the struggle for survival. In his analysis, however, Mikhailovskii did not abandon his skeptical attitude toward current efforts to make the struggle for survival not only a law of the living world but also a law of human society. He recognized the wide scope of the struggle for existence in human society, but he also claimed that this struggle often played a "regressive" role by victimizing the more advanced social groups and by perpetuating the social power of "degenerate groups."[104] In making this statement, Mikhailovskii was obviously moved by ideological motives.

A contributor to the prestigious *Herald of Europe* presented Darwin's biological ideas as a dominant current of modern thought. He told his readers that these ideas brought Darwin "his much deserved fame on both hemispheres."[105] He added that Darwin's theory served as a basis for Herbert Spencer's grand effort to construct "a full system of philosophical thought by relying exclusively on the facts of science."[106] The time had come to make the new theory a source of models for the systematic study of the succession of civilizations. He reviewed the work of Sir Henry Maine and Max Müller to illustrate the great promise of the new evolutionary thought as a basis for the comparative study of human society and religion.[107]

At the same time, L. K. Popov, a conservative writer of popular articles on scientific themes, published a long piece in the journal *Observer* (*Nabliudatel'*) in which he surveyed Darwin's basic contributions to science. He took the opportunity to refute Darwin's view of natural selection as the primary agent of organic evolution. Obviously not well versed in the substance and in the style of Darwin's argumentation, he depended heavily on Kölliker and other contemporary critics of evolutionary theory. He praised Darwin for having been the first naturalist to make the notion of evolution the central topic of biological research strategy and to carry this notion far beyond the limits of the scientific community. The grandeur of Darwinism, as he saw it, was in giving modern science a historical orientation. This contribution, applicable to the inorganic world no less than to the organic world, would have been sufficient, he thought, to ensure Darwin a permanent place in the annals of science.[108]

In the journal *Delo*, A. Moskvin provided the culminating point for the procession of Darwinian accolades. He surveyed the wide scope of Darwin's scientific output from the *Voyage of the Beagle* (1840) to *The Formation of Vegetable Mould* (1881), in which he saw the pieces that made up the grand idea of evolution, the key notion in the scientific study of living nature. Darwin's main achievement was not in producing an errorless theory, or in answering all the questions he set out to answer; his undeniable contribution lay in inaugurating a new system of biological inquiry based on a number of general principles that were as simple as Newtonian gravitation and equally open to empirical examination and testing. He cited Wallace's statement that Darwin's greatness was much more in showing how to study living nature than in drawing irrefutable conclusions.[109] Nor did he overlook Darwin's personal characteristics that made him a person of "vast knowledge," "tireless industry," "clear thought," "powerful analytical skills," and "extreme simplicity and unpretentiousness."[110]

This sample of commemorative writings shows that Russian writers viewed Darwin not only as a giant in a specific branch of scientific endeavor but also as the creator of an epoch in the growth of human thought. He did much more than formulate the pivotal principles of organic evolution. By establishing the reign of natural causation—and the rightful jurisdiction of science—in the study of the vast domain of life, he reunited man with nature, contributed to the triumph of a cosmic view dominated by historicism, added immeasurable strength to the rationalist pillars of modern culture, and served as an impeccable model of pure dedication to the methodological and ethical standards of science.

The developments during the 1870s showed also the gradual emergence of a national style in Russian interpretations of evolutionary thought, in general, and of Darwin's scientific legacy, in particular. Four features of this style were particularly notable.

First, a general allegiance to Darwinian thought did not prevent a large group of Russian biologists from having a strong inclination to look favorably at the Lamarckian principles of the direct evolutionary role of the external environment and the inheritance of acquired characteristics. Of the Lamarckian evolutionary principles, only the idea of an innate drive for progress, categorically rejected by Darwin, found little support in the Russian scientific community. Despite the enormous popularity of individual principles of Lamarck's evolutionary theory, Lamarckism did not exist in Russia as a comprehensive and codified system of theoretical principles.

Second, at the beginning of the 1880s the seeds were sown for a unique tradition in Russian evolutionary thought which considered cooperation and mutual aid a strong factor in the evolution of plants and animals. Subsequently codified and popularized by Petr Kropotkin and A. F. Brandt, this orientation played a strong role in both pro-Darwinian and anti-Darwinian ideologies. In the scientific community it received primary support from individuals who, responding to external ideological pressures, chose to address themselves not only to professional peers but also to the general reading public. Despite the popularity of the idea of mutual aid, the Darwinian struggle for existence continued to be the stock in trade for most theoretical and empirical biologists. Unlike mutual aid, prominently displayed in popular writings, the struggle for existence was generally hidden in technical literature. The strong ideological interest in mutual aid, whether it was expressed by the ideologues of autocracy or by the spokesmen for the radical intelligentsia, contributed to the general unpopularity of Social Darwinism in Russia.

Third, Russian evolutionary biologists made no comprehensive or systematic effort to combine residues of transcendentalism or vitalism with Darwin's principles. Keeping metaphysics out of biology, they were generally consistent in their criticism of the holdovers of *Naturphilosophie* in German evolutionary biology. This did not mean, however, that they did not depend heavily on the stream of challenging and fertile comments or elaborations on Darwin's ideas that originated in the German scientific community.

Fourth, tight—and rapidly becoming tighter—censorship, more than any other factor, accounted for the unusually weak interest of Russian biologists in the basic idea of *The Descent of Man:* the origin and evolution of the human species. No scientist rated the notion of the anthropoid origin of man higher than a hypothesis far removed from adequate empirical support. It was not until the turn of the century that the serious possibility of an anthropoid origin of man was discussed in the scientific literature. The evolutionary aspect of physical anthropology was clearly one of the less developed academic interests in Russia.

CHAPTER THREE

Waves of Criticism

During the 1870s and early 1880s, Darwinism became firmly entrenched in a wide spectrum of natural and social sciences, as well as in philosophical thought. It also became a target for relentless attacks by critics representing the major categories of intellectual endeavor, ranging from science to religious philosophy. The critics were united in a determined effort to expose the flaws in both the substance and the logic of Darwin's evolutionary theory. They shared a belief that Darwinism was an ally of positivism and materialism and that it represented an antithesis to the dominant values of Russian society. The critics ranged from Karl von Baer, a giant among scientists, to marginal journalists, always ready to translate autocratic values into ideological slogans.

A Synthesis of Anti-Darwinian Arguments: Karl von Baer in the 1870s

The most powerful criticism of Darwin's theory originating in Russia came from Karl von Baer, who added new logical and substantive arguments and a heightened sense of urgency to the war on the new evolutionary theory.

Von Baer retired from the St. Petersburg Academy of Sciences in 1862 at the age of seventy. To pay homage to one of its most honored members the Academy established the Karl von Baer Prize to be given periodically for outstanding work in biology by Russian scientists. In 1867 he moved to Dorpat, Estonia, where his scholarly activity continued unabated. A concerted effort to deepen and consolidate anti-Darwinian arguments clearly dominated his research activities.[1] He died in 1876, the year of the publication of "Über Darwins Lehre," his most noted and most comprehensive critique of Darwin's theory on scientific, philosophical, and moral grounds. This study impressed contemporaries as an effort to point out the unexplored complexities of biological evolution and to broaden the scope of evolutionary philosophy. Von Baer did not belong to the group of scientists, typified by Kölliker, who both criticized Darwin and advanced their own theories of evolution. He was too busy criticizing Darwin to carry his own theory of organic evolution beyond a clearly articulated, but sketchy, analysis.

In 1873 the *Memoirs* of the St. Petersburg Academy of Science published von Baer's long and unrelenting attack on A. O. Kovalevskii and German embryologist C. Kupffer, who had written about the ascidians as a species linking the invertebrates with the vertebrates. Von Baer expressed bitter resentment over Darwin's enthusiastic endorsement, in *The Descent of Man,* of the great promise of Kovalevskii's embryological research. Aware of the current criticism of Kovalevskii's and Kupffer's views of ascidians as "forefathers of man," his arguments relied much more on logical deductions than on substantive analysis. Clearly angered about the ascidian affair, he unleashed a bitter attack on the growing ranks of "dilettantes," whose transmutationist ideas "had no basis in science." As on several other occasions, von Baer made it known that he did not criticize organic transmutation "as a general principle." He limited his criticism to unsupportable claims by transmutationist "dilettantes."[2]

In the same year, von Baer summed up his anti-Darwinian arguments in an article published in *Augsburger Allgemeine Zeitung.*[3] Addressing himself to the general public, he relied primarily on logical arguments in expressing his ideas about the nonsimian origins of man, the teleological functioning of the living world, and the legitimacy of religious explanations of the mysteries of nature inaccessible to science. He was particularly disturbed about the Darwinian view of evolution as a "blind force," free of predetermined "goal-directedness" (*Zielstrebigkeit*). This time von Baer was much more interested in articulating a generalized argument against the possibility of a full scientific explanation of evolution than in lodging a scientific attack on the basic conceptions of Darwin's theory. This article showed clearly that von Baer's assault on Darwin's theory was part of a general war against scientific materialism as a fountain of atheistic ideas.

"Über Darwins Lehre," taking up 245 pages in volume 2 of von Baer's *Addresses*,[4] showed that von Baer was fully aware of contemporary criticism of Darwin's transformist ideas; but it also showed that in discussing the anti-Darwinian arguments, he depended much more on his own constructions than on mechanical summations of borrowed ideas. But whatever he did, he produced one of the first systematic and comprehensive critiques of Darwin's thought. He made it easier for philosophers, theologians, and free-lance contributors to popular journals to select and elaborate antievolutionist themes and give the anti-Darwinian campaign inner unity and common purpose. Jane Oppenheimer stood on firm ground when she asserted that in "Über Darwins Lehre" von Baer directed his heaviest guns at three components of Darwin's theory: the explanation of the general nature of evolutionary processes; the ethical implication of the disregard of teleology; and the uniformitarian orientation in the explanation of the causes of transformation. There was a strong possibility, von Baer argued, that in the distant past "a much stronger formative force must have prevailed on earth than we know now."[5]

In the Introduction to the *Origin of Species* Darwin acknowledged von Baer's pioneering contribution to evolutionary thought in biology. Without mentioning its title, Darwin referred specifically to von Baer's "Über Papuas und Alfuren," published in the fall of 1859 by the St. Petersburg Academy of Sciences, a few months before the publication of the *Origin of Species*. He gave von Baer credit for expressing "his conviction, chiefly grounded on the laws of geographical distribution, that forms now perfectly distinct have descended from a single parent-form."[6] In fact, this was the article von Baer had mentioned in a letter to Thomas Huxley in which he stated that he "expressed the same ideas on the transformation of types or origin of species as Mr. Darwin."[7] In *The Descent of Man* Darwin repeated his reference to von Baer as a supporter of the common origin of animal forms. In his 1876 essay von Baer was eager to make it clear that he was never an evolutionist in the Darwinian sense, particularly that he never abandoned the idea of transformation taking place only within the preset limits of each of the four types of animals. He analyzed his own earlier work for the purpose of showing that he was never close to the idea of monogenetic evolution.[8]

Von Baer is recorded in the annals of science as a leading anti-Darwinist of his day. But his own noted contributions to science showed that Darwin was absolutely correct in treating him as one of his precursors. In several of his major works von Baer elaborated the transformist

idea in much more than a casual manner. Although some of his suggestions had a clear Darwinian ring, he advanced a theory of the origin
of species which avoided a radical break with creationism. He stuck
closely to three general principles: no species can evolve beyond the general limits of the type to which it belongs; within every type there are
two categories of species, those that are independent creations and those
that are the results of evolution; and all transmutations are caused by
the interaction of living forms and the environment—geography is the
only factor of evolution.[9]

With his record cleared of suggested Darwinian admixtures, von
Baer undertook the task of dismantling the theoretical edifice of his eminent foe. He proceeded with the wrecking job only after making it clear
that he did not question Darwin's qualifications as an established scholar.
In addition to recognizing the contributions of the *Voyage of the Beagle*
to natural history, he noted its author's sagacity in formulating a feasible
theory of the origin of coral islands.[10] He made no effort to examine
every salient ramification of the general theory Darwin had advanced.
The idea of natural selection did not attract much of his attention. His
main criticism was directed at the treatment of all existing species as
transitional stages in the infinite succession of the forms of life, and at
the suggestion that all these forms stemmed from common ancestors,
with the human species as no exception.[11] Von Baer assembled three
types of arguments: those that, in his opinion, invalidated the idea of
common origin and the unity of the evolutionary process; those that
helped him refute Darwin's claim that his theory had nothing in common with atheism; and those that worked against Darwin's alleged
effort to become the Newton of biology by extending the mechanical
principles of the physical world to the domain of life. Newton solved the
riddle of the physical world by explaining the motion of celestial bodies
as the work of "a mathematical-physical law" that brings "mass" and
"force" into causal relationship. The problems of life, which Darwin
tried to answer in the Newtonian spirit, differ fundamentally from the
problems of physical reality. The riddles of heredity and adaptation
are problems of a completely different order, for their understanding re-
quires a concern with teleology rather than with causality.[12] Von Baer
also argued that Darwin had made the unpardonable error of trying to
explain organic evolution at a time when science was not in a position
to explain the origin of life, the starting point of the transmutation
process.

In marshaling arguments against Darwinian transformism, von Baer injected compounded uncertainties into his own theoretical edifice. He did not want to accept a single cardinal argument Darwin had advanced; yet he did not want to abandon the idea of transformism altogether. In at least one place, he said that transformism was not only a universal law of organic nature but also a process free of supernatural interference.[13] To abandon transformism, he said, would mean to abandon a legitimate area of scientific inquiry. In general, von Baer was much more successful in pointing out specific flaws in the elaborate structure of Darwin's theory than in presenting an adequate and scientifically promising substitute for it. Throughout his long essay he followed an unwavering line of attack: the idea of transmutation was most probably correct, but it continued to be an unfathomable riddle of nature. Darwin's "solution" of this riddle must be fully rejected, for it depended on the blind materialism of contemporary natural science rather than on a scientifically rigorous empirical analysis. Darwin's theory, according to von Baer, violated the principles of scientific methodology, the inductive-empirical orientation of modern natural science, and verification standards and procedures. He ended his essay with this advice to scientists:

> I want to offer only one thought to scientists: a hypothesis may be necessary and valuable only if it is treated as a hypothesis, that is, if one takes its basic premises as topics of special inquiry. But a hypothesis may be unnecessary and harmful if, by disregarding proofs, we treat it as an end product of our search for knowledge. Our knowledge is fragmentary. Some persons may find satisfaction in filling in the gaps in scientific knowledge by relying on presuppositions, but that is not science.[14]

Although von Baer's essay lacks precision, consistency, and theoretical clarity, it is a work of notable historical value. As B. E. Raikov has pointed out, it is a basic document for an understanding of von Baer's complex and extensively ramified world view, which reflected a dedicated search for a middle ground between the new ideas in biology and the echoes of the old science born in the early decades of the nineteenth century.[15] It is also the most thorough synthesis of early anti-Darwinian arguments advanced by the representatives of various branches of biology. Even though it opened the doors for a systematic and thorough criticism of Darwin's transformist ideas, its Russian contemporaries received it with inexplicable silence, a fact that caused much grief to Strakhov.[16] Written in German and not translated into Russian, von Baer's major anti-Darwinian arguments were actually accessible only to a small segment of interested writers.

It was for his past achievements, particularly in comparative embryology, that both Darwinists and anti-Darwinists recognized von Baer as a leading figure of nineteenth-century science—and that he was awarded the Copley Medal by the Royal Society of London. The idea of teleology that dominated von Baer's criticism of Darwin was not part of a surrender to metaphysical speculation and theological dictates; it was part of an earnest search for an empirical account of "concrete purposiveness" in the dynamics of living nature. Von Baer's teleology was a rightful ancestor of twentieth-century teleonomy. His faith in the power of science was pure and undeviating. Science, he wrote, is built on an eternal fountain; its authority is not limited in space and time, its full compass goes beyond the reach of measurement, and its goal is unachievable.[17] In Stephen Jay Gould's general assessment:

> Despite shifting emphases, von Baer's general opinion changed very little during his long life. He was a teleologist: he disliked the mechanistic aspects of Darwinian theory. He allowed for limited physical evolution within types, but no transformation among them. His early words on general advance in the universe refer not to physical descent, but to the same ideal progress that Schelling and other anti-evolutionists took as the universal law of nature.[18]

Timothy Lenoir is correct in considering Darwin and von Baer the leading representatives of the two main nineteenth-century theories of evolution, each built upon unique philosophical suppositions: whereas Darwin's theory is firmly fastened to the mechanistic orientation of contemporary natural science, von Baer's theory is dominated by a clearly postulated, logically argued, and vehemently defended teleological principle.[19]

Von Baer's writing quickly became recognized as a rich source of ideas that helped the critics of Darwinism bolster their arguments. Particularly in Russia, very few scholars tried to challenge von Baer's anti-Darwinian crusade. Russian scientists preferred to limit their references to the pre-Darwinian work that made von Baer a most illustrious leader in nineteenth-century embryology. They were unanimous in considering von Baer an eminent member of the Russian scientific community and a great national asset. Vladimir Vernadskii carried this attitude into the twentieth century when he stated that von Baer had played a significant role in the development of modern Russian culture.[20] Even D. I. Pisarev, the iconoclast leader of the nihilists and one of Darwin's most uncritical admirers, referred to von Baer with the utmost respect.

Among the rare scientists who responded to von Baer's criticism by taking Darwin's side, Georg Seidlitz occupied a most prominent position. He was a professor at Dorpat University at the time when von Baer resided in the town of Dorpat—present-day Tartu—after retiring from the St. Petersburg Academy of Sciences. Born in St. Petersburg to a Baltic-German family—his father was a professor at the Medical-Surgical Academy—Seidlitz graduated in zoology from Dorpat University in 1862. As a student he developed a strong interest in Darwin's theory and corresponded with Haeckel. After intensive study in Germany, he earned a doctorate at Dorpat University in 1868. The dissertation dealt with the morphology and systematics of beetles and showed a clear and profound influence of Darwinian thought. In 1869 he was appointed a privatdocent at his alma mater. His research branched out in many directions, but his general emphasis was on the Baltic fauna. In 1871 he published *The Darwinian Theory,* a broadly conceived effort to present Darwinism as a system of evolutionary principles, a culminating point in the history of transformist ideas, and a most satisfactory method for an integrated study of life.[21] The book offered the first Darwinian bibliography to appear in Russia. For its time, this was generally one of the most systematic and extensive bibliographies of studies related to Darwinism.[22] It mentioned only a few contributions by Russian biologists. Omitting the embryological work of I. I. Mechnikov was one of its major flaws. No studies in the Russian language were cited. The book made no mention of the Russian forerunners of Darwin.

Von Baer's essay "Über Darwins Lehre"—a tightly woven assemblage of arguments against Darwin's evolutionary conception—appeared in 1876. In the same year Seidlitz published a 160-page essay entitled "Baer and Darwinian Theory."[23] Just as von Baer was convinced that Darwin's theory was built on shaky foundations, so Seidlitz was determined to show that von Baer's attack was scientifically unfounded and philosophically misdirected. Certain components of von Baer's criticism seemed to him to have applied less to Darwin than to Oken, Schopenhauer, and Hartmann, or, in some instances, to Lamarck, É. Geoffroy Saint-Hilaire, and most of all, Haeckel. Relying on natural history facts and on logic, Seidlitz was particularly eager to show that purposiveness in living nature was a "product" rather than a "cause" of organic evolution. He also tried to show that von Baer gave his individual notions contradictory definitions and that he deliberately distorted Darwin's ideas to make them easier targets for attack. Von Baer's anti-Darwinian effort convinced Seidlitz that there was a pressing need for a "dictionary of the theory of descent" that would help in making Darwinian studies more precise, systematic, and productive.[24] Seidlitz himself contributed to a clarification of the lines of reasoning that separated Darwin's evolutionary ideas from Lamarck's views, particularly with regard to the evolutionary role of external environment and the meaning of "adaptation." He agreed with August Weismann's claim that only adaptation related to natural selection can throw light on the process of organic transformation.[25] Von Baer, he said, agreed with Haeckel in accepting the Lamarckian idea of "adaptation" as "accommodation." Seidlitz observed that the elimination of organisms that did not pass the test of environmental adaptation took place in conformity with the laws of nature and could be expressed mathematically with the help of the calculus of probability. His prophetic suggestion found its first solid expression in the twentieth century—in the work of S. Wright and R. A. Fisher.[26]

The conservative university administration brought a quick end to Seidlitz's course on Darwin's theory. A few years after he published his essay on von Baer's criticism of Darwin's theory, the besieged scholar found it advantageous to take a minor university position in Germany. In *The Descent of Man*, Darwin cited Seidlitz's comment on reindeer antlers as sexual characters.[27]

Criticism from Many Sides

During the 1870s the attacks on Darwin's theory were very much on the upswing. They came from many quarters and in many forms. The publication of *The Descent of Man* in 1871 brought new recruits to the anti-Darwinian camp and gave new breadth and added vigor to the general war on transformist ideas. Inspired by von Baer and Western criticism, Russian antievolutionists made their campaign against Darwin a sustained and elaborate effort. Attacks on Darwin came from professional defenders of autocratic ideology, conservative scientists, idealistic philosophers, alert theologians, and commentators on the current dilemmas of Russian society. Most attacks on Darwinism at this time were much stronger in emotional appeal than in methodical and documented analysis.

Mikhail Pogodin, without peer in his loyalty to autocratic values and institutions, was among the first to react. He did not read Darwin, nor did he have to: all the anti-Darwinian arguments he needed were provided by the conservative journal *Russian Herald*, currently engaged in publishing selected English antievolutionist essays in Russian translation. One of these essays contained John Tyndall's attack on the theory of pangenesis and on the idea of the origin of all species from a single

"primordial embryo."[28] He also depended on his ideological sensitivities and instinctive repugnance for the growing "materialism" of
contemporary science. The simultaneous publication of three Russian
translations of *The Descent of Man* within a year after the appearance
of the English original helped only to heighten his apprehension about
the persistence of the pernicious role of Darwinism in shaping the
thought of the rebellious intelligentsia. Pogodin was particularly disturbed by the deliberate efforts of "nihilists" and "positivists" to make
Darwinian evolutionism a vital part of antiautocratic ideology.

Spread over 152 pages in a collection of Pogodin's essays collectively
entitled *Simple Words about Wise Things,* "Darwin's System" is a biting
and rambling paper made up mainly of running comments on numerous
excerpts from *The Descent of Man,* and of sarcastic attacks on "materialists" and "positivists." His analysis of Pisarev's lengthy comment on
the *Origin of Species* consists mainly of casual, and often petty, comments on selected excerpts. Pisarev, he said, was so preoccupied with
"minute modifications" in nature that he fully ignored such macroscopic
phenomena as the origin of life.[29] Not equipped to handle the subtleties
and intricacies of scientific analysis, he relied on sarcasm as the main
weapon in his war on Darwin and his allies. He also concentrated on
Darwin's work as a classic example of science as a distinct expression of
national character. Darwin's ideas were both incorrect in substance and
non-Russian in their soul.

Pogodin's acrimonious attack concentrated on two features of Darwin's alleged identification with positivist philosophy: the treatment of
science as a culminating point of man's intellectual endeavor, and the
belief that every advance in science represented a corresponding decline
in religious and metaphysical thought. Pogodin found Darwin guilty
not only of giving the wrong answers to the questions he had raised but
also of leaving many questions both unraised and unanswered. Neither
Darwin nor science in general was in a position to answer the fundamental ontological questions related to the creation of matter, the origin
of life, and the nature of space. Most contemporaries, particularly the
growing ranks of scientists and the more astute leaders of the intelligentsia, read Pogodin's acrid writings with much trepidation. They interpreted them as a declaration of war on science as a free, autonomous,
and critical study of nature and society. They knew that the campaign
against Darwinism had become a campaign against science and the scientific world outlook. Few contemporaries showed signs of readiness to

accept Pogodin's manner of argumentation, built on rhetorical bombast and saturated with sarcasm.

In preparing the manuscript for publication, Pogodin asked an un-named naturalist for a critical appraisal of his main arguments. The anonymous naturalist found little in the paper to agree with and was particularly unhappy about Pogodin's vicious and reckless attack on sci-ence.[30] He commented that science was not incompatible with religion, and that it could develop only in societies that protected it from attacks by malicious and destructive forces outside the scientific community. Pogodin welcomed the response only because it gave him another chance to attack Darwin. And again he concluded with the statement that Darwin had not come anywhere near to answering the crucial question of the origin of life. Darwin had failed, he said, because he did not have the vaguest idea where to start his study of organic evolution. Pogodin did not respond to the statement of the naturalist-critic that "the early Russian enthusiasm for Darwin has passed, and, at the present time, even the young people have begun to view Darwinism in a more sub-dued and sober manner."[31]

Pogodin did not have to look far to find out that the enthusiasm for Darwin among the representatives of critical thought was very much alive, even though it was tempered by the rising intensity of the anti-Darwinian campaign. He was no doubt familiar with A. P. Shchapov's new book—*Social and Educational Conditions of the Intellectual De-velopment of Russia,* published in 1870—which lamented the painfully slow development of rationalist tradition in Russia and argued in favor of natural science as the only salvation for Russia. The progress of Rus-sian society, he said, depended on substituting a forward-looking in-volvement in natural science, which alone could solve the grave prob-lems of the present day, for the "archeological" orientation, and its overcommitted involvement in reconstructing and romanticizing the medieval past. The supporters of "archeological" orientations were in-terested in preserving the ideology of autocracy rather than in creating a basis for modern institutions and scientific technology. In Shchapov's nomenclature, Pogodin's activities provided a graphic example of a thorough commitment to the archeological style of thought. Shchapov never failed to mention Darwin among the towering figures of natural science, who showed the path to a higher level of human existence and social values.[32]

For a variety of reasons, a goodly number of erstwhile Darwinists began to voice criticism about the major postulates of evolutionary thought. S. A. Rachinskii, who became famous as the first translator of the *Origin of Species* into Russian, and who used his lectures in botany at Moscow University to popularize the evolutionary ideas, became a determined foe of Darwinism.

N. N. Strakhov, one of the first writers to popularize Darwin's theory in Russia, was now a confirmed and uncompromising anti-Darwinist. He criticized the evolutionary theory as a clear illustration of the intellectual myopia of natural science and fought for a revival of philosophical idealism, untrammeled by the rigidity and formalism of the scientific method. He was more than Darwin's adversary: he organized and led a general antievolutionary movement. In his anti-Western crusade, which did not subside until the 1890s, he treated Darwin's theory as a sweeping and reckless negation of the ideals of humanity built into Russian culture. In Darwin's ideas he saw a triumph of "Western nihilism" that combined a materialistic position in philosophy with a rapid erosion of moral principles, particularly after the irreversible collapse of Cuvier's antitransformism.[33] As he saw it, Darwinism was a unique expression of skepticism bequeathed by Bacon, Hobbes, Locke, and other masters of the English theory of knowledge and moral philosophy. Darwinism, he noted, was a child of modern science, based on Newtonian mechanical models. Even if it did meet the standards of science, it did not ascend the heights of philosophical vision. In 1878 Strakhov wrote to Leo Tolstoy that he was sufficiently familiar with science to grasp its limitations and to avoid becoming one of its "superstitious worshippers."[34]

In Strakhov's opinion, Darwin's view of variation as a random occurrence was not an empirical fact in science but a logical deduction from a materialistic philosophy that denied the role of a supreme intelligence in upholding the harmony and inner order of nature and in ensuring the regularity and predictability of natural processes. Darwinism was a pseudoscience and a materialistic conspiracy, for it eliminated the role of divine powers in maintaining the regularity of natural processes. "The weakness of Darwin's theory," he wrote, "is in that, like any theory that places the primary emphasis on random change, it cannot encompass natural phenomena in their full magnitude and explain their very essence."[35]

Strakhov's attack on Darwin was part of a broader attack on the scientific community—and on science. Strakhov fought what he believed to be the two major "superstitions" of the scientific community: the belief that science alone was in a position to resolve the greatest mysteries of nature, and the habit of certifying scientific knowledge by relying on the consensus of scholarly opinion rather than on the weight of empirical proof.[36] He also attacked nihilist ideology, which, in his view, supported the intellectual imperialism of science. Strakhov contended that Darwinism belonged to the currents of modern scientific thought responsible for a massive decline of idealistic metaphysics.

Strakhov's attitude toward the scientific community was full of animosity and cynical negation. He conducted bitter attacks on science as an archenemy of the sacred values of Russian culture. It came as no surprise to him that naturalists accepted Darwin's theory in such large numbers; after all, he said, it was well known that the scientific community produced more "materialists, pantheists, and individuals who rejected every notion of supernatural interference with the order of nature" than any other group. Cuvier's nontransformist views, in Strakhov's judgment, were rejected, not under the pressure of accumulated scientific facts, but by a sudden shift in the "beliefs" of the scientific community unrelated to true science.[37] It was not enough to point out that Darwinism was moored in the contaminated waters of "materialism" and "pantheism"; Strakhov found it important to note that, among its many other frailties, Darwinism was non-Russian. To prove his point, he relied on the recently published *Russia and Europe*, a much-noted book by N. Ia. Danilevskii, which argued in favor of a national character of science. Danilevskii made it known that in order to reach the inner depths of Darwin's mystique it was necessary above everything else to note that his works, including the *Origin of Species,* were products and expressions of English cultural and social values, scientific traditions, and methodological preferences. "Every nation and every epoch," Strakhov wrote, "prefers certain theories not because of the power of their logical development but because of their moral appeal."[38] Darwin's theory, no less than Hobbes's views on the state and Adam Smith's orientation in political economy, is a reflection of the moral makeup of the English nation. Russia must reject Darwinism on two grounds: it does not agree with "established" (Cuvier's) science, and it is the product of a moral atmosphere that placed strong emphasis on "skepticism" and "empiricism," totally alien to the moral matrix of Russian thought.

Strakhov and Pogodin briefed the high officials in the Ministry of Public Education about Darwinism as an open attack on the sacred values of tsarist autocracy. In 1871 D. A. Tolstoi, Minister of Public Education, responded to their warnings by promulgating new statutory regulations for gymnasiums. The classical gymnasium was now recognized as the only direct institutional path to university education. Forty percent of the curriculum concentrated on classical languages. Biology was not represented in the new curriculum. The new curriculum was one of the most potent mechanisms the government had employed in its effort to stem the tide of "natural science materialism," interpreted as a major enemy of autocratic institutions. Pogodin and Strakhov were the leading interpreters of Darwinism as one of the chief pillars on which the new materialism was erected.

At first the criticism of Darwin's theory was a mere trickle. The writings of Strakhov and Pogodin helped accelerate the growth of the anti-Darwinian movement and make it a force of great consequence. The growing criticism came from many sources: from disgruntled naturalists and the new breed of theological scholars, from the pioneers of idealistic metaphysics, and from self-styled patriots eager to ward off the corrupting influences of Western materialism. The towering figure of von Baer stood in the background, providing a constant flow of comments on which the growing breed of critics relied to strengthen or embellish their anti-Darwinian arguments.

I. F. Tsion was one such critic. A graduate of Berlin University and a professor of physiology at the Medical-Surgical Academy and, on a part-time basis, at St. Petersburg University, he distinguished himself by work in the physiology of blood circulation and the nervous system. He wrote on the experimental methodology of physiology and produced the first Russian textbook in physiology. Provocative expressions of conservative political views brought him into constant conflict with both students and peers. A student demonstration in 1875 led to his dismissal from the medical academy.[39] From 1875 to 1891 he lived in Paris, representing the Russian Finance Ministry.

In 1878 Tsion chose a unique way to vent his anti-Darwinian views: instead of attacking Darwin directly, he chastised Ernst Haeckel, at that time generally considered one of the most enthusiastic—and the most uncritical—of the supporters of the new evolutionary theory. The choice of target was different, but the effect was the same, the more so because his article appeared in the *Russian Herald,* a widely read "thick journal," and was written in dynamic and lucid prose. Tsion made sure to create an air of objectivity by pointing out both the greatness and the weakness of Darwin's theory. Darwin, he said, deserved credit for transforming a whole array of descriptive and diffuse biological disciplines into theoretically integrated bodies of knowledge. Darwin's success in making "the theory of transformism" a solid part of modern science earned him "a place among the greatest naturalists."[40] Even if Darwin had done nothing else but collect the vast stores of empirical material, his name would have been enshrined among those of the leading masters of natural science.

After paying homage to Darwin's scientific stature in a few short passages, Tsion devoted the remainder of his 66-page article to a harsh criticism of Haeckel's interpretation of the basic principles of Darwin's theory. The criticism of Haeckel, however, was only a ploy; the main intent of the article was to undermine the very foundations on which Darwin built his theoretical edifice. Combining his own logical constructions with critical hints supplied by the writings of von Baer, de Quatrefages, Kölliker, and other contemporary critics, he concluded that natural selection had no empirical support. Even worse, it was a sterile hypothesis, for it had no chance of receiving empirical verification of any kind.[41] "No scholar," he said, "ever thought of placing Darwin's theory on the same plane with the law of gravitation and the law of the conservation of energy." It was not Darwin, according to Tsion, but Haeckel who transformed a dubious hunch about the struggle for existence into a general law of nature.[42]

Criticism emanating from theological quarters became more common than during the 1860s, and it also became less dependent on outright translations or summaries of Western anti-Darwiniana. During the 1860s the theologians concentrated on marshaling religious and moral arguments against the more disagreeable components of Darwin's theory. At this time they were not ready to make a systematic use of arguments raised by Darwin's critics in the scientific community. The publication of the three Russian translations of *The Descent of Man* gave theological criticism a new sense of urgency and a more precise challenge. During the 1870s theological criticism adopted two distinct modes of operation. One group of theologians, depending mainly on Western translations, worked on rewriting natural history to bring it in tune with scriptural explanations and to make it a source of authoritative arguments against Darwin's theory. The second group concentrated on analyzing and systematizing anti-Darwinian arguments advanced by such eminent scientists as von Baer, Wigand, and Kölliker. Both groups depended almost exclusively on theological journals for publication of critical essays.

The first group—the contributors to a new natural theology—categorized its activity as "scientific-theological research," or as an effort to advance that part of theology that was "particularly close to natural science."[43] The "scientific" branch of theology dealt with the biblical history of the independent creation of individual forms of life. It directly opposed Darwin's theory, which dealt with the history of continuous transformation of species and which treated the origin of new species as a result of long evolutionary activities. Darwin's categorical claim of the anthropoid origin of *homo sapiens* gave theological writers both a most critical target to attack and a most difficult generalization to refute. Although theological naturalists made extensive use of suitable citations from the scientific literature, syllogistic constructions, and philosophical arguments, they depended on scriptural teachings as the highest and most reliable source of information on the structure and the dynamics of the universe. This type of criticism proved ineffective and compelled theological scholars to search for other modes of operation. The most learned theologians quickly gave up on the idea of creating a theologically acceptable natural history as a systematic body of knowledge. They did not give up on expanding their interests in, and their comments on, the burning theoretical developments in modern science. The attack on Darwin's theory did not lose momentum; it merely moved to another ground.

The second group of critics, who shifted the emphasis from theological to scientific grounds, relied on a strategy that called for protecting science, rather than theology, from the Darwinian plague. These critics kept in touch with the proliferating anti-Darwinian literature produced by recognized scientists. A. P. Lebedev, professor at the Moscow Theological Academy, was one of the better-known early representatives of this group. In 1873 he published a lengthy article in the *Russian Herald* which summarized the antievolutionary views of the leading contemporary critics of Darwin's theory. Presenting his essay as a philosophical discourse based on logic and science, Lebedev dwelt on the "weaknesses" of Darwin's "naturalist deism," the narrow use of empirical data, and excessive dependence on philosophical materialism. But the main thrust of his attack depended on anti-Darwinian statements by recognized biologists. Lebedev advanced his arguments calmly and methodically, with all the trimmings of an objective academic discourse. His conclusions, however, stayed completely within the realm of expectations: he relegated the theories of both the *Origin* and the *Descent* to the realm of fictitious knowledge. He saw no reason to challenge von Baer's assertion that Darwin's transformist thesis was "a product of phantasy without any basis in real observation."[44]

Lebedev's critique marked a turning point in the history of Russian theological criticism of Darwin's theory. His predecessors limited their criticism primarily to refuting Darwin's atheism by recounting the scriptural interpretation of the creation of the world. Lebedev, by contrast, concentrated on two novel lines of attack: the incompatibility of Darwin's arguments with the established facts of modern science—or, rather, the facts he chose to regard as established—and the flaws in Darwin's theoretical reasoning. Although he directed his main arguments at *The Descent of Man* and its specific concern with the origin of moral norms, he relied heavily on von Baer's attack on the *Origin of Species,* which avoided a discussion of human evolution. By a careful selection of citations from the works of Charles Lyell and Thomas Huxley, he tried to create an impression that these two naturalists had serious doubts about the new theory. The two groups of theological writers relied on different strategies in examining Darwin's biological legacy, but they produced identical results: without exception, both groups rejected every part of Darwinian thought.

The mushrooming criticism of Darwin's evolutionary theory was not unexpected. Unexpected was the tangible increase in the number of theologians who commanded well-grounded knowledge about the scientific arguments centered on evolutionism. The emergence of theological criticism of Darwinism marked the beginning of a concerted and systematic effort of religious scholars to make the major developments in modern science topics of learned theological commentaries and to involve the church in the discussion of modern knowledge. Theologians preferred to write about scientific developments that threatened the orthodoxy of theological thought, and showed no interest in a synthesis of scientific and theological ideas. A reconciliation of scientific and religious thought was outside the realm of their intellectual concerns.

Theological writers did not limit their activity to attacks on the arguments Darwin had used in building his theoretical edifice. They also disseminated information about noteworthy developments in the West which provided a background for anti-Darwinian criticism. In 1880, for example, the journal *Christian Readings* reported extensively on the bitter dispute between the Darwinist Haeckel and the anti-Darwinist Virchow at the Congress of German Naturalists and Physicians, held in Munich in 1877.[45] While reporting the details of the argument, the journal did not hide its full concurrence with Virchow's views. Dispassionate readers, however, benefited from the report's survey of ideological questions related to Darwin's theoretical structures.

Mainly in a roundabout way, Darwin came under attack also from

representatives of religious philosophy, which at this time was expe-
riencing a strong revival. In 1874 the twenty-one-year-old Vladimir
Sergeevich Solov'ev completed his magisterial thesis at Moscow University, an event that marked a turning point in the history of philosophy
taught in Russian universities.[46] An insightful and forceful analysis of
Western philosophy, particularly as represented by Schopenhauer and
Hartmann, led Solov'ev to announce the beginning of a new era in the
growth of philosophical thought, which he interpreted as a meeting
point of "rational knowledge" in the West and the spiritual legacy of the
East.[47] The new philosophy represented a fusion of "the logical perfection of Western forms" with the "substantive fullness of spiritual contemplation" in the East. In this work, as well as in his doctoral dissertation, "The Abstract Principles of Philosophy," he attacked positivism as
a philosophy closely associated with science. The future, as he saw it,
belonged to a unity of philosophy and religion, not to a unity of philosophy and science. In science he saw a repository of shallow and fragmented knowledge. Metaphysics alone can lead to fundamental and absolute truth and to a universally integrated body of knowledge.[48]

Under Solov'ev's strong influence, Russian professors of philosophy,
without exception, followed a strategy of playing down the role of
science in the intellectual conquests of modern man. Their attacks concentrated on the "mechanistic" and "materialistic" foundations of
Newtonian science, which, of course, included Darwinism. The philosophers showed a strong inclination to interpret the spreading crisis in
Newtonian science as a rapid and irretrievable decline of the intellectual
authority of science in general. In their eagerness to point out the narrow horizons of scientific knowledge, Solov'ev and his university followers were actually engaged in a war against the ghost of nihilism, a
philosophy that denigrated all nonscientific modes of inquiry and types
of knowledge.

The writings of Lebedev and Solov'ev were the fountainheads of
two notable anti-Darwinian traditions anchored in religious thought.
Lebedev was a progenitor of the theological tradition that made extensive use of arguments advanced by anti-Darwinian scientists. Theological writers helped in diffusing the ideas of such astute critics of Darwin's
theory as Wigand, Kölliker, Nägeli, de Quatrefages, and von Baer.
Solov'ev, by contrast, was the chief progenitor of a philosophical-
religious tradition that contributed to the anti-Darwinian movement by
concentrating on the cognitive inadequacies of science in general, rather
than on the contestable specifics of Darwin's theory. Solov'ev's followers

operated on the assumption that, by its very definition, science cannot avoid a close alliance with the guiding principles of "mechanism" and "materialism."

Boris Chicherin, a Hegelian who defended the superiority of idealistic metaphysics over science, lost no time in adopting von Baer's anti-Darwinian arguments and in making them the subject of a general public discourse. In *Science and Religion* (1879) he concentrated on two major "flaws" in Darwin's theory. First, he thought that Darwin's method was illogical: it tried to explain life, a purposive natural phenomenon, by relying on mechanistic models that implied a rejection of every kind of purposiveness. The element of purposiveness, in Chicherin's view, falls completely outside the competence of science; it is a problem that can be handled successfully only on philosophical grounds. Second, Darwin based his theory on inadequate empirical data; for example, he did not marshal sufficient evidence in support of adaptation as a mechanism of organic evolution. This deficiency, according to Chicherin, forced many Darwinists to resort to fabricated ideas that often had nothing in common with reality and confused "science" with *Naturphilosophie*.[49]

The main criticism of Darwinism came from three sources: from the scientific community, from an expanding group of theological commentators on the main currents of scientific development, and from the representatives of new stirrings in idealistic philosophy. But it also came from individuals who did not belong to any one of these easily recognizable groups. Among these individuals the most noted were general commentators on the state of Russian society, who defended autocracy from three real or imaginary enemies: democracy in politics, materialism in science and philosophy, and utilitarianism in ethics. N. N. Novosel'skii, the author of *Russia's Social Problems*, published in 1881, was a typical commentator of this order. He wrote about local government, the *zemstvo* movement, incipient industrialization, labor, and many other topics, always with an eye on the developments he considered inimical to the growth of the nation. He was also interested in the intellectual stirrings of the day and in the compatibility of their general orientations with the spirit of Russia.

Evolutionism in general and Darwinism in particular did not escape Novosel'skii's attention. He devoted forty pages to an analysis of Rudolf Virchow's attack on Darwinism—particularly on Haeckel's brand of it—at the Congress of German Naturalists and Physicians, held in Munich at the end of September 1877. His book represented the first systematic effort to alert the government to possible ties between Darwinism and socialism. Novosel'skii devoted much attention to arguments showing the fundamental incompatibility of the evolutionary theory with Christian beliefs. Biological evolutionism, he said, is based on a mechanistic orientation in science and a materialistic orientation in philosophy, the leading enemies of religious beliefs and moral principles. He differed from other antievolutionists in one important respect: he openly suggested that the Russian government undertake sweeping police and censorship measures to wipe out the new heresy.[50]

As Darwinian ideas and attitudes became a recognized topic of public debate, the literary figures of the age could not escape taking notice of the new source of heretical ideas. In *At Daggers Drawn*, an antinihilist novel published in installments in the conservative *Russian Herald* in 1870–71, the novelist Nikolai Leskov had room only for a contemptuous attitude toward the moral code of the *negilisty*, a lingering offshoot of the rapidly waning nihilist movement. Unlike the *nigilisty* of the early 1860s, the *negilisty* of the late 1860s believed in the Hobbesian principle of *homo homini lupus* as built into Darwin's theory of the struggle for existence. "Gobble up others lest they gobble up you. . . . Living with wolves, act in a wolflike fashion and hang on to what you hold in hand."[51] The aim of Leskov's antinihilist novel was clearly to lend support to the conservative view of Darwinism as an attack on the humanistic base of moral principles.

F. M. Dostoevsky—in the novel *The Possessed*, among others—made use of Darwinian metaphors. He took note of the rapid spread of Darwin's ideas. In 1876 he remarked—in *The Diary of a Writer*—that there was a fundamental difference between Western and Russian attitudes toward Darwin's contributions to science; while in the West Darwin's theory was viewed as a "brilliant hypothesis," in Russia it quickly acquired the authority of an "axiom."[52] Perhaps Dostoevsky's extensive, and often biting, use of naturalist allegories owed some debt to Darwin's suggestive ideas.

Echoes of Philosophical Criticism
in Germany

German philosophers wasted no time in responding to the intellectual challenges of Darwin's theory. The most noted comments came from Eduard Hartmann and Friedrich Albert Lange, who advanced distinct philosophical interpretations that helped shape contemporary thought in and outside the community of philosophers. The main works of both went through several editions and reached an unusually broad cross section of the reading public. Both were noted in Russia and added potent fuel to the fire of Darwinian debate. They helped give sharper focus to the philosophical inquiry into the theoretical structure of evolutionary thought and to the criticism of various general aspects of Darwinism.

In the *Philosophy of the Unconscious* (1869) Hartmann undertook the arduous task of showing that the alliance of science and spiritualistic metaphysics was an absolute necessity for both modes of inquiry. He pointed out the intellectual futility of the reigning alliance of science and materialistic—mechanistic—metaphysics. In a sweeping attack on the "materialism" of nineteenth-century science, Hartmann did not over-look Darwin. He thought that Darwin's principles could be accepted on the most general level but that "the philosophy of the unconscious"—a metaphysical amalgam of Schopenhauer's "Will" and Hegel's "Reason"—could give them more depth and a much broader perspective. Darwin's theory, he argued, recognized the teleological aspects of the evolutionary process, but it limited teleology to an empirical and induc-tive level. He stressed the need for a metaphysically comprehended tele-ology, not bound by the narrow limits of empirical knowledge. To achieve this goal meant to bridge the chasm between idealistic meta-physics, with its traditional unconcern with reality, and the empirical sciences, and their distaste for immaterial absolutes. It was clearly for this reason that Nietzsche labeled him an "amalgamist."

This orientation found a particularly clear expression in the *Truth and Error of Darwinism,* an effort to find a middle ground between Darwin's ideas and the views of Darwin's principal opponents, particu-larly von Baer, Wigand, Kölliker, and Nägeli. Hartmann developed a strategy that subsequently became particularly attractive to the lead-ers of neovitalism: the mechanistic orientation of general biology, par-ticularly of Darwinian theory, should not be rejected but should be recognized only as the beginning of a more intensive search for deeper meanings of evolution. This orientation made it necessary to com-bine "mechanical causality" (a scientific concept) and "teleology" (a metaphysical notion) into a "higher principle of logical necessity" that explains both the organic and the inorganic universe.[53] In all this, Hartmann clung steadfastly to the view that "teleology," rather than "mechanical causality," provided the safest path to the innermost depths of the evolutionary riddle. His world outlook was dominated by a belief in the superiority of idealistic metaphysics over empirical science. He directed his basic criticism at the widely held opinion that Darwin's major contribution was in freeing biology from metaphysical speculation and in making natural science the only legitimate source of knowledge on the infinite complexities of organic evolution.

Hartmann's philosophical position evoked varied reactions in Russia. Vladimir Solov'ev, at the dawn of his illustrious philosophical career, praised Hartmann for his assault on the "narrow" vision of modern science and his sustained effort to revive idealistic metaphysics as a source of the most sublime wisdom. Despite all the flattery, however, Solov'ev could not accept Hartmann's mode of philosophical thinking, particularly his ambitious—and unrealistic—effort to build a metaphysical system upon empirical foundations and inductive method. Solov'ev resented Hartmann's effort to create a system of philosophical thought without resorting to mysticism. He heartily approved of Hartmann's merciless attack on "science materialism."

In 1878 N. N. Flerovskii, the well-known author of the *ABC of the Social Sciences,* published the *Philosophy of the Unconscious, Darwinism, and Realism,* an unwieldy and hastily written effort to draw the main lines of a modern world view. His reference to the work of Darwin and Hartmann had only one purpose: to help him give his own philosophical views sharper focus and more depth. His own philosophical position was built on two premises describing the unity of the universe. The first principle emphasized knowledge—"information" in the more modern parlance—as the basic regulative force embracing the entire universe. He gave this force extensive philosophical and scientific treatment. The second principle emphasized solidarity as a centripetal force operating on the cosmic level. Whereas Darwin's work, in his opinion, suffered from a lack of clear and articulate philosophical vision, Hartmann's work, by contrast, represented a paradoxical effort to build a metaphysical system by relying on the inductive method of the natural sciences.[54]

Flerovskii thought that both Hartmann and Darwin erred in placing too strong an emphasis on the struggle for existence as a universal law of nature. He contended that this emphasis justified immorality as a norm of human behavior. Not competition, fostered by Western industrial societies, but cooperation, deeply ingrained in the inner fabric of the Russian *obshchina,* he saw as the sound basis for social progress.[55] Darwin, according to Flerovskii, erred also in placing too much emphasis on the evolutionary role of adaptation. As a mechanism of evolutionary progress, adaptation is contradicted by the "known fact" that the simpler forms of life achieve more nearly perfect adjustment to the environment.[56] Nor did Darwin's theory explain man's involvement in the dual process of adaptation to nature and to society. To bring organic evolution in tune with his principle of cosmic solidarity, Flerovskii went so far as to recognize an "internal impulse" as "the main and primordial cause of organic development."[57]

With the exception of a few introductory remarks, Flerovskii made no effort to compare Darwin's science and Hartmann's philosophy. He showed clearly that his familiarity with and reference to Hartmann's ideas exceeded by far his familiarity and concern with Darwin's contributions. In Hartmann's philosophical ideas he saw a deliberate effort to blunt the intensity of Darwin's influence on modern thought and to ensure a triumph for German philosophy over English and French science.[58] While disagreeing with individual principles built into Darwin's theory, he judged Darwin's general contribution to science equal to Newton's greatest achievements. Abstruse and unstructured, Flerovskii's study attracted an unusually small number of readers.

N. K. Mikhailovskii, the intellectual leader of populism, agreed with Flerovskii's contention that Hartmann's "philosophy of the unconscious" represented an effort to replace Darwin's theory of evolution as a guiding force in modern science. Hartmann's effort, he thought, was part of a rising movement, particularly in German philosophical circles, to reaffirm the intellectual supremacy of metaphysics over science. Hartmann's metaphysical notion of the "unconscious" was intended to be a response to Darwin's scientific notion of "evolution." History has borne out Mikhailovskii's prophetic statement that "the philosophy of the unconscious" was only a passing attraction. During the first ten years after its publication, he said, Hartmann's book caused much excitement among philosophers, but despite its primary intent, it made no impact on scientific thought. In Darwin's theory of evolution, by contrast, Mikhailovskii saw a development of epochal significance. He thought that Darwin's theory was built on solid scientific foundations and exercised a strong influence on all the sciences. It not only precipitated a revolution in biology but also exercised a strong influence on psychology and linguistics. "It received support from physics, gave birth to such new disciplines as comparative study of culture, and, in a most resolute fashion, it knocked on the door of all the social sciences."[59]

Hartmann, as Mikhailovskii saw him, created a system of sterile thought: his philosophical system neither achieved an effective synthesis of contemporary thought nor opened new vistas for research. Darwin, by contrast, created both a new scientific orientation and a solid foundation for philosophy. Mikhailovskii disagreed with individual principles of Darwin's theory, but he did not challenge Darwin's revolutionary role in broadening the vision of both science and philosophy. He showed no reticence in criticizing various aspects of Darwin's theory—particularly the excesses of Social Darwinism—but he never failed to acknowledge the immense intellectual resources of evolutionary thought. In his opinion, the differences between Darwin's and Hartmann's views had deep roots in the unique features of English and German cultures. While Darwin tied his science to the English tradition in empirical philosophy, Hartmann limited his "science" to the German tradition in metaphysical thought of an idealistic orientation. By expressing a preference for the Darwinian method, Mikhailovskii reaffirmed the antimetaphysical stance of Russian populism.

F. A. Lange commented on Darwinism in his classic *History of Materialism,* particularly in the second edition, published in 1875. The publication of the Russian translation of this work in 1881–83 acquainted the general reading public with a German interpretation of Darwin's thought which differed markedly from Hartmann's criticism. Lange and Hartmann took radically different positions. Hartmann thought that the future belonged to the unity of science and metaphysics, with metaphysics serving as the chief and most competent judge of scientific contributions. Philosophy, in Hartmann's view, should serve as a clearing house for scientific ideas.[60] Lange, by contrast, believed that science and metaphysics are separate modes of inquiry and that they have no point of contact. Science is knowledge; metaphysics is a specific form of poetic expression. Metaphysics has no right to oversee or interfere with the development of science. Hartmann declared war on materialism in both science and metaphysics; both materialisms, he thought, have the same makeup. Lange had no argument with materialism in science; in philosophy, however, he considered materialism a crippling force of major magnitude. He accepted materialism in science as long as it appeared as an empirical rather than as a metaphysical phenomenon. In the principle of causality, the backbone of Newtonian "materialism," he saw an empirical phenomenon and a legitimate domain of the scientific method.

The Russian translation of the *History of Materialism* was of the third German edition, published in 1877. It retained all the elaborations and revisions that went into the second edition, including a sizable expansion of the chapter on Darwin. The Russian translator was Nikolai

Strakhov, who at that time had firmly established himself as a leading Russian anti-Darwinist. In 1899 the book appeared in a second Russian edition, a clear indication of persisting interest in its contents.

Lange operated on two levels. On one level, he tried to draw a sharp line separating science from metaphysics and teleology and to show that scientific knowledge and metaphysical learning are incommensurable. On another level, and quite independently of his philosophical stance, he pointed out what he considered to be the weak points in Darwin's theory of evolution. On both levels, he discussed the relevant problems with calmness and persevering thoroughness, depending on both documentary support and logical deduction.

As a philosopher, Lange thought as a true Kantian. Science, he said, deals with particular creations of the human mind; it does not deal with the thing-in-itself, the reality unpierceable even by the sharpest instruments of scientific inquiry. Metaphysics deals with the thing-in-itself, but it depends on poetic constructions rather than on empirically grounded methods of inquiry.[61] It deals with unverifiable products of imagination, not with tested knowledge. With one sweep, Lange acknowledged the legitimacy of metaphysical speculation and denied the right of metaphysics to interfere with the work of the scientific community. He viewed Hartmann's unrealistic effort to establish a mastery of metaphysics over science as "a national philosophy of the first rank," but also as a sure symptom of the intellectual decadence of his time.[62] He responded clearly and significantly to Hartmann's effort to construct a metaphysical system by relying on a flagrant abuse of the inductive method. "It would be difficult," he said, "to find another modern book with so much natural-scientific material that stands in sharp contrast to all the essential principles of the scientific method."[63]

The Russian translation of the *History of Materialism* appeared at the time of a rapidly growing confrontation of science and philosophy. It did not influence the university professors of philosophy who, to a man, chose to go along with Hartmann's effort to make science a secondary force in the organized search for knowledge. It helped reinforce the growing tendency of the scientific community to seek isolation from all metaphysical currents of the day and to make epistemology its major concern. At no time in history did Russian scientists show more interest in philosophy than during the last three decades of tsarist rule, and at no other time did they show so much distaste for metaphysical exhortations.

How did the change in the relations of science to philosophy affect the Darwinian studies? The scientific community learned quickly to disregard the metaphysical criticism of the evolutionary theory. The rare scientists who lent support to neovitalism—and its attitude toward organic evolution—made a deliberate and strenuous effort to dissociate themselves from spiritualistic metaphysics, usually by defining teleology as an empirical—rather than as a transcendental—force. Lange did not create this shift; he merely helped give it a stronger impetus and a clearer vision. He showed the Darwinian scholars a way to disregard the speculation of metaphysics and still maintain an intensive contact with philosophy. The scientific community learned to appreciate both the right of metaphysical scholarship to conduct a war on Darwinism, and its own privilege to ignore metaphysical criticism.

Despite its title, Lange's philosophical treatise dealt with both materialistic and idealistic metaphysics. In the chapter on Darwin he dealt extensively with the pitfalls of idealistic metaphysics. While attacking teleology, as vitalistic metaphysics interpreted it, he joined the Newtonian camp by praising Darwinism for a "thorough application of the principle of causality"—and by considering the principle of causality the most rational prescription for an understanding of the work of nature.[64] Nor did Lange eliminate teleology altogether. He discarded the "anthropomorphic form of teleology"—which relied on human attributes in interpreting the purposiveness of natural processes—and accepted the natural or empirical form of purposiveness, detached from human predilections and intellectual and emotional biases. Whereas Hartmann made causality a weak ancillary of "anthropomorphic teleology," Lange made teleology, trimmed of metaphysical excesses, a ramification of causality. The problem of teleology versus causality, particularly in its relation to the theory of evolution, in general, and Darwin's transformist views, in particular, became one of the central philosophical topics that engaged the attention of Russian naturalists. Such Russian biologists as S. I. Korzhinskii, I. P. Borodin, A. Ia. Danilevskii and A. S. Famintsyn, who entered the field of philosophy during the late 1880s or the 1890s, did not necessarily accept Lange's specific suggestions for separating teleology from metaphysics, but they did accept Lange's claim that the relationship of causality to teleology represented a key epistemological problem of contemporary evolutionary biology. They also accepted Lange's warning that their scientific explanations should under no condition take refuge "in a mystically interfering teleological force."

Lange accepted the Newtonian base of the Darwinian theory; he had no serious quarrel with the materialism and mechanism of Darwin's scientific outlook. This did not prevent him, however, from challenging individual propositions that became the building blocks of this theory. While essentially correct, Darwinism needed much refinement and additional ramifications. Lange objected to the tendency of Darwinian scholars to treat the species as a precisely defined group of animals, disregarding considerable overlapping of key characteristics, particularly among the lower forms. He thought that the Darwinian notion of "imperceptible slowness" did not give a full explanation of variation and that "rapid change" should also be recognized. In all this criticism, Lange relied on scientific arguments and metaphors and avoided metaphysical involvement. He tried to provide examples illustrating the working of the time-honored rule that only the scientific community has the methodological tools and the moral authority to certify scientific knowledge. More than any previously published book, Lange's study gave Russian scientists a clear warning that the future of Darwinism was not in a defense of orthodoxy but in the ability to adjust to the onrush of new developments in evolutionary biology.

CHAPTER FOUR

N. Ia. Danilevskii: Codification of Anti-Darwinism

Science Under Attack

During the 1880s the forces of reaction were in the ascendant. The government broadened and intensified the process of Russification in the peripheral regions, particularly in Poland, passed anti-Semitic laws, encouraged the harassment of Catholics and Lutherans, and sharpened the claws of censorship. The university charter of 1884, the most telling symbol of the age of escalated oppression, put a wedge between students and professors by instituting special academic controllers responsible exclusively to the Ministry of Public Education, reversed the process that made higher education accessible to all social classes, dismissed professors known for their liberal views, made it a crime for students to belong to organizations outside the legal framework of the university, and transferred the administration of student examinations from professors to specially appointed government commissions.[1] The universities suffered heavy losses of traditional privileges to act as autonomous bodies. Ministerial fiat, rather than academic consensus, became the basic mechanism of university administration. Deans and higher university officials ceased to be elected representatives of the faculty.

Konstantin Leont'ev, the ideological leader of the revived forces of reaction, defended the "new" spirit of Russia as an antithesis to Western thought built on "the rationalist delusion of democratic and utilitarian progress."[2] He built his philosophy on the premise that science was an evil force, a source of "eudaemonism and utilitarianism," the main reason for the reckless rapidity of the development of modern society. To preserve the stationary *obshchina* as the sacrosanct base of Russian society, he proposed to slow down the development of science as the mechanism of social progress. He thought that "the great rapidity of modern life," a product of scientific achievement, led to unnecessary and injurious social mobility and to the eradication of differences between social estates, nations, customs, and religions which could not but exercise an injurious influence on "the mental state of mankind." The rapidity of modernization gave "vagueness, feebleness, uncertainty, and instability to our mental life."[3] As a cure for the ills of the day, he suggested an outright recognition of "the powerlessness of our intellect," "the subordination of rationalism to mysticism," and a full submission of all "raw and accessible knowledge to the secret principles of a higher order."[4] He revived the old argument that science was contrary to the spirit of Russia because it encouraged materialism in thought, utilitarianism in ethics, and democracy in politics.

It was in this atmosphere that Nikolai Iakovlevich Danilevskii published his *Darwinism*, a monumental study with one goal: a full demolition of Darwin's theory of organic evolution. He died a few days before volume 1—published in 1885—had reached the public. This volume appeared in two separate books, totaling over a thousand printed pages. Volume 2, much shorter and incomplete, appeared in 1887. Known as a conservative thinker deeply steeped in Slavophile ideology, Danilevskii made his attack on Darwinism part of a general and relentless war on Western "materialism." Like Leont'ev, Danilevskii looked on the triumphs of Western science as a major threat to the spiritual values that kept the Russian soul together and gave it a distinct cultural and historical individuality. Both fought the spirit of nihilism and its prophets, who expressed an optimistic view of science as the only true source of designs for the salvation of Russia. They gave ideological support and philosophical comfort to the architects of one of the most oppressive periods in Russian history.

In his youth Danilevskii belonged to the ill-fated Petrashevskii circle, a group of young intellectuals immersed in the ideas of Fourier and Proudhon and concerned much more with self-education than with the diffusion of socialist ideas. In 1849 the police, agitated by the revolutionary waves in western Europe, disbanded the circle and exiled its leading members to various isolated places in eastern Russia and Siberia. The novelist F. M. Dostoevsky found himself in Siberia. Danilevskii was sent to Samara, where he was kept under the watchful eye of local police.

In 1853 Karl von Baer organized the first in a series of expeditions to the Caspian basin for a comprehensive study of fishing resources. He chose Danilevskii—whom he did not know personally—as a member of his team. Designated as the "statistician" of the expedition, Danilevskii quickly took on many additional activities. By the time the expedition completed its work in 1857 he had covered the full gamut of relevant challenges in research ranging from ichthyological observations to economic, ecological, and technological analyses. His reports are also rich in ethnographic descriptions of fishing communities along the Caspian rivers and contain valuable suggestions for the conservation of fishing resources. He quickly established himself as one of the expedition's most productive members and as von Baer's chief assistant. Contributions to von Baer's expedition made Danilevskii a leading expert in the biological and ecological aspects of Russian fishing. It was for this reason that the Russian Geographical Society elected him a full member in 1858 and awarded him one of its most prestigious medals for scientific achievement. Karl von Baer was one of those who recommended him for this honor.[5]

After completing his work as a member of the von Baer Caspian expedition, Danilevskii joined the Ministry of State Properties, first as a member of the Scientific Committee and then as a member of the Ministerial Council, a chief advisory body. These high positions did not prevent him from organizing several expeditions to study the fishing resources of various river systems and lakes. Death prevented him from carrying out a study of fishing resources at Lake Gokcha in the Caucasus. Contemporaries admired his unmatched familiarity with technical literature, vast reservoirs of personally collected and recorded observations, and ability to transform masses of raw data into precisely arranged and logically rigorous reports.

Danilevskii was a man of many talents and research interests. In addition to valuable studies on the fishing economy and technology, he wrote about the Russian national character, the relations of Russia to the Franco-German War of 1871, Russian demography, the Ice Age, Russian geographical terminology, the struggle against phylloxera in southern Crimea, Russian monetary policy and protective tariffs, and the evils of Russian nihilism. He displayed enviable patience and adequate skill in handling statistical data; but he was equally proficient in philosophical analysis. He was clearly one of the most original Russian philosophers of history. In one respect his publications fall into two categories: some were products of his professional activity, usually as a member of government research teams; others were products of his personal concern with the critical intellectual and ideological issues of the day, particularly as these affected the spirit of Russia.

Danilevskii's major works—*Russia and Europe* and *Darwinism*—belonged to the second category. They were united by a deep concern with the cultural, particularly the philosophical, roots and manifestations of an irreconcilable conflict between western Europe and Russia. The sociologist Pitirim Sorokin interpreted the antagonism between Russia and Europe as a reality with deep roots, reinforced by unconscious historical instincts and by conscious considerations and historical conflicts. Danilevskii, he said, wanted to trace the sources of the antagonism between the Russian and the Teutonic-Romanic tribes, and to explain why it had persisted and whether it was "a sort of temporary misunderstanding destined to disappear in the future" or something perennial.[6] This conflict was at the bottom of Danilevskii's philosophy of history.

In 1869 Danilevskii published *Russia and Europe,* a widely noted treatise in the philosophy of history and a literary landmark.[7] Intricate and erudite, the volume was clearly dominated by a single idea: the incommensurability of the dominant values of Russian culture and other—particularly western European—"historical-cultural types." Danilevskii agreed with the Slavophiles who elaborated the idea of the cultural uniqueness of Slavic civilization, but he resented their exaggerated reliance on Western philosophical metaphors. Like the Slavophiles, he tried to show the full autonomy of Slavic culture and to refute claims of the advocates of the relative superiority of Western civilization. Topical priorities of *Russia and Europe* showed convincingly that he did not consider science a particularly strong expression of the inner springs of Russian culture. He talked about the national strength of English science, of German science, and of French science, but the national strength of Russian science did not attract his attention. It was no accident that he did not place a single Russian name among the most illustrious and creative contributors to science. In the West, as he saw it, science represents a central cluster of values; in the Slavic world it is merely borrowed material unaffiliated with the primary values. It was exactly at the time when he wrote *Russia and Europe* that, unbeknown to him, D. I. Mendeleev announced his discovery of the periodic law of elements, the most comprehensive law in chemistry formulated in the nineteenth century, and that N. I. Lobachevskii's non-Euclidean geometry received full recognition as one of the most challenging developments in modern mathematics.

Dostoevsky was impressed with *Russia and Europe* as a synthesis of current arguments against all major theories that regarded the Russian way of life and values as an extension of Western culture.[8] He was under the spell of Danilevskii's philosophy of history when he asserted that science may be a source of great wisdom but that it is not Russian. The great novelist acknowledged the social usefulness of science and said that Russia should take full advantage of modern developments in experimental science, but he insisted that the true genius of Russian culture lay not in secular wisdom and positive knowledge but in religious contemplation.

Danilevskii did not forget to remind his readers that English science, a particularly strong component of English culture, stressed competition and struggle as the social and cosmic forces of the first order. The most typical representatives of the English national character in the pursuit of science made competition the central theme of their inquiry.[9] In the middle of the seventeenth century Hobbes formulated a general political theory that emphasized the role of the state in regulating competition (*bellum omnium contra omnes*) as a mechanism of social dynamics. In the second half of the eighteenth century, Adam Smith formulated the theory of free competition as a source of economic progress. In the middle of the nineteenth century, Darwin made the principle of natural selection, based on the struggle for existence, the backbone of organic and social evolution. The English national character both encouraged the development of science and made science a one-sided activity.

Danilevskii's *Darwinism*

In *Darwinism*, Danilevskii presented an elaborate critique of the major principles built into the *Origin of Species*. To make his criticism appear as a dispassionate discourse, he made occasional and rather perfunctory references to the strengths and commendable qualities of Darwin's work. In this effort, however, he was very careful to praise Darwin for his style of work and general mode of scientific thinking, rather than for his contributions to evolutionary theory. He called Darwin the first true biologist—the first to look at plants and animals within the context of their natural environment. Darwin recognized the value of the microscope in embryology, of the herbarium in botany, of special theaters in anatomy—and of natural history museums in general—as indispensable sources of pertinent scientific information. To the skillful use of research instruments he added careful and thorough observation of organisms in places where "they live and influence each other."[10] This kind of approach, according to Danilevskii, made biology a true and full science.

Danilevskii praised Darwin also "for adducing scientific arguments in favor of nationalism over cosmopolitanism." After all, what else was nationalism if not an accumulation and transmission of combinations of physical, intellectual, and moral characteristics that make up the distinctive features of national groups? These characteristics, in turn, influence the development of political, industrial, artistic, and scientific activities, and are the basis of national progress. He did not state it explicitly, but Danilevskii acknowledged the scientific value of comparing the notion of "species" with the notion of "nation." Darwin, he said, enriched the theory of education by showing why schoolteachers should shift their goal from cultivating individual characteristics to helping preserve group characteristics acquired and transmitted by many generations.[11] Danilevskii was quick to add that all these were only lateral contributions and should not be allowed in any way to cover up the "fundamental falseness" of Darwin's evolutionary principles.

In praising Darwin, Danilevskii relied heavily on Karl von Baer's technique of making perfunctory and unillustrated references to Darwin's "great erudition," "profound sincerity," and "noble spirit."[12] In Darwin he saw "a precise observer, accomplished experimenter, and keen synthesizer." Carefully worded and unambiguous, the accolades were only a tactical move to strengthen Danilevskii's hand in the ambitious task of dismantling the complex structure of Darwinian evolutionism.

Darwinism can best be described as an elaborately structured synthesis of all anti-Darwinian arguments in circulation at the time it was written. Too dependent on random changes, Darwin's book, as Danilevskii saw it, violated the absoluteness of the laws built into Newtonian science. Darwin's theory was outside the realm of science, for it transgressed the ground rules of predictability. By substituting "pseudoteleology" for von Baer's principle of organic purposiveness, Darwin created a biological orientation that ignored one of the basic attributes of life. According to Danilevskii, the principle of adaptation as formulated by Darwin was a "pseudo-teleological" principle, for it denied that all transmutations in the organic world were predetermined by a higher intelligence. Danilevskii defined organic transmutation as the work of a "superior intelligence" that gives living nature meaning and direction. Darwin's emphasis on the randomness of organic change was a denial of the supreme authority of divine reason.[13] Overgeneralizing the role of natural selection as the prime mover of organic evolution, Darwin failed to place due emphasis on the accumulated effects of environmental influence, on hybridization, and on "sudden leaps" as mechanisms of evolution.[14] To accept natural selection as the prime factor of evolution meant to view evolution only as a result of "external" influences on individual organisms and on species; Danilevskii, who preferred "development" as a neutral notion to "evolution" as an unrealistic and emotionally tinted category, made "internal" factors the real source of ordered and directed organic change and of the universal harmony of living nature. He devoted an entire section to pointing out the "flaws" in Darwin's logic.

Danilevskii's treatise did not add new ammunition to the war on Darwinism, nor did it open new vistas that promised to lead to previously untapped anti-Darwinian arguments. But it did give the old arguments a more ordered form and an augmented sense of urgency. It drew the attention of philosophers to the broader compass of the new heresy; it heralded the coming of an accelerated growth of theological attacks on Darwinism; and it contributed to the emergence of a new wing of critics of Darwin's theory in the scientific community. Danilevskii returned persistently to philosophical, ethical, and ideological fallacies he thought he had detected in Darwin's reasoning.

In his criticism of the substance of Darwin's theory, Danilevskii relied on standard western European works, particularly on A. J. Wigand's *Der Darwinismus und die Naturforschung Newtons und Cuviers* (3 vols., 1874). He contended that the refutation of Darwin's theory was an accomplished fact at the time he undertook to write his copious synthesis. He noted that the rejection of Darwinism came from "the leading naturalists of our time." In his view:

> The great naturalist-philosopher Karl von Baer was the first critic. Immediately after him came the outstanding disciples of Georges Cuvier: Louis Agassiz and Milne-Edwards. These were followed by the eminent comparative anatomist Richard Owen, the famous paleontologists A. Brongniart, H. R. Göppert, H. G. Bronn, and J. Barrande, whose opinion was of particular relevance. Then came the phytogeographer A. Griesebach, the botanists J. Decaisne and A. J. Wigand, the famous contemporary histologist A. Kölliker, and the physiologists A. de Quatrefages, H. Burmeister, and E. Blanchard.[15]

In criticizing the formal structure of Darwin's theory, Danilevskii depended primarily on his own intellectual resources. He showed remarkable skill in immersing himself in the intricacies and minutiae of

Darwin's tedious efforts to link empirical data to theoretical propositions. His lengthy argumentation produced one general conclusion:
Darwin's theory must be rejected in its entirety, for it contradicted empirical evidence accumulated by modern biologists and was marred
by inconsistencies in theoretical reasoning. In Danilevskii's judgment,
Darwin made four erroneous interpretations of evolution. First, he ignored "sudden leaps" as a factor of organic development, even though
they alone accounted for the transition from one species to another.
Second, by emphasizing the randomness of variation, he overlooked the
purposeful—and predetermined—direction of evolution. This view—
named orthogenesis—found a strong spokesman in the German evolutionist Theodor Eimer, particularly in his works after the publication
of Danilevskii's *Darwinism*. Third, Darwin did not go beyond what
Danilevskii labeled the mosaic nature of variation; he did not relate the
changes in individual organs to the adaptive changes in the organism as
an integrated whole. Fourth, the idea of the struggle for existence, as
Darwin used it, did not have a basis in empirical facts. Danilevskii devoted most of his study to the refutation of this notion. He stated:

> The struggle for existence undoubtedly exists and Darwin deserves great
> credit for attracting the attention of scholars to it; this phenomenon, how-
> ever, does not have the power of selectivity—it is a *biogeographical principle*
> determining to a great extent the distribution of organisms on the face of the
> earth, but it does not have, and cannot have, a *biological significance*.[16]

Danilevskii's criticism centered on the general assumption that "Darwin's theory is not so much a zoological and botanical theory as it is a
philosophical theory." Darwinism, he said, "changed not only our com-
monsense and scientific ideas but also our world view."[17] Darwinism
profoundly affected both idealism and materialism. Before Darwin, materialism was based on vague and uncertain scientific generalizations
and theoretical views. It was Darwinism, according to Danilevskii, that
"ensured" victory for "a universal materialistic view of nature"; it
achieved this by strengthening the belief in the progress of science and
by eliminating a supreme intelligence from the work of nature.[18] Guided
by his own diagnosis, Danilevskii concentrated on exposing Darwin's
main philosophical sin: a blind commitment to narrow empiricism and
to materialistic naturalism. Since, in reality, Darwin avoided philosophical analysis, Danilevskii accused him of sneaky efforts to hide his general ideas behind endless arrays of empirical data.

Danilevskii resented particularly Darwin's rejection of supernatural interference with the origin and the dynamics of organic nature. He firmly opposed the view of organic nature which had no room for divine authority and for spiritual forces veiled in mysticism. He wrote: "Darwinism removes the last traces of what has come to be known as mysticism; it even removes the mysticism of the laws of nature and the mysticism of the rational structure of the universe." [19] To add strength to his criticism, Danilevskii made occasional attacks on Darwin's alleged failure to meet the logical and methodological rigor of Newtonian science. He preferred the universality of Laplacian causality to the causal randomness of Darwinian variation. In general, however, he gave prominence to every anti-Darwinian argument in circulation.

Danilevskii's treatise was not without realistic and defensible challenges to specific components of Darwinian thought. What alienated most serious experts from *Darwinism* was that, in general, it was an emotionally charged attack on every ramification of Darwin's theory. Darwin's evolutionary thought attracted Danilevskii's attention only as a set of diabolical schemes that must be totally rejected. To compromise with one Darwinian idea meant to compromise with the entire theoretical structure of Darwinian evolutionism—to allow the devil to sneak in through the back door. Nor did he show the least inclination to look critically at the evidence on which the critics had depended in fashioning their anti-Darwinian arguments.

Danilevskii's anti-Darwinian arguments rested on two pillars: criticism borrowed from a long list of well-known anti-Darwinists, and logical constructions of his own making. Relying on logical constructions, Danilevskii concluded that Darwin's theory was a pseudoscience contradicting two of the objective criteria of validating scientific conclusions. In the first place, Darwin's theory did not have an objective and identifiable object of inquiry: in Danilevskii's view, the problem of the origin of species was so complex and so indefinite that it could not be made part of science. In the second place, it violated the rules of logic: relying on syllogistic maneuvers, Danilevskii claimed to have shown that Darwin's premises did not support, but actually eliminated, natural selection as a prime mover of organic evolution. He also claimed that Darwin's comparison of artificial and natural selection rested on faulty premises.

Marshaling logical arguments against Darwin, Danilevskii tried to convince his readers that he relied on the Newtonian models of scientific reasoning. He clearly tried to convey a message to his readers that Darwin was not a true scientist, because he had failed to conform to the standards of Newtonian scholarship. All this did not mean, however, that Danilevskii was an admirer of Newtonian science. As a strong supporter of mystical experience and supernatural teleology, he had little use for the world outlook based on rigid determinism, mechanistic philosophy, and natural causation. Relying on Newtonian norms only when he could use them to bolster his arguments against Darwin, he praised Wigand's *Darwinismus und die Naturforschung Newtons und Cuviers* because it made Cuvier, rather than Darwin, a true scientist—a genuine supporter of Newtonian natural philosophy.[20]

Theologians and philosophical critics of materialism were quick to greet Danilevskii's work as a most thorough and devastating critique of Darwin and his supporters. For the first time, theologians concentrated much more on the "empirical" and "theoretical" weaknesses of Darwinian thought than on the incompatibility of the notion of anthropoid origin of man with the moral and intellectual aspects of humanity. Danilevskii provided them with an extensive catalogue of anti-Darwinian arguments that could be easily woven into larger complexes of criticism. He supplied them with "scientific" arguments revealing the pitfalls and shortcomings of Darwin's "pseudo-teleological" stance and pointing out the objectionable features of his materialist philosophy. Like Danilevskii, theologians tried to save science from Darwin's "anti-scientific" manipulations. As indicated by P. Ia. Svetlov's article in *Pravoslavnoe obozrenie*, theological writers were particularly eager to spread the word about Danilevskii's study as a grand summation of anti-Darwinian arguments acceptable to the church.[21] They made it a policy, however, to take no part in the bitter feud between the supporters and enemies of Danilevskii outside the church community.

I. Chistovich wrote in *Belief and Reason*, a leading theological journal published in Kharkov, that Danilevskii was an ideal man to dissect the Darwinian menace because he combined extensive knowledge in natural science with a philosophical vision and critical mind. According to Chistovich, Danilevskii succeeded on three levels of discussion: on the scientific level, he showed the total absence of empirical support for the notion of natural selection; on the philosophical level, he adduced strong arguments against the Darwinian notion of random variation as a denial of the purposiveness of natural phenomena; and on the moral level, he helped unveil the false premises of the Darwinian effort to "lower man in his own eyes by reducing him to the simian level."[22]

It would be a mistake to treat theological criticism of Darwin's theory as a monolithic structure built on identical arguments and on a unitary line of reasoning. Nikanor, bishop of Kherson and Odessa, for example, fought Darwinism on two fronts. First, he saw Darwin's theory of evolution as a heresy that contradicted the biblical explanation of the creation of the living world. Second, he considered Darwinism a most thorough embodiment of Western materialism, a philosophy that had conquered and corrupted natural science. Nikanor did not hesitate to count Kantian, Schellingian, and Hegelian philosophy among the pillars of Western materialism.

Relying heavily on Danilevskii, P. Ia. Svetlov concentrated on "logical flaws" that made Darwin's theory far removed from true science.[23] Darwin, he said, was guilty of drawing diametrically opposite conclusions from the same empirical data, of disregarding information that did not support his theoretical constructions, of "logical inconsistencies," and of "analytical shallowness." In addition to the "flaws" in logic, Svetlov pointed out two major "weaknesses" in Darwin's research strategy. Darwin, he claimed, concentrated on the transmutation and origin of species, but he could not show a single example of transitional forms. He did not prove that he had a legitimate topic for scientific inquiry. By disregarding the purposiveness of life processes and by assuming the absolute reign of random variation, he abandoned one of the most essential characteristics of organic processes.[24] In Svetlov's view, Darwin's emphasis on the randomness of natural processes was a unique surrender to materialistic philosophy.

There was one notable difference between Danilevskii and his theological followers. Danilevskii's rejection of Darwinism did not imply a rejection of the idea of transmutation in general. Most of his anti-Darwinian sources were actively engaged in constructing evolutionary schemes. He inferred that science was not yet ready to elevate evolution above the level of a hypothesis. The theologians, at least at this time, operated on the assumption that transmutation was not a fact of nature in the first place. All this, however, did not prevent them from making an extensive, but selective, use of Danilevskii's bountiful argumentation.

Nor did the government waste time on giving Danilevskii's anti-Darwinism its blessing. In January 1886 the Scientific Committee of the Ministry of Public Education ordered that *Darwinism* be included in all basic libraries of the higher classes of secondary schools, as well as in the libraries of teachers colleges.[25] At the same time, the official journal of the ministry gave the study a most favorable review. The reviewer commended the book as a rich source of antievolutionary thought and as an exemplar of "irrefutable logic," "extraordinary wit," and "correct analysis." Danilevskii's work, the reviewer noted, had shown that Darwin's theory, in its entirety, did not belong to the realm of truth.[26]

N. N. Strakhov's Defense of Danilevskii

Soon after the publication of the first volume, *Darwinism* received a most glowing review from the conservative gazette *New Time*. El'pe, pseudonym of L. K. Popov, the author of the review, thought that Danilevskii was eminently successful in proving that Darwinism lacked a solid scientific base and was erected on a faulty philosophical doctrine.[27] In El'pe's view, Danilevskii's main contribution was in removing one of the most dangerous modern challenges to the scriptural ["ideal"] explanation of the origin of "organic forms." Since *New Time* normally echoed the views of the St. Petersburg government, El'pe's feuilleton gave an impression of the readiness of the authorities to give wide circulation to their views and sentiments about the mounting "threat" of Darwinism.

Writing in the *Russian Herald* in 1886, P. Semenov found Danilevskii's scientific and philosophical criticism of Darwinism much more thorough and fundamental than any similar criticism in the West. Even Albert Wigand's three-volume *Darwinismus und die Naturforschung Newtons und Cuviers* was no match for the depth and the compass of Danilevskii's magnum opus.[28] Semenov was particularly impressed with Danilevskii's own observations on the origin of variation among cultivated fruit trees. He noted that Danilevskii, his close friend, had planned to complete his study with a special volume entitled *Metaphysical Deductions,* but that his sudden death on November 7, 1885, prevented him from carrying out this project. Most informed contemporaries would not have agreed with Semenov's claim that in writing *Darwinism* Danilevskii acted as a judge without "preconceived ideas."

Church and government supporters of Danilevskii's anti-Darwinian crusade received particularly strong support from the writings of N. N. Strakhov, a secular commentator on the broader philosophical aspects of science, a noted contributor to the current revival of idealistic metaphysics, and a staunch defender of the originality and purity of Russian thought. In a long review of Danilevskii's *Darwinism,* published in the *Russian Herald* in January, 1887, he argued that Danilevskii was eminently successful in his effort to refute the principal ideas built into Darwin's evolutionary theory. In his analysis he stayed close to Danilevskii's text and relied heavily on direct citations summarizing and systematizing the major categories of criticism. At this time Strakhov was widely known as the first thorough, persistent, and uncompromising critic of Russian nihilism—an ideologically saturated philosophical movement that extolled the secular and critical wisdom of science as the only sound path to a better future for Russia. Fighting Darwinism as a scientific theory close to the materialistic and positivistic bent of nihilism, Strakhov fought not only the radical intelligentsia but also the more outspoken members of the scientific community, who were prone to emphasize science as the motive force of modern civilization.

Nihilism, in Strakhov's view, was a grave Western disease transmitted to an otherwise healthy Russian body. Darwinism was the most pernicious component of nihilist pathology. Strakhov found no difficulty in identifying the broader intellectual context of Darwinian heresy:

> How could it happen that such a mirage deluded and has continued to delude a large segment of the scholarly world and of educated people in general? The answer lies undoubtedly in the *spirit of the time* and in the inclination of human beings to believe in everything they find pleasant to believe in. During its early decades, the nineteenth century represented a miraculous upsurge in thought, science, and poetry. At the middle of the century, however, signs of an unexpected and sharp decline in the brilliance of intellectual creations and in expressions of sentiment became clearly visible. . . . At this time the doctrines of theoretical and practical materialism became preeminent. From England, the classical land of skepticism, utilitarianism, and all kinds of base ideas, the materialistic orientation began to influence the intellectual world of continental Europe. But no idea was received with so much enthusiasm as that of Darwin, obviously because it untied a most difficult knot, solved a puzzle that defied simple explanation, and stood before the human intellect as a huge sphinx.[29]

Although Danilevskii's tomes were saturated with emotional outbursts, rhetorical exaggerations, and translucent syllogistic maneuvers, Strakhov showed no hesitancy in treating the work as one of the rare "scientific" studies destined to be read and cherished by many generations. He wrote:

> The richness of its coordinated information, the depth of its logic, its inordinate brilliance, its scientific rigor, and fullness of the questions it raises compel us to conclude that N. Ia. Danilevskii's work is one of the most extraordinary phenomena in world literature. It may be stated emphatically that this book has brought honor to all of Russian literature and that it will forever link the author to the most important and most profound questions of natural science. N. Ia. Danilevskii deserves eternal honors for his fight against natural selection, one of the greatest delusions of our century.[30]

Strakhov did not mention, however, that Danilevskii was not an impartial judge of Darwin's scientific contributions. Searching through the sprawling empire of Darwinian literature, Danilevskii looked only for—and made use of—negative criticism. Nor was he moved by the interests of science; he relied on ideology, translated into selected tenets of morality, as the primary source of criteria for selecting arguments and directing analysis. It was primarily as a systematic compilation of all major anti-Darwinian arguments that his book became the focal point of one of the most bitter polemics in the late 1880s.[31] Danilevskii's attack on Darwinism was a special attack on "nineteenth-century science materialism" and an elaborate expression of the cosmic mission of Russian culture heralded by *Russia in Europe*, published in 1869.

Critics of Danilevskii and Strakhov

As might have been expected, K. A. Timiriazev took it upon himself to write a major refutation of Strakhov's glowing review of Danilevskii's work. Eager to disprove Strakhov's claim that Danilevskii's work presented "a conclusive refutation of Darwinism," Timiriazev made his argument more polemical than academically rigorous. Expansive in loquacity and rich in allegory and metaphor, his prose did not have what it needed most: straightforwardness and lucidity. Numerous detours made his essay unwieldy and asymmetrical. The essay offered a discourse on scientific objectivity as an ethical norm. It presented a tedious recounting of the logical and empirical ramifications of natural selection as the prime mover of organic evolution. Just as Danilevskii and Strakhov found nothing in Darwin's theory to recommend, so Timiriazev found nothing to reject. He even defended the Malthusian core of natural selection, which was particularly unpopular among the radical intelligentsia.

Timiriazev's response to Strakhov was part of an effort to tie Darwin's science to a democratic ideology, a truly brave effort at a time of mounting government efforts to eradicate the last vestiges of academic autonomy. By bringing in ideological issues, Timiriazev merely responded to Danilevskii's claim that Darwin's theory presented "a special world view," and "a higher explanatory principle" transcending the limits of legitimate academic concern and covering the entire structure of the universe and every form of existence. According to Timiriazev, Danilevskii aimed his arrows at Darwinism, not because it expressed a scientific view, but because it supported a materialistic philosophy. Danilevskii was not interested in protecting science from its ideological enemies; he wished to protect a political ideology that imposed crippling limitations on the freedom of scientific inquiry. In refuting Danilevskii, Timiriazev recapitulated the history of papal opposition to Galileo's Copernicanism to show that in science the ultimate victory belonged to objective truth.[32]

Timiriazev opened his criticism with favorable comments on current Western scholarship—as represented by Nägeli, Romanes, and Weismann—which built upon the Darwinian legacy even when it modified some of its principles. He did this in an effort to refute Danilevskii's—and Strakhov's—claim that Darwin's theory could in no way serve as a basis for an evolutionary theory. He ended his article by presenting himself as an unwavering supporter of orthodox Darwinism. Relying on the polish of his rhetoric, he made it clear that he had no reservation about treating natural selection and the struggle for existence as the basis upon which the grand process of evolution is built. He wrote:

> All organic forms are capable of change, sometimes at infinitesimal rates. These changes may be heritable. At the same time, all organisms increase in geometrical progression. . . . A large majority of these forms disappear in the struggle with enemies and with environment—as well as in various kinds of competition. All this falls under the general category of the struggle for existence. Only the best adapted forms are preserved. These forms transmit their new characteristics to their offspring, and the process of selection is again repeated. In such a way useful characteristics are preserved and accumulated. This is the principle of natural selection.[33]

Timiriazev's biting criticism of Strakhov's anti-Darwinian exhortations did not bring the Danilevskii affair to an end. At the end of 1887 Strakhov published a blistering response to his critic, adding more pungency to his anti-Darwinian rhetoric and more acrid sarcasm to his attack on Timiriazev's argumentation. Concentrating much more on Timiriazev's rhetoric than on substantive questions, he attacked Timiriazev as a "typical professor," uncompromisingly critical of all those who held opposite views. Strakhov chastised Timiriazev for labeling Danilevskii a mere dilettante in science and for a servile attitude toward Western scientific scholarship. Moreover, he took this opportunity to point out the major sins of the world of Russian scholarship. Russian scholars, he said, reminded him of religious fanatics, the only group that surpassed them in the intensity of prejudice and warlike intolerance toward opposing views. They belonged to the groups that were "most inflexible in defending their authority and least capable of abandoning preconceived ideas." To them science was a creed—a dogma. They were "contaminated with scientific fanaticism." [34] At this point in his long career as a free-lance writer, Strakhov was convinced that natural science was the main source of the moral and intellectual crisis that troubled the world.

Strakhov was correct in claiming that Timiriazev's essay on Danilevskii could not be viewed by any stretch of the imagination as a model of lucid writing and logical rigor and that it was richer in distracting loquacity than in substantive thought. The same, however, applied to his answer to Timiriazev. In an outburst of sentimentality, he thought that Timiriazev's sweeping attack on a dead man (Danilevskii died in 1885) was both inappropriate and uncalled for. Short on facts and unaware of finer points in the current developments in evolutionary biology, Strakhov depended mainly on logical maneuvers, emotional outbursts, and endless citations.

Strakhov drew the attention of his readers to K. von Nägeli's monumental *Mechanical-Physiological Theory of Descent*, published shortly before Danilevskii's death, which he viewed as the last in the series of masterful refutations of Darwinism. He was sure to add, however, that in the logic and the depth of his argumentation Nägeli was no match for Danilevskii; in his view, no work, not even Karl von Baer's, emulated the sweep, the logical precision, and the intellectual power of Danilevskii's magnum opus. Strakhov's essays had one central task: to present Danilevskii's anti-Darwinism as a high point in the ongoing war on "scientific materialism." In Strakhov's thinking, Danilevskii's work was not only a most accomplished assemblage of anti-Darwinian arguments but also a unique personification of the Russian soul. *Darwinism,* as he saw it, brought a sure victory for modern Slavophilism.

Timiriazev was ready for Strakhov's offensive. In the "Powerless Anger of an Anti-Darwinist," published in 1889 in the journal *Russian Thought,* he answered Strakhov on two levels. On the scientific level, he gave a new recapitulation of Darwin's theoretical views and at several places pointed out Darwin's awareness of empirical difficulties in drawing his generalizations. He resented the widespread effort on the part of critics to exaggerate the discrepancies between Darwin's generalizations and their empirical base. Timiriazev was particularly eager to point out Strakhov's extensive practice of misinterpreting the theories of Western biologists in order to present them as more anti-Darwinian than they actually were.

On the philosophical level, Timiriazev carefully pointed out that

Strakhov, not aware of differences between philosophical and scientific metaphors, wavered between a sweeping philosophical rejection of the scientific world view and a rhetorical allegiance to "the solid ground of science." Despite his profuse criticism of the scientific community, Strakhov, according to Timiriazev, tried to present himself as a person whose argumentation adhered closely to scientific standards; at the same time, he accused Timiriazev of indiscriminate excursions into philosophy. Timiriazev, of course, did not agree with this description; he contended that Strakhov made philosophical statements when he thought he was speaking at the level of science.[35] Indeed, he viewed Strakhov's anti-Darwinian arguments primarily as articulations of a philosophical view of nature. Timiriazev was correct in stating that both Danilevskii and Strakhov had not the least interest in defending the welfare of science: their criticism of Darwin's theory was motivated by ideological, philosophical, and religious reasons. They attacked "variation," "struggle for existence," and "natural selection" because these concepts offered only a "crude" explanation of the harmony of nature and because they disregarded the purposive dynamics of the organic world based on supernatural predetermination of the forms of life.[36]

In his criticism of Darwinism, Strakhov found himself in a paradoxical position. On the one hand, he rejected the Laplacian bias of Darwin's thought, which placed the theory of organic evolution into a framework of mechanistic determinism. On the other hand, he rejected the Darwinian notion of the randomness of variations as a violation of the regularity and predictability of mechanical laws. He attacked Darwin for his close adherence to Newtonian mechanism, but he also attacked him for committing serious digressions from the deterministic regularity of Newtonian causality.[37] As the occasion demanded, he found Darwin guilty either of a blind adherence to Newtonianism or of serious digressions from the canons of the Newtonian mode of scientific explanation.

Strakhov was determined not only to erase Darwinism from modern thought, but also to spread the word about the narrow perspectives of science as a general source and depository of wisdom. At the same time, Timiriazev, in his emotional defense of Darwinism, was blind to the demands of new developments in biological theory and methodology and to extensive changes in the treatment of the evolutionary process. No doubt Strakhov was more philosophical—and more animistic and anthropocentric—than he was willing to admit. But Timiriazev was also primarily philosophical—except that his philosophy was much closer to the spirit of science. The major contribution of the Strakhov-Timiriazev feud was much more in articulating the philosophical questions raised by Darwinism than in giving a new direction to Darwinian science. Vagueness and lack of clarity prevented the papers of both writers from exercising a stronger influence on contemporary thought. The papers of both writers suffered from imprecise organization, strained logic, and heavy reliance on rhetorical exaggeration.

The Strakhov-Timiriazev debate went far beyond the limits of Darwinian controversy. The question whether or not Darwin's theory had solved the riddle of organic evolution became part of the much larger question of the place of science in modern culture. This was a time of rising government oppression, which encouraged attacks on "natural science materialism" as a pernicious ideology. Populists, anarchists, Marxists, and a wing of academic liberals were unyielding in their determination to present scientific knowledge as the only sound path to the salvation of Russia. At the same time, the government received unlimited aid from idealistic philosophers, led by V. S. Solov'ev and Boris Chicherin, who concentrated on the intellectual narrowness and materialistic underpinnings of scientific thought. K. A. Timiriazev was the most ardent and the most uncritical champion of the unlimited powers of science. Nikolai Strakhov, by contrast, was the most determined and the most vociferous critic of science.

Relying on his accomplished oratorical skill, Timiriazev extolled science as a panacea for all human ills and as an unerring guide to a better future. He was eager to show the primary role of science in creating a healthy base for the moral code of modern Russia. To protect science as a depository of secular wisdom, the scientific community, he thought, must concentrate on eliminating the last vestiges of metaphysics from the process of scientific cognition. He was not always sure in drawing a line between "science" and "metaphysics"—for example, he had no reservations about labeling the biological theories of August Weismann as metaphysical aberrations. In defending science, Timiriazev fought not only the resurgence of idealistic metaphysics but also the stirrings in the scientific community that challenged the inviolability of the mechanistic—or Newtonian—view of the universe. He found no good reason for abandoning his allegiance to nihilist "materialism" and its worship of the method and the content of natural science. As a champion of academic autonomy and a proponent of democratic ideals in politics, Timiriazev was one of the most eminent and influential representatives of the liberal professoriat, ever ready to challenge the oppressive measures undertaken by the Ministry of Public Education. His consistent and stubborn defense of Darwinism was tightly woven into the intricate texture of a philosophy that combined positivism and materialism.

N. N. Strakhov was at the opposite pole. His scorching attacks on Darwinism—and his enthusiastic endorsement of Danilevskii's anti-Darwinism—were part of a general campaign to point out the assumed intellectual limitations of science, in general, and of the "materialism" of nineteenth-century science, in particular. Of all the modes of inquiry, Strakhov argued on many occasions, science has the narrowest horizons: it suffers from the rigidity of "rational forms" in which it is cast, it lacks internal unity, and it is fastened to a myopic philosophy.[38] Darwinism is a product, rather than a creator, of "scientific materialism." In his *Struggle against the West* Strakhov saw the intellectual future of Russia in a system of thought dominated by religious teachings and only peripherally affected by science.

Strakhov flattered himself for having played a decisive role in attracting the attention of the general reader to Danilevskii's anti-Darwinism. In his first answer to Timiriazev, he lamented the virtual silence with which Danilevskii's work had been received during the first two years after its publication.[39] He implied that his review, published in February 1887, was the first serious effort to give Danilevskii's ideas the public exposure they so well deserved. In an expression of modesty, he admitted that even this effort did not do full justice to the grandeur of Danilevskii's arguments. In November 1889, Strakhov, no longer worried about the fortunes of Danilevskii's work, announced joyfully that "N. I. Danilevskii's books have attracted universal attention."[40] No doubt he was convinced that his writings deserved the main credit for making Danilevskii's *Darwinism* a literary landmark of the age. Strakhov erred in estimating the popular appeal of Danilevskii's work. Iu. A. Filipchenko, a geneticist, came closer to the truth when he stated in the 1920s that Danilevskii's *Darwinism* received very little attention primarily because the scientific community had chosen to emphasize only its negative aspects.

Known for his strong sentiments, perfunctory handling of complex philosophical questions, superficial knowledge about the burning scientific questions of the day, and torrid Slavophile bias, Strakhov was held in high esteem only by a small group of close friends dedicated to autocratic values. A younger contemporary noted that Strakhov was a dilettante not only in science but also in philosophy and that the shallowness of his antievolutionary arguments was the main reason for the dismal failure of his effort to marshal public support for Danilevskii's crusade against Darwinism.[41]

Other Views

The vast proportions of the Strakhov-Timiriazev controversy prompted the editor of the journal *Russian Wealth* to ask N. A. Kholodkovskii, a young St. Petersburg professor and a confirmed Darwinist, to write an essay appraising the current state of Darwin's theory. In response, Kholodkovskii wrote "Darwin's Theory: Its Critics and Future Development," published in 1888 in two installments. He admitted that Danilevskii's work had created "something of a sensation" and noted that the Strakhov-Timiriazev controversy was more a general polemic than a scientific discussion.[42] His study, written in a subdued tone and lucid style, concentrated on the evolutionary theories of Kölliker and Nägeli. Kölliker's criticism of Darwin formed the backbone of Danilevskii's work.

Albert Kölliker, who began to develop his evolutionary theory four years after the publication of the *Origin of Species*, tied organic transformism to three general ideas, all contrary to Darwin's views: the multiple origin of living forms, the internal causes of variation, and "sudden leaps" (heterogenesis) in the evolutionary process. After a detailed analysis of these ideas and a comparison with Darwin's views, Kholodkovskii concluded that Darwin's theory was immeasurably more advanced than Kölliker's theory because "it gives a more satisfactory explanation of many aspects of organic development and does not allow for fanciful digressions of any kind."[43] Darwin's theory, he said, presented a full picture of the cosmic position of organic evolution and made a major contribution to the understanding of the harmony of nature.

Kölliker offered a fully crystallized theory of evolution in 1872, expressing a hope that it would quickly supersede Darwin's "hypotheses." By 1888, when Kholodkovskii published his essay, Kölliker's followers were small in number and weak in influence, and his name continued to be recalled primarily in nonscientific anti-Darwinian literature. At the same time, Kholodkovskii argued, Darwin's theory had continued to preoccupy the scientific world, and the number of its supporters had grown at a steady pace. Although Darwinism continued to harbor many weak points that demanded vigilant study and major revisions, it showed more promise and had stronger appeal than any competing evolutionary theory. Kholodkovskii noted a glaring inconsistency in Danilevskii's attitude toward Kölliker. Danilevskii, he wrote, relied heavily on Kölliker's anti-Darwinian arguments, which did not stop him from asserting that Kölliker's basic theory was not science but guesswork.[44]

In 1884 Karl von Nägeli published his capital work on the "mechanical-physiological" theory of descent, which drew the attention of theoretical biologists to two general evolutionary ideas. First, it recognized natural selection as a factor in the evolution of species, but treated it only as a secondary factor. Natural selection, as Nägeli saw it, plays no role in the origin of new species; its role is limited solely to eliminating the living forms incapable of adapting themselves to environmental changes.[45] Second, it represented an effort to place the theory of evolution on a physiological basis, the only way, according to Nägeli, to make use of the laws of Newtonian mechanics in biology—and to make biology a true science. Physiology, as he saw it, is the mechanics—the scientific backbone—of biology.

Kholodkovskii rejected the major ideas von Nägeli had advanced. He turned down Nägeli's—and Kölliker's—view of evolution as a natural process dependent on internal, rather than on external, conditions of existence. Nor did he favor Nägeli's effort to tie the internal causation of evolution to a universal force that gives unity and direction to organic change. With equal vehemence he criticized Nägeli's categorical refusal to regard natural selection as a primary factor of evolutionary change. He criticized Nägeli's physiological orientation on two grounds. In the first place, Nägeli's study of the idioplasm was more "morphological" than "physiological." In the second place, Kholodkovskii contended that any effort to create a physiological theory of evolution was premature.[46] All this did not stop him from expressing an optimistic view about the future development of the physiology and morphology of evolution as distinct disciplines, the former dealing with the physical and chemical foundations of life and the latter with the forms of life in their "coexistence and succession." Kholodkovskii thought that Nägeli and Danilevskii had one attribute in common: both presented Darwin's arguments by omitting those ramifications that would invalidate their criticism.[47]

Kholodkovskii's paper marked an important step in the history of evolutionary thought in Russian biology. While recording the Russian scientific community's firm adherence to Darwin's theoretical legacy, it also showed clearly and categorically that Darwinian scholars had been aware of the urgent need to examine the new flow of suggestions for filling the gaps in evolutionary theory and for eliminating obvious errors from Darwin's major works. Kholodkovskii criticized Kölliker and Nägeli for overemphasizing internal causation as a factor of evolution. Although he favored external causation, in the form of natural selection, he urged Darwinian scholars to place much more emphasis on internal causes than they had done in the past. He supported Darwin's heavy concentration on random variation as the source of evolution, but he also noted that future Darwinian scholars would be required to pay more attention to the mechanism of heredity than their predecessors were inclined to do. Although Darwinism was much more advanced than any other evolutionary theory, its ultimate survival depended on its ability to absorb—or to complement—new advances in biological theory.

In 1889 A. S. Famintsyn, professor of plant physiology at St. Petersburg University and a member of the St. Petersburg Academy of Sciences, entered the debate evoked by Danilevskii's work. His verdict was clear:

> Danilevskii's *Darwinism*, as stated by the author, is intended for educated persons without special training in biology. In this category of works, devoted primarily to the diffusion of scientific knowledge, particularly important is the scientific accuracy of presented scientific material. . . . In my judgment, Danilevskii's work has not obeyed this rule; throughout the work the foundations of Darwin's theory have been presented incorrectly. No other person has made the social influence of Darwin's work appear so pernicious and destructive, and none has been so distraught and bitter about it. Spreading such views about the significance of Darwin's work—views that contradict the judgment of all experts without exception—is, in my opinion, most regrettable, undesirable, and harmful.[48]

Famintsyn introduced himself as a person who did not belong to the "group of absolute worshippers of Darwin." He admitted that on some essential points he parted company with Darwin. At all times, however, he considered it his duty to protect Darwin from absurd criticism and to treat him as "one of the greatest naturalists of our time."[49] He thought that Darwin was a target of Danilevskii's angry attacks not because of intrinsic flaws in his theory but because all kinds of "materialists" have made extensive use of his ideas in shaping the philosophical foundations of their ideologies.[50] Famintsyn cited several scientists, including Wigand (one of Darwin's most severe critics), to show that Darwin had no intention of turning his theory against religion. All this did not prevent him from paying tribute to Danilevskii as an "outstanding Russian" and from asserting that "zoologists and botanists" will find useful information in *Darwinism,* which presents "all the objections to Darwin's theory and scattered factual material of great interest, for which we will always be indebted to him."[51]

Danilevskii lost his anti-Darwinian war even though some of his criticism was both astute and difficult to refute. By staying strictly within the realm of anti-Darwinian arguments, Danilevskii gave up any legitimate claim to objectivity and a real chance to be taken seriously by the scientific community. Many years later, Iu. A. Filipchenko, a modern critic of Darwinism as a general evolutionary theory, stated that "Danilevskii's work, like that of Wigand, was totally unsuccessful, primarily because the critics, led by Timiriazev, concentrated exclusively on its weak features."[52] He said that the members of the scientific community were united in refuting Danilevskii's sweeping denunciation of Darwin's theory. A. Borozdin, who wrote the biography of Danilevskii for the *Russian Biographical Dictionary,* stated that the members of the scientific community were much less belligerent than Timiriazev, but that, without exception, all expressed generally unfavorable opinions.[53]

Nor was the St. Petersburg Academy of Sciences favorably disposed toward Danilevskii. On the basis of an appraisal by A. P. Karpinskii and A. S. Famintsyn, two of its more distinguished members, the Academy turned down an external recommendation that *Darwinism* be awarded a special prize in recognition of the originality of its ideas. Criticizing it from a geological point of view, Karpinskii noted that *Darwinism* was seriously marred by inaccurate details and major flaws in logic. Worst of all, Danilevskii had denounced Darwin before he undertook to write a book about him.[54] Famintsyn claimed that Danilevskii had misrepresented Darwin's ideas. Both critics noted that Danilevskii did not add new strength to the criticism of Darwin's thought and that he limited his work to mere systematization of widely circulated anti-Darwinian arguments.

The 1880s ended with Famintsyn's sober warning that the scientific treatment of Darwin's theory should be kept apart from the ideological uses to which it was put—that the scientists should follow the internal logic of the development of biological theory, untrammeled by ideological considerations. He also announced that the biologists who defended Darwin's theory in its original form were a rapidly dying species, and that a new Darwinism, sensitive to developments in experimental biology, was very much in evidence. More than any other person in Russia, Famintsyn felt the coming of a new era in evolutionary studies. Opening

Darwinism to serious reexamination, he also opened Timiriazev's orthodoxy and Danilevskii's anti-Darwinism to scientific criticism.

Danilevskii's *Darwinism* attracted little attention among general
readers because it did not receive direct support from a single spokesman for the scientific community. Biologists were either committed to
the cause of Darwinism, almost always in a modified form, or chose to
keep silent on all the issues of evolutionism. Had Danilevskii published
his study ten years later he would have had no trouble in finding influential supporters in the scientific community. Although Danilevskii was a
man of great erudition and penetrating logic, his book was neither a literary success nor a dispassionate historical analysis: he wrote to con-
demn Darwinism rather than to subject it to a careful, cautious, and fair
analysis. All this, however, does not deny the orderly synthesis and rich
insights that made *Darwinism* a notable work. Danilevskii made a tactical error: he did not follow the example set by Kölliker and Nägeli,
who criticized Darwinian evolutionism unsparingly, but who carefully
pointed out and elaborated the points of Darwiniana that were fully
compatible with their own scientific thinking. They chipped away at the
undesirable parts of the complex edifice of Darwin's theory, but they
contributed to the preservation of the grandeur of its triumphant position in modern scientific thought. The members of the scientific community who found much to criticize in Darwin's legacy made no effort to
seek help from Danilevskii. His work appealed only to theologians and
conservative philosophers, who used it as an inexhaustible source of information on the negative and heretical features of Darwin's view of
evolution.[55]

There was another reason for Danilevskii's failure to attract serious attention: the leaders of metaphysical thought, who were particularly encouraged by the rising tide of tsarist oppression, were far from
united in lending support to Danilevskii's anti-Darwinian crusade. V. S.
Solov'ev, the most erudite and influential Russian philosopher, considered Strakhov's role in the defense of Danilevskii a strong reason for the
unenthusiastic reception of *Darwinism*. In the bitter duel between Strakhov and Timiriazev, Solov'ev, who was involved in a feud with Strakhov
over other issues related to Danilevskii as well, had no reservations
about accepting Timiriazev's—and Famintsyn's—verdict that Strakhov
used poor judgment in presenting Danilevskii's *Darwinism* as one of the
greatest works of the nineteenth century. He, too, thought that *Darwinism* was scarcely more than a compilation of anti-Darwinian arguments

advanced by other writers. He accepted Timiriazev's argument that
Danilevskii, by his own admission, wrote his book for the sole purpose
of attacking Darwin. The idea of a dispassionate and well-rounded
analysis of Darwin's arguments was completely alien to Danilevskii,
who wrote on a scientific theme by relying on a method that was totally
unacceptable to the scientific community.[56]

Strakhov was not without loyal and dedicated allies. One of them
was Prince V. P. Meshcherskii, who noted resentfully that at the end of
the 1880s it was difficult to find a high school student unfamiliar with
Darwin's ideas. He quickly added that it was equally difficult to find a
professor who had a "clear idea" about Danilevskii's criticism of Darwinism. The prince added that it was most unfortunate that Danilevskii's
Darwinism—the book that "completely destroyed" Darwin's theory—
had attracted only a handful of readers. In response to these and similar
statements, N. V. Shelgunov, a writer with liberal views, noted that
Meshcherskii had neither the "intellect" nor the "talent" to debate the
issue of Darwinism. He also noted that Danilevskii, Strakhov, and
Meshcherskii were engaged not only in "destroying" Darwinism but
also in writing articles against the freedom of the press.[57]

V. V. Rozanov, a commentator on modern developments in science
and philosophy and a noted religious thinker, echoed the views of
Danilevskii and Strakhov.[58] The profound crisis in Western thought—
which had a strong echo in Russia—came, in his view, on the wings of
three massive "absurdities": "Darwinism in the study of organic nature,
positivism in the sphere of logic, and mechanism in the search for an
understanding of the entire universe."[59] At long last, he said, Darwinism
had ceased to exist as part of living science: while von Baer, Agassiz,
and Kölliker had not been successful in their war on the new heresy,
Danilevskii and Strakhov were eminently successful in laying bare the
monstrous flaws in Darwinian thought and in removing it from the
realm of acceptable knowledge. He wrote in his review of Strakhov's
Philosophical Essays: "Together with Danilevskii's monumental work,
Strakhov's articles have made a singular contribution to the spread of a
correct interpretation of [Darwin's] theory in our society. Earlier it was
difficult to encounter a person who questioned Darwin's theory; now it
is difficult to find a person who does not question it."[60]

Rozanov concentrated on one major "weakness" of Darwinism: its
close and indefensible alliance with the mechanistic orientation in science which had reigned supreme since the time of Galileo and Kepler.
Every mechanical explanation in biology, according to Rozanov, is a

causal explanation, and every causal explanation, as part of a mechanical orientation, is an "external explanation" that does not go below the surface of organic reality. Not "external causality" but "internal purposiveness," irreducible to mechanical explanation, holds the key to the understanding of the dynamics of living nature. "Causal" explanation is acceptable only insofar as it recognizes the existence of a transcendental "plan" that predetermines the development of nature. In the world of life, causality is only a specific manifestation of teleology. Darwin's "temporary" success came from the simplicity of his theory and the wealth of illustrative material he drew from natural history, rather than from a deep analysis of experimental data.

In 1899 the *Russian Herald* published a review of the second volume of Danilevskii's *Darwinism*, of which only the initial chapters were completed. Rozanov, the author of the review, repeated the old argument that Darwin's natural selection explained the preservation of existing forms rather than the emergence of new forms. Darwin, he added, rejected every teleology as a mode of scientific explanation, but was unsuccessful in his effort to replace it with a precise and logically rigorous reliance on causality. Rozanov obviously felt that the Darwinian emphasis on the random nature of variation went against both teleology and causality as tools of science. He claimed that particularly the disregard of teleology forced Darwin to limit his research to the surface of life processes.[61]

Outside the Danilevskii Controversy

The Danilevskii crusade did not encompass all attacks on Darwinian thought during the 1880s. Significant criticism of Darwin's theory came from individual members of the scientific community, who were guided more by the inner logic of the growth of evolutionary thought than by ideological considerations and who made no references to Danilevskii's *Darwinism*. Two scientists deserve special attention because their criticism anticipated the anti-Darwinian movement of the 1890s. P. F. Lesgaft, at this time employed by a military gymnasium as an instructor in physical education, advocated a synthesis of Lamarckism and Darwinism, giving the former a central position in the evolutionary scheme. S. I. Korzhinskii, professor of botany at Tomsk University, heralded the coming of neovitalism as one of the major forces opposed to Darwinism.

In 1889–90 the journal *Russian Wealth* carried in long installments Lesgaft's comprehensive study of the nature and evolutionary role of he-

redity. The essay surveyed the history of embryological studies of heredity from Caspar Wolff to Darwin. It concentrated on the relative merits
of Lamarck's and Darwin's contributions to the theory of heredity as
part of a broader theory of organic evolution. In Lesgaft's judgment,
both Lamarck and Darwin explained the origin of plants and animals
by relying on the inductive method—the method of empirical science.
Their theories received strong support from the facts gathered by paleontology, comparative anatomy, the theory of rudimentary organs,
plant and animal geography, and developmental or ontogenetic studies.[62] While Darwin, according to Lesgaft, offered a generalized idea of
organic transformation as a continuous process, Lamarck "explained"
the mechanism of transformation by stressing the direct influence of the
environment and the paramount role of physicochemical explanation in
biology.[63] Lamarck, he thought, was much more successful in preserving
the full effectiveness of the law of causality in the scientific study of organic evolution. In brief, the future of Darwinism lay in taking over the
very essence of Lamarck's theoretical principles. Lamarck's approach,
according to Lesgaft, was genuinely scientific because it conformed to a
mechanical model of the living world. The same could not be said about
Darwin's approach, which did not clear the ground for physicochemical
analysis of living phenomena, did not present an unambiguous picture
of causality, and did not give exclusive primacy to the environment as a
factor of evolution.

In 1888 S. I. Korzhinskii, a noted expert on the flora of eastern Russia, with a primary interest in plant geography and systematics, read
his inaugural lecture, entitled "What is Life?" at the newly opened
Tomsk University.[64] The lecture focused on the limitations of the mechanistic orientation in biology, which stressed physicochemical analysis of
physiological processes and the Darwinian interpretation of organic
evolution. Once a moderate Darwinist, Korzhinskii now became the
main catalyst of an anti-Darwinian orientation in the community of biologists. Although Korzhinskii's attack did not echo Danilevskii's anti-
Darwinian argumentation, it received strong impetus from the same
currents of thought that encouraged Danilevskii to attack Darwin in
the first place. Its major roots were, however, in the academic community's growing dissatisfaction with Newtonian mechanism as the scientific key to all the mysteries of nature. At the end of the 1880s the
antimechanistic orientation appeared for the first time as a clearly articulated movement. Most biologists, however, were not ready to make
public statements against Darwinism and to join academic philoso-

phers, theological scholars, and professional ideologues in a crusade against evolutionary thought.

Just as Danilevskii refueled the anti-Darwinian campaign conducted by metaphysicists and theologians, so Korzhinskii ignited the first sustained anti-Darwinian campaign in the scientific community. There was one basic difference in the strategies of the two campaigns. Whereas Danilevskii declared a total, direct, and uncompromising war on Darwin's theory, Korzhinskii, who made no overt effort to be identified with Danilevskii, started only a partial and indirect war. In the first place, he criticized Darwin only by inference—by criticizing the mechanistic orientation of contemporary biology. In the second place, he did not rule out the use of mechanical models and physicochemical analysis in biological studies; he merely attacked their role as preeminent methodological tools in the study of life. By drawing a conclusion from the arguments set up in "What is Life?" Korzhinskii stated:

I can formulate my views in the following manner: life is an inordinately complex phenomenon, made up of heterogeneous types of processes. Some of these types bear a purely chemical or physical character. Such are the processes of nutrition, building and destroying organic substance, and energy transformation. These activities, however, do not exhaust the entire spectrum of life processes. They play only an auxiliary role. The essence of life consists, first, of activities coming from the ability to respond to external stimuli in a purposeful manner, and second, of processes related to the development of organisms. These purely vital phenomena are based on something that is common to, and specific for, organisms and has no place in the inorganic world. This quality—this principle—belongs to the plasm. It should not be related to chemical or physical qualities, because it creates phenomena that have nothing analogous in the inorganic world. It cannot be split into constituent elements and it resists exact inquiry. Conditionally, this quality may be termed *vital energy*. This is not a *vital force;* it is not a distinctive and inexhaustible source of force peculiar to organisms. Living energy is not an exception to the law of the conservation of energy. It appears in organisms as a result of the disintegration of complex organic substances and of the continuous oxidation process in organisms.

All this may lead us to conclude that the recognition of a special vital quality of organisms—of a special vital principle—represents a backward step, returning us to vitalism and rejecting the mechanical theory as erroneous. This conclusion would not be fully correct. As a scholar examines daily experience he becomes convinced that there are no views that are totally erroneous, just as there are no views that are absolutely correct. Every view contains a degree of truth, and every conclusion represents only an approximate solution of a problem. If this is true, then my view represents only a new approximation to the truth.[65]

Korzhinskii's views found few, though strong and influential, supporters. His Tomsk pronouncements helped inaugurate a new orientation in the Russian Darwinian studies—an orientation that sought a compromise between mechanism and vitalism in the scientific study of life processes. The dispute over the philosophical foundations of biology opened the door for a fresh examination of the Darwinian legacy. The crisis in Darwinism became a special case of the crisis in Newtonianism, spurred by the streams of fresh and "unexpected" knowledge produced by laboratory experiments in physics. It was no accident that at the very end of the nineteenth century Korzhinskii's concern with the inadequacies of mechanical models in biology became a direct attack on Darwin's theory.

N. G. Chernyshevskii's Critique of Darwin's Theory

Nikolai Chernyshevskii, like Danilevskii, operated from an ideological position. Unlike Danilevskii, however, he represented and articulated the views of a strong liberal wing of the Russian intelligentsia. Presenting himself as "Old Transformist," he was among the writers who both defended the idea of organic evolution and sought a critical reexamination of Darwin's theory, particularly on the basis of modern studies of the physiological aspects of organic change. He made no reference to Danilevskii, nor did he give any indication of familiarity with his anti-Darwinian work. In 1889 the journal *Russian Thought* published his article on the origin of the struggle for existence.[66] The editor of the journal introduced the article with a brief note underlining the main thrust of Chernyshevskii's argument. He wrote:

> *Russian Thought* has published several articles explaining the evolutionary theory and defending it from one-sided attacks. In our society, however, there has been a tendency to narrow the meaning and the significance of evolution (transmutation)—to show that it depends exclusively on the struggle for existence, presented as an unfailing source of progress. Because of this confusion, the editors of *Russian Thought* consider it their duty to publish the article of the author who has chosen to go under the name of Old Transformist.[67]

Chernyshevskii's attack on Darwin's heavy reliance on the struggle for existence followed two lines of reasoning. In the first place, Darwin interpreted the struggle for existence as a mechanism for eliminating species not contributing to organic evolution as a progressive process.

Chernyshevskii could see no connection between the struggle for existence and the growing physiological complexity of organisms, the most significant index of progress in living nature. "If the transformation of species were a product of natural selection then it would be limited to a degradation of organisms. . . . There would be no progress beyond unicellular organisms."[68] To save evolution as the cornerstone of modern biology, it was necessary to widen the base of its study far beyond the Malthusian limits.

In the second place, Chernyshevskii criticized Darwin and his uncritical followers for their effort to transform the theory of organic evolution into a system of orthodox thought and unchallengeable authority. He blamed Darwin for disregarding many clues that both contravened his theory of evolution and helped make biology an open system of scientific knowledge, propelled by fresh empirical evidence and new theoretical challenges.

Chernyshevskii's argumentation was more an exercise in logic than an exhibition of scientific erudition. His paper presented an orderly sequence of tight arguments relying on personal judgment and unadorned by customary citations from Western sources. Chernyshevskii tried to save the place of honor for Darwin in both modern science and modern ideology. He saw Darwin's main contribution in his giving a historical orientation to both biology and ideology. Darwin faltered, however, in his close reliance on Malthus's identification of the primary mechanism of evolution. His errors were minor in comparison with the grandeur of his achievement and scholarly character. Chernyshevskii wrote about Darwin as a human being and as a scholar: "Darwin combined the qualities of a very good, perfectly honest, and unusually noble man with the attributes of a great scholar: powerful intellect, enormous command of knowledge, and unremitting search for new wisdom."[69] All this, however, did not mean that Darwin was a man of impeccable scholarly habits: for example, he indulged in digressions that took him away from the questions for which he sought answers, and he did not always draw a clear line between essential and trivial facts.[70]

In 1888 *Russian Wealth* published a rebuttal of Chernyshevskii's article. The anonymous critic was obviously unaware of Chernyshevskii's authorship. He thought that by devoting two-thirds of his article to refuting Malthus's theory of different ratios in the growth of animals and their food supplies, the author had succeeded only in showing his ignorance of the theoretical base of Darwin's evolutionary views.[71] He was satisfied to note, however, that the Old Transformist was concerned only with the naturalist bias of Darwinian views on the origin and evolution of moral precepts.

Chernyshevskii's arguments helped bring an end to the uncritical worship of Darwin as a pillar of academic liberalism. The publication of the article coincided with the emergence of a new generation of biologists, no longer satisfied with a mere recounting of the high points of Darwin's achievement on both scientific and ideological fronts. These biologists began to point out the research areas and theoretical riddles in which Darwin's legacy needed not only expansion and refinement but also major retreats before the onslaught of fresh insights, theoretical challenges, and experimental discoveries. Moreover, it became clearly evident that criticism of Darwin's theory had ceased to be a monopoly of conservative thought. The new criticism, however, was not part of a campaign to dismantle the intricate structure of Darwinian thought; it was part of an effort to integrate Darwinism into the changing mainstream of modern biology. The rising criticism did not gather enough strength—and sufficient numbers of strong supporters—to present a serious threat to the firmly entrenched authority of Darwin's theory. In a short biography of Darwin written for the Brockhaus-Efron *Encyclopedic Dictionary,* M. Engel'gardt gave a fitting description of the towering position of Darwinism at this time:

> It would be difficult to find a nineteenth-century scholar whose ideas have had a broader scope of influence than those of Darwin. Relying on the theory of natural selection in explaining the processes of the development of the organic world, he has made the long-heralded idea of evolutionism a part of science. The future will show whether the factors he has introduced—the struggle for existence, variation, and heredity—are sufficient to explain all the phenomena of organic development or whether there will be a need for a consideration of additional factors, but whatever happens biology will continue to be evolutionary. Many branches of knowledge—the social sciences, anthropology, sociology, and ethics—have been or are in the process of being transformed in the spirit of evolutionism. Darwin's book on the origin of species has opened a new era not only in biology but also in the history of human thought.[72]

Criticism: An Overview

The years 1885–1890 marked a turning point in the annals of Russian Darwinism. This period produced the first Russian effort to consolidate anti-Darwinian arguments on four intellectual fronts. On the philosophical front, Strakhov and Rozanov led a strong crusade against Darwinism as an expression of "natural science materialism" and sought to justify the new education policy aimed at a drastic reduction of science subjects in university and gymnasium curricula. University professors of philosophy—all imbued with the spirit of mysticism—accepted Danilevskii's work as a crushing blow to Darwinism. On the theological front, this was a period of concerted effort to establish closer working ties with the philosophy of new conservatism and to build defensive guards around the values built into autocracy. It was at this time that theological journals intensified and systematized their effort to deal with the threat of materialism represented by Darwinism. The *Theological Herald* and *Belief and Reason* became important tribunes for a running evaluation of the relationship of new developments in science to religious values. Theological criticism became more comprehensive and sophisticated. On the scientific front, Korzhinskii started a movement that considered Darwinism too closely tied to mechanistic views to serve as a guide to the future development of biology. Korzhinskii's criticism gained support from the unorthodox discoveries in the laboratories of experimental biology and from new theoretical insights. The scientific community showed increasing unwillingness to defend extremist views: it expressed a clear tendency to ignore Danilevskii's exhortations and to avoid Darwinian orthodoxy, which Timiriazev had cultivated so diligently. On the ideological level, Chernyshevskii's criticism of the Malthusian bias built into Darwin's theory was particularly significant. From now on, the criticism of Darwinism could count on support from a strong and constantly growing wing of liberal thought.

This period was marked particularly by a consolidation of conservative thought engaged in a relentless war on Darwinism. The leaders of the new conservatism—Danilevskii and Strakhov, Rozanov and Leont'ev, K. P. Pobedonostsev and Meshcherskii—made their attack on Darwin part of a more general attack on English values. They exhibited remarkable consistency and persistence in attacking the two pivotal characteristics—in their opinion—of English culture: a commitment to political democracy, and a firm belief in the intellectual superiority of science and of a philosophy built upon the riches of empirical science. The architects of the new conservatism treated Darwinism as a specific and particularly pernicious expression of this dual threat to the sacred values of autocracy. They wanted a "science" that did not encourage critical thought and did not stand in the way of mysticism and philosophical irrationalism. Rozanov went so far as to lay the "logical" and "epistemological" foundations for a science that would be fully compatible with the "Russian soul."[73]

The horror of the 1880s became an indelible part of Timiriazev's memory. He wrote subsequently:

> The polemic [evoked by Danilevskii's *Darwinism*] mirrored the full field of our recent cultural history. It expressed the oppressive atmosphere of the eighties, when the triumphant forces of reaction thought it possible to make a decisive break with the legacy of the recent past and, acting on behalf of a primordial, national-Byzantine principle, to declare war on the whole of Western culture. Danilevskii's campaign against Darwinism was one of the characteristic episodes of that struggle. He received help from all reactionary spheres. Ministers, influential St. Petersburg circles, obliging capital (without whose help Danilevskii's thick volumes would not have been published in the first place), literature . . . , leading dailies, philosophers, and official science . . . all were on the side of Danilevskii.[74]

While the challenges of conservatism grew in compass and intensity, they were not strong enough to become a serious threat to the scientific power and intellectual resilience of Darwinism. The time did come for Darwinism, however, to search for accommodation to the new scientific and philosophical challenges. The university community, for the most part quiet and invisible to the lay public, continued to be the main and most resourceful guardian of the massive gains of Darwinian thought.

In the meantime, the opposition to Darwinism received help from a powerful and influential source not allied with the ideologues of autocracy: the great novelist Leo Tolstoy. In 1887 Tolstoy told George Kennan that he considered Darwinism a "great deception" and that he "was told" that Danilevskii "has written a book which will completely demolish the Darwinian theory."[75] Kennan gained the impression that Tolstoy did not have an "adequate conception of the cumulative strength of the mass of evidence which now supports the theory of development," and he decided not to pursue this particular discussion. A year later Tolstoy labeled Chernyshevskii's critique of Darwin's theory "beautiful" and "powerful."[76]

CHAPTER FIVE

Mechanism and Vitalism

New Challenges and Controversies

In 1894 M. M. Filippov wrote in *Science Review* (*Nauchnoe obozre-nie*), the journal he edited, that at the time of his writing the main theoretical interests of Russian biologists were not the same as those of Western biologists. While Western biologists were deeply involved in a feud between neo-Darwinism and neo-Lamarckism, the Russian biologists continued to be split into Darwinists and anti-Darwinists. Obviously, he took the Spencer-Weismann controversy over the inheritance of acquired characteristics as an apt representation of the Western situation. At the same time, he took the writing of K. A. Timiriazev, who divided all biologists into Darwinists and anti-Darwinists, as representative of the prevalent thinking among Russian biologists. Filippov contended that a typical Russian biologist had not yet been accustomed to thinking that the future of his science was neither in inflexible Darwinism nor in blind anti-Darwinism but somewhere between the two extremes.

Filippov gave forthright advice to Timiriazev: instead of wasting energy in fighting "third-rate anti-Darwinists," the time had come to take note of new developments in biology and to make these developments the base for a scientific reevaluation of Darwin's ideas. It was for this purpose that he made *Science Review* a vehicle for the diffusion of new biological ideas, particularly in the disciplines concerned with the evolutionary process. The journal went so far as to publish, in installments, a Russian translation of A. de Quatrefages's *Les émules de Darwin*, which presented the Russian readers with rich information on two kinds of naturalists: those who viewed evolution from non-Darwinian positions and those who "tried to perfect the doctrine of the master." Darwin's friends Huxley, Haeckel, and Romanes received as much attention as Darwin's sworn enemies Richard Owen and George Mivart. In summing up his appraisal of the achievements in evolutionary thought, de Quatrefages, relying on du Bois-Reymond's labels, preferred the term *ignoramus* to the term *ignorabimus*.[1] While welcoming the critical spirit of the "new biology," Filippov wanted to make it clear that evolutionary biologists had made enviable progress because they stood on the shoulders of Darwin, a giant among scientists.[2]

During the 1890s Russian Darwinism was a deeply rooted and consolidated theory of organic evolution. It was the reigning power in biological theory at university centers, and it commanded the primary attention of the popular mind. Moscow and St. Petersburg universities were Darwinian fortresses. They were the primary institutional factors in assuring Darwinism of general supremacy in and outside the scientific community. All this did not mean that Darwinism did not encounter powerful challenges from new developments in biology. The state of Darwinism during the 1890s and the main currents of biological thought challenging Darwinian orthodoxy are the central topics of this chapter.

Darwinism in the 1890s: New Heights and Expectations

In 1895, in an encyclopedia article on Lamarckism, N. M. Knipovich spoke for a majority of Russian biologists when he noted that Darwinism was clearly the strongest orientation in contemporary biology. Regardless of differences in the explanation of the origin of variation, a "vast majority" of biologists, he wrote, accepted natural selection as "the main factor of evolution."[3] A year later, the journal *Russian Wealth* offered an important clue for the general assessment of the current state of Russian Darwinism:

> During the last year two publishers have announced simultaneously their decisions to bring out new translations of Darwin's *Origin of Species, Descent of Man, Expression of the Emotions, Voyage of the Beagle,* and *Autobiography.* These announcements testify to a significant interest among Russian readers in the works of the famous thinker and biologist. Although the *Origin of Species* does not, and cannot, account for all the progress after 1859 in the study of the evolution of organic forms, it is, and will continue to be, a work of extraordinary significance. To an amateur it offers the path to familiarity with the basic elements of organic transformation; it even offers a

brief historical review of the evolutionary theory. . . . The specialist sees in it a work that has laid the foundations for an entire school in biology. The *Origin of Species* contains a great number of facts that will forever constitute a solid foundation for evolutionary theory.[4]

In 1897 the geologist A. P. Pavlov noted that the recent decades represented an important epoch in the history of science, distinguished by brilliant undertakings, unceasing struggles, and solid achievements in the study of evolutionary processes. He identified the four decades following the publication of the *Origin of Species* as an epoch of the triumphant victory of evolutionary thought in many disciplines. Evolutionary theory, he noted, transformed paleontology from an auxiliary of geology to a distinct discipline, paleobiology, expected to provide the key for the understanding of some of the most fundamental questions of biology.[5] Vladimir Vagner, an expert in zoopsychology, observed that Darwinism had taken deep roots in "philosophy, history, ethics, and aesthetics."[6]

In 1890 F. A. Brockhaus (a German publisher) and I. A. Efron (a Russian publisher) undertook the publication of *Encyclopedic Dictionary*, a cultural event of monumental proportions and a magnificent achievement. The content of the new encyclopedia was actually a combination of articles translated from the German edition of *Conversations Lexicon* and new articles contributed by Russian experts. The encyclopedia was particularly strong in the coverage of Russian topics, all subject to freshly written articles. All Darwinian topics—including natural selection and the struggle for existence—were covered by new articles written by Russian experts. The articles related to the general topic of "transformism" made the clear impression that contemporary biologists were preoccupied with adding the finishing touches to the edifice of Darwinian thought. They agreed that Darwin deserved the main credit for the full demise of the static view of living nature, and for building an indestructible base for the scientific study of the evolution of life.

With the exception of new vitalism, the encyclopedia articles did not negate the new theoretical and methodological orientations in evolutionary biology. This did not prevent their authors from placing the primary emphasis on Darwin's thought and from taking serious note of the growing number of disciplines affiliated with Darwinian evolutionism. At first the evolutionary idea, according to V. M. Shimkevich, the author of the article on transformism, was an abstract idea, a product of sheer speculation. Then, thanks to Darwin, it found a secure home in biology, where it received precise theoretical formulation and rich empirical backing. Finally, it spread to the other sciences and became a new philosophy. In essence, the triumph of the theory of natural selection was the triumph of Darwinism in modern biological thought. The triumph of Darwinism, in turn, was the triumph of two great evolutionary ideas: the idea of the gradual transformation of species and the idea of the common origin of all forms of life.

As might have been expected, the strongest and most eloquent description of the triumphs of mature Darwinism came from Kliment Timiriazev. In a paper read before the Eighth Congress of Russian Naturalists and Physicians in 1890, Timiriazev asserted that Darwin's theory not only continued to dominate evolutionary biology but had ascended new heights of achievement. In his estimate, it had also gained strong recognition in the countries, like France, that were at first openly antagonistic to it.[7] Aware of the growing criticism of Darwin's theory, Timiriazev found himself constrained to respond to the claim of philosophers Hartmann and Spencer and scientists E. D. Cope and Karl von Nägeli that Darwinism dealt with the conservation, but not the origin, of the forms of life best adapted to external conditions.

Timiriazev noted that modern biology handled two major problems: the step-by-step development of organisms which holds the key to unveiling the secrets of the emergence of new forms of life, and the adaptation of existing forms to the changing conditions of life. While the first problem required an experimental method and a "physical" explanation, the second problem depended exclusively on a "historical" approach. The first problem belonged to the domain of "experimental morphology," or "experimental embryology"; the second problem was the rightful and exclusive domain of Darwinism. He saw no point in criticizing Darwinism for ignoring a problem that was outside its competence and interest in the first place. At no time did he try to explore the close relation between experimental morphology (or embryology) and Darwinism. Unperturbed by the gathering storm, he asserted boldly that Darwinism had grown into a full body of theoretical and empirical knowledge without encountering serious opposition.

Darwinism, Timiriazev wrote at this time, was not only a legitimate division of science but also a consistent and far-reaching philosophy built on mechanistic principles. On philosophical grounds, it received strong support from current studies based on the ideas of the unity of the inorganic and organic worlds and of the superiority of the physicochemical analysis of biotic phenomena. This kind of orientation en-

abled V. V. Dokuchaev to study types of soil as unique historical config-
urations of inorganic and organic constituents, helped M. V. Nentskii, a
Pole by origin, produce challenging data on the basic similarity between
chlorophyll and hemoglobin, and made it possible for N. S. Vino-
gradskii, subsequently a member of the Pasteur Institute in Paris, to explain the role of microorganisms in supplying soil with various inorganic nutrients.[8]

Another group of scientists pointed out the possibility of making the
chemical analysis of vegetal matter a tool in the study of the evolution of
plant species. E. A. Shatskii showed that the chemical characteristics of
alkaloid extracted from plants that belonged to the same family were so
similar that they could be taken as an indicator of genetic affinity. S. L.
Ivanov pointed out in 1914 that genetically related plant families produced chemically similar oily substances.[9] This and related information
prompted V. L. Komarov to assert that biochemical characteristics un-
derlie the pure morphological features that help identify and classify
plant species on genealogical grounds. Chemical analysis became an important tool in the study of phylogenetic affinities in the plant universe.
All this, in turn, supported Timiriazev's contention that the future progress of biological disciplines would come from a heavier and more diversified reliance on mechanical models and guides.

As Darwinism grew in breadth and depth so did anti-Darwinism.
The power and prestige of Darwinism compelled the critics to exercise
special care in advancing their arguments on substantive, logical, and
philosophical levels. Although there were exceptions, anti-Darwinists
could no longer depend on moral pathos and on blind adherence to autocratic ideology. The case of I. A. Chemena made this change very
clear. In 1892 he published a tome on Darwinism with the sole purpose
of assembling and coordinating anti-Darwinian statements of the leading critics of the evolutionary theory.[10] He went from one discipline to
another—from ethnography, anthropology, and sociology to psychology and mathematics—with the exclusive purpose of distilling anti-
Darwinian arguments. While scientific journals chose to ignore the
book, popular journals devoted much attention to it and, without exception, attacked it as a product of careless and incompetent scholarship. The journal *Russian Thought* accused the author of fabricating
"information" and of peddling biased ideas. It attacked his disregard for
the rules of grammar. Writing for *Theological Herald,* S. S. Glagolev, a
theological scholar, wrote that Chemena's book contributed to Darwinism by showing the ignorance of anti-Darwinists.[11]

New Competition

During the 1890s five distinct currents competed with Darwinian orthodoxy for a preeminent position in evolutionary biology. The representatives of all these orientations paid homage to Darwin's contributions to the triumph of the evolutionary idea in biology and also digressed—sometimes extensively—from his scientific views, usually on a selective basis.

The first orientation went under the generic name of neo-Lamarckism, divided into specific modes of expression, depending on the French evolutionist's ideas selected for special emphasis and elaboration. The supporters of neo-Lamarckism ranged from down-to-earth toilers in "folk science" to highly erudite metaphysicists propelled by cosmic vision.

The second orientation, known as neo-Darwinism, was guided by the powerful figure of August Weismann. It eliminated the role of external environment from the factors of evolution, but it gave prominence to a modified version of natural selection. It served as a historical link between Darwinism and modern genetics.

The third orientation was known as neovitalism, a label used more by antagonists than by advocates. It was preoccupied with metaphysical excursions into the realm of secrets that belonged exclusively to the living world. It emphasized the aspects of the biological order not open to experimental study. In the view of its champions, the very essence of life, and the evolution of the forms of life, belonged to a reality inaccessible to the analytical models of inductive science. The historical role of neovitalism during the early decades of the twentieth century was to provide criticism of, rather than a substitute for, the heavy dependence of many branches of biology on the methods and styles of physics and chemistry.

The fourth orientation may be termed behavioral evolutionism. It differed from neovitalism and various metaphysically oriented branches of panpsychism in groping for a rigorously scientific approach to the evolutionary role of mental activity. It was this tradition that laid the foundations for the subsequent emergence of ethology as a distinct discipline. Some of the early representatives of this orientation distinguished themselves more as advocates than as practitioners of experimental research in plant and animal behavior.

Haeckelism, the fifth orientation, was unique. Named after the German scientist-philosopher Ernst Haeckel, it recognized the basic principles that gave Darwin's theory a distinct place in evolutionary biology, but it built these principles into a metaphysical system that had little in common with Darwin's scientific views and aspirations.

The five orientations became fully formed during the 1890s. Each appeared not only as a unique commentary on the Darwinian revolution in science and world outlook, but also as a distinct reaction to rapid advances in experimental biology. Collectively, they echoed both the shifts in the strategy of scientific research and the rising vigor of the philosophical controversies of the day.

Neo-Lamarckism

During the 1890s a revival of Lamarckian ideas became a strong component of the general upsurge in biological thought. Characteristically, the conditions and expressions of the revival varied from one country to another. In Great Britain the Lamarckian scene was dominant by Herbert Spencer's thunderous attack on August Weismann's equally thunderous rejection of the Lamarckian notion of the inheritance of acquired characteristics as a motor of organic evolution. In the United States the situation was dominated by Edward Drinker Cope's heavy reliance on Lamarck's ideas in a frontal attack on the randomness of Darwin's idea of evolution. In France, Edmond Perrier represented a strong group of biologists eager to show that Lamarck's evolutionary theory should be treated not as a rival of Darwin's theory but as an "indispensable base" upon which Darwin built his theoretical edifice.[12] Alfred Giard made Lamarck's views the basic component of an elaborate evolutionary theory in which the environmental determinants occupied the primary position, and in which Darwin's natural selection was recognized only as one of "the secondary factors."[13] He pointed out that Bossuet, Montesquieu, and Buffon, the illustrious French thinkers in the eighteenth century, had recognized climate and other physical features of the environment as a direct and primary factor of organic transformation.[14] Y. Delage and M. Goldsmith noted that many biologists were actually "shy Lamarckists."[15] Felix le Dantec, the Sorbonne biologist-philosopher, went back to Lamarck to help Darwin's theory become a legitimate part of the Newtonian picture of the world.

In no country was the revival of Lamarckian thought more widespread and intensive than in Germany. Among many strains of resurrected Lamarckism, two stood out because of the unusual strength of their influence and the clear delineation of their interests. One group concentrated on the experimental study of the direct influence of external environment on the evolutionary process and on the inheritance of acquired characteristics. This orientation reached the culminating point in the experimental work of Max Standfuss, E. Fischer, Hans Przibram, and Paul Kammerer.[16] The Russian commentators on modern developments in biology preferred to identify this orientation as mechanical Lamarckism or mechano-Lamarckism, a term used by Ludwig Plate in his classic *Selektionsprinzip,* primarily because of its goal to reduce vital processes to physicochemical explanations.[17] The second group elaborated on Lamarck's idea of an internal impulse for perfection as the basic motor of organic evolution. Whereas the first group emphasized the external causation of evolution, the second group attributed the primary significance to internal causation. Whereas the first group sought a physicochemical explanation of evolution, the second group was concerned primarily with psychological explanations.[18] For this reason it earned the title psychological Lamarckism.

The mechanical orientation in Russia was for a long time inspired by the wide publicity given to Shmankevich's experimental study of the transformation of *Artemia salina* into *Artemia Mühlhausenii.* Shmankevich interpreted this unique "transformation" as a result of the direct influence of changes in the salinity and temperature of water in the Odessa lagoons. The experimental research of this kind became so popular that two eminent observers—A. N. Beketov and A. N. Severtsov—went so far as to claim that Lamarckism was the dominant orientation in evolutionary biology during the 1890s.[19] Beketov and Severtsov did not document their claims, and in the specific reference to Russia, they could not have documented them. During this period Russia did not produce a single Lamarckian scholar capable of attracting the attention of the scientific community. Intense and far-reaching, most Lamarckian thought was submerged in Darwinism.

During the 1890s Shmankevich's reputation waned rapidly. The results of accumulated experimental research contradicted the results of the Odessa studies and helped undermine one of the prize claims of experimental Lamarckism. Challenges to Shmankevich's claims came first from the West. Alfred Russel Wallace argued that Shmankevich's study did not explain "the cumulative effects" of external influences on "the very lowest organisms."[20] A true Darwinist, he knew that to give in to the Lamarckian emphasis on the "direct influence" of environmental conditions would mean to give up on natural selection as the key factor of organic evolution. As early as 1894 William Bateson claimed that *A. salina* and *A. Mühlhausenii* were not different species, as Shmankevich had claimed, but varieties of the same species.[21] In 1898 V. P. Anikin, curator of the Zoological Museum of Tomsk University, reported that the "transformations" Shmankevich claimed to have observed were actually pathological effects of a drastic change in the environment. He also found that *A. Mühlhausenii* appeared in an unusually large number of varieties. Three years later P. N. Buchinskii reported to the Eleventh Congress of Russian Naturalists and Physicians that his long and tedious experiments with the same species and in the same environment did not induce anything resembling the transformation Shmankevich had reported.[22] Shmankevich came under attack not so much because he was preoccupied with the direct influence of the environment on the transformation of species but because his claims were too extravagant. The rise of Weismann's neo-Darwinism encouraged experimental ventures that contradicted Lamarckian claims. The collapse of the Shmankevich "school" made it much more difficult for the Russian Lamarckists to become a distinct community of scholars.

V. V. Zalenskii, a member of the St. Petersburg Academy of Sciences, noted that the direct influence of the environment on the process of evolution continued to be an open and unanswered question. After surveying a fair sample of contemporary Western studies of the influence of specific environmental factors on the transformation of species, he concluded in 1896 that the paucity of experimental data made it impossible to draw conclusions of a more general nature. He could not point out a single scientific contribution of this type of research which found a practical application in animal husbandry.[23] He commended Shmankevich for the precision and originality of his research, but he continued to be generally skeptical about relating experimental findings of environmental studies to the laws of evolution.

Psychologically oriented Lamarckism was dominated by the German botanists-philosophers August Pauly and Raoul Francé.[24] L. Plate stated that the views of these two scholars could be termed "psycho-Lamarckism."[25] Yves Delage and Marie Goldsmith named this orientation "vitalistic or teleological neo-Lamarckism."[26] The psycho-Lamarckians built their evolutionary theories on the idea of an innate "striving for progress," characteristic for all animals regardless of the complexity of their organization. It made no difference to them that this claim did not originate with Lamarck but was attributed to him first by Cuvier and then by Darwin.[27] No doubt, psycho-Lamarckians were attracted to the idea of "inherent impulse" because it invited and legitimated a psychological approach to evolution. They also preferred

Lamarck over Darwin because he presented, they thought, a picture of the living world that was ordered, purposeful, and harmonious with the universal aspects of human society and culture. Lamarck lived in an age in which the line separating biology from metaphysics was neither precise nor fixed.

Theodor Eimer, a noted German zoologist, was generally recognized as an immediate ancestor of the psychological branch of neo-Lamarckism. His studies made many appreciative references to Lamarck's and Cope's statements on the external environment as the primary factor of evolution. He linked a minute empirical study of the wing patterns of two large groups of butterflies to a general theory of evolution. Instead of Darwin's random variation, Eimer presented a picture of organic transformation as a one-directional process, which he named orthogenesis. This theory assumed the existence of an "internal constitution," responsible for making the transformation of organic forms an ordered and purposeful process. It lent support to the cosmology of the day built on two pillars: evolution and progress.

In Russia, Eimer's ideas attracted the attention of popular commentators on the current state of evolutionary theory much more than they attracted the attention of the members of the scientific community engaged in research. They pleased anti-Darwinists without offending Darwinists. Anti-Darwinists appreciated Eimer's firm criticism of Darwin's random variation and close affiliation with the Newtonian world view. Darwinists appreciated the massive proportions of his collection of empirical data sustaining the notion of evolution.[28]

Psycho-Lamarckians represented one of the best-known wings of a large aggregate of schools of thought that went under the general name of "panpsychism"—an effort to treat every plant and every animal as a full system of intertwined and intricate mental acts. Adaptation, a vehicle of transformation, results, in the eyes of psycho-Lamarckians, from mental activities of living beings. It comes about as a result of the innate ability of organisms to act in accordance with a "principle" expressing definite "purposes" of evolutionary processes.[29] Behavior makes plants and animals active participants in their own evolution. A typical supporter of psycho-Lamarckism argued that Darwin attributed to plants and animals only a passive role in the process of organic evolution.

In the opinion of Raoul Francé, Darwin's recognition of evolution as a key to the full scientific understanding of the dynamics of life was a great and indelible contribution to modern thought. Darwin's explanation of the inner workings of evolution, however, was inadequate and did not belong to the realm of science. It was too closely tied to mechanistic philosophy—and to philosophical materialism—to penetrate the bewildering complexity of life processes. One of the basic tasks of twentieth-century biology was to modernize Darwinism by freeing it from mechanistic and materialistic controls and by making it an integral part of modern idealism. In the *Current State of Darwinism*, Francé noted that science had reached the heights of perfection by recognizing the primacy of the psychological factor in living nature. He referred to his ideas as animation theory (*Beseelungslehre*), which he presented as the twentieth-century replacement for the nineteenth-century science built on mechanistic principles.[30] The modern study of evolution shifted the emphasis from external (physicochemical) to internal (psychological) causation.[31]

Psycho-Lamarckism found little support in the Russian scientific community. A typical biologist viewed the theories presented by Francé, Pauly, and other psycho-Lamarckians as unwholesome mixtures of Lamarckian science and vitalist metaphysics. The defenders of both orthodox and moderate Darwinism in the scientific community were obviously convinced that psycho-Lamarckism was an eclectic movement with little promise for crossfertilization with other orientations in evolutionary biology. Vladimir Vagner argued that all efforts to equate the innate "impulse for progress" with the "vital force" postulated by idealistic philosophers produced a morass of metaphysical speculation and contradicted Lamarck's evolutionary theory in the first place. Metaphysical intrusion, in his view, produced one major result: it exaggerated the difference between the evolutionary views of Lamarck and Darwin. Psycho-Lamarckism was much closer to the vitalistic leanings of *Naturphilosophie* than to Lamarck's theoretical views.[32] According to V. V. Lunkevich, Francé's and Pauly's version of panpsychism was no less one-sided and "primitive" than the extreme materialism of Vogt, Büchner and Moleschott.[33]

While the experimentalist and psychological branches of Lamarckism showed only faint signs of existence in the Russian biological thought, the presence of classical Lamarckism was all too clearly noticeable. In Russia classical Lamarckism had three general characteristics: it found expression in scattered and fragmented ideas rather than in systematic theory and elaborate experimental research; it showed a more conciliatory attitude toward Darwinism than was the case in Germany; and it placed much stronger emphasis on the direct influence of the environment on the transformation process and on the inheritance of acquired characteristics than on "the internal impulse for perfection."

There was a strong tendency among Russian biologists to minimize the differences between classical Lamarckism and Darwinism. Quite often a Lamarckian scholar worked within a modified framework of Darwinism. Nor was it uncommon to encounter Darwinian scholars who were eager to point out fundamental similarities between Darwin and Lamarck, particularly in interpreting the evolutionary role of the external environment and the inheritance of acquired characteristics. Before becoming distinct entities, the theories of Lamarck and Darwin, according to Menzbir, were made out of similar parts. Menzbir wrote:

> Like Lamarck, Darwin finds it impossible to set a limit on variation in plants and animals. He, too, explains variation by resorting to heredity which is responsible for transmitting inborn and acquired characteristics from parents to children. Again like Lamarck, Darwin recognizes the possibility of strengthening individual characteristics in cases where they are equally shared by parents. He, too, looks at variation as a product of changes in the external conditions and of use and disuse of organs, which Lamarck viewed as key factors of evolution. In everything else, however, Darwin appears as a fully independent scientist and follows a path that is completely his own.[34]

M. M. Filippov, a mathematician who wrote popular articles on scientific, literary, and philosophical themes, was also firmly convinced of the fundamental similarity in the views of Lamarck and Darwin. Both scientists, he wrote, clung steadfastly to the idea of the purposiveness of vital processes as products of natural causes, even though they provided different specifics in explaining it. Neither Darwin nor Lamarck made use of vitalistic conjectures. Both believed in the inheritance of acquired characteristics. And both recognized the direct role of the environment in the evolutionary process, even though not with the same degree of firmness.[35] Despite these similarities, Filippov did not overlook the critical differences between the two theories. In general, however, he believed that the future of evolutionary biology lay in a firmer and more comprehensive synthesis of Lamarck's and Darwin's views.

V. V. Lunkevich, a careful and informed observer of current developments in biology, diagnosed the situation correctly when he stated that to understand Darwinism and classical Lamarckism was to understand them as complementary, rather than as mutually exclusive, biological orientations.[36] The botanist V. L. Komarov, with strong Lamarckian leanings, observed that in the study of the origin of species the question of the hereditary transmission of acquired characteristics was not resolved. He also noted that it did not make much difference whether the biologists emphasized direct or indirect influence of the environment on the evolution of species. Nor did it make much difference whether they placed the primary emphasis on the emergence of new characteristics or on natural selection.[37] In their views on the evolutionary role of the environment, Mikhail Menzbir and Aleksei Severtsov, the staunch defenders of Darwinian orthodoxy, showed strong Lamarckian tendencies but only on a most generalized level.

There were also biologists who accepted the idea of a fundamental affinity of Darwinism with neo-Lamarckism, even in its psychological variety. After an analysis of the psychological branch of neo-Lamarckism, M. M. Novikov drew two general conclusions. First, unlike other neovitalist orientations, neo-Lamarckians agreed with Darwinism in accepting evolution as the pivotal notion of biology. A typical neovitalist, by contrast, thought that the time had come to remove evolution from the strategic heights of biological research. In his limited reference to evolution, Driesch, for example, relied exclusively on paleontological and zoogeographical data. Second, neo-Lamarckians and Darwinists accepted the idea of the inheritance of acquired characteristics. In this respect, neo-Lamarckism and Darwinism were united against Weismann's neo-Darwinism, their common enemy.[38]

All this, however, did not mean that revived Lamarckism was not exposed to growing attacks from varied sources. Neo-Darwinism, the archenemy of environmental determinism in organic evolution, became the leading force in the anti-Lamarckian movement. Kholodkovskii, a moderate Darwinist, wrote in 1891 that it was much easier to show the influence of a specific environmental factor on the emergence of new characteristics than to prove the transmissibility of these characteristics by the mechanism of heredity.[39] Shmankevich's study of the *Artemia* and *Branchipus,* he said, presented a conclusive picture of the active role of the salinity and temperature of Odessa lagoons in inducing morphological changes, but it presented an inconclusive picture of the relationship of these changes to the system of heredity.

Kliment Timiriazev, in his persistent defense of Darwinian theory, did not take the trouble to attack individual branches of neo-Lamarckism: he limited his attack to classical Lamarckism. Lamarck erred, he said, in making generalizations that had no empirical base, particularly in claiming the environment as the direct source of changes in races and species and the inheritance of acquired characteristics. In 1909 Timiriazev went a step further by rejecting Lamarck's notion of adaptation.

Darwin, he wrote, opposed Lamarck's tendency to identify every morphological change as an adaptation to the environment and as a "progressive step" in the evolution of species.[40] Nor did Lamarck succeed in
giving a precise explanation of the work of purposiveness in nature, an
effort in which Darwin, according to Timiriazev, was eminently successful. Darwin criticized Lamarck's "unfortunate effort" to treat "mental"
or "volitional" activities of animals as a driving force of evolution; he
did not criticize Lamarck's claim that the environment is the primary
cause of modifications in living forms. Timiriazev saw the future of evolutionary biology in a combination of Lamarck's views on the evolutionary role of the environment and Darwin's views on natural selection.[41]

In the search for the unity of Lamarckism and Darwinism, A. N.
Beketov occupied a particularly strong and appealing position. In this
effort he never digressed from the idea that the environment was the primary and direct cause of variation in species.[42] At the end of the nineteenth century, he wrote, the ideas of Lamarck and É. Geoffroy Saint-
Hilaire had taken a "preeminent place" in the theory of evolution. In
Beketov's view, only the study of the external causes of variation can
raise the theory of evolution to the level of exact science. In other words,
only a Lamarckian approach could lead to a "mechanical" (physicochemical) study of living nature, the only one science admits. Beketov
recognized natural selection as a factor of evolution but placed a stronger
emphasis on "mutual aid."[43] Despite his extensive and consistent identification with Lamarck's theory, Beketov made no effort to comment on
current developments in various neo-Lamarckian schools. He clearly
wanted to be known as a Darwinist with strong Lamarckian leanings.
More than any other biologist, he helped nurture a strong Lamarckian
tradition in Russia—not as a fully crystallized school of thought but as
a reservoir of suggestive ideas that formed a distinct enclave in the mainstream of Darwinian tradition. The rapidly unfolding developments in
experimental biology that led to the rise of genetics as a special science
and a key discipline concerned with organic evolution found no echo in
his scientific contributions.

It was not until 1911 that Lamarck's main work, *La philosophie zoologique,* was published in a Russian translation. In his introduction
V. Karpov made an earnest effort not only to give an unbiased analysis
of the translated work but also to discuss its significance within a
broader intellectual and historical perspective. He thought that Lamarck
viewed organic evolution as a specific ramification of the cosmic evolution, a self-evident process that required no special explanation. A *Pri-*

roda reviewer noted that "despite the important role Lamarckism has played in modern biology, very few biologists have read Lamarck in the original." [44] By this time Darwin's major works were available in several Russian translations.

The plant physiologist Ivan Borodin was correct when he asserted in 1910 that the widespread influence of Lamarckism at the turn of the century owed much to the dual appeal of Lamarck's theoretical legacy. The Lamarckian ideas of the direct influence of the environment on morphological modifications and of the inheritance of acquired characteristics appealed to the defenders of mechanistic views in biology. The Lamarckian emphasis on the evolutionary role of an inner—mental— impulse for perfection appealed to many adherents of vitalism and panpsychism. [45]

Neo-Darwinism

As it was understood during the 1890s, neo-Darwinism differed from neo-Lamarckism in one important respect: it was the creation of one man—August Weismann. Ernst Mayr found no problem in identifying neo-Darwinism as Weismannism. [46] Weismann's cytological theory of heredity rests on two general principles: "the continuity of the germ plasm," and "the all-sufficiency of natural selection." [47] All inheritable variation, as he saw it, is a result of recombinations of microscopic "determinants" contained in the germ plasm; adaptations to the environment do not result in heritable characteristics. Natural selection takes place in the germ plasm itself and is an outcome of the struggle among "determinants" of various characteristics for the limited supply of nourishment. It is the main factor that advances or suppresses variation caused by recombinations of "determinants." The first principle made Weismann a forerunner of genetics; the second principle made him one of the staunchest defenders of Darwinism.

M. A. Menzbir, a leading Russian defender of Darwinian orthodoxy, was on the mark when he stated that "after Ernst Haeckel, August Weismann occupied the first place among the popularizers of Darwinism in Germany." [48] Another writer gave a somewhat different interpretation. In his view, a notable difference separated Darwin from neo-Darwinists: while Darwin regarded natural selection as a primary factor of evolution, neo-Darwinists, led by Weismann, regarded it as an exclusive factor of evolution. [49] In this specific case, "neo-Darwinists were more ardent Darwinists than Darwin himself."

Few Western biologists at this time had attracted more attention than Weismann. A contributor to the journal *Russian Thought* noted in 1896 that Weismann's neo-Darwinism was the most widely recognized biological orientation of the day because it answered a "much larger number of questions" than any other theory.[50] In his essay on the achievements of nineteenth-century biology, V. I. Taliev noted in 1900 that Weismann's theory was so well known that it required no detailed discussion for a Russian audience.[51] N. A. Kholodkovskii observed that Weismann's theory "not only does not contradict the theory of natural selection but goes further than Darwin's theory in emphasizing its primary role" in the evolutionary process.[52] He was pleased to record that Wallace, in *Darwinism*, gave full support to Weismann's theoretical views.[53] Weismann's heated debate with Spencer in the early 1890s—in which he defended the idea of the noninheritance of acquired characteristics—found a strong echo in Russia. In assessing Weismann's scientific work, a typical Russian commentator had something to agree with and something to disagree with. There was no general consensus on the scientific merit of Weismann's scientific work. Taliev contended that Weismann's rejection of Lamarck's idea of the inheritance of acquired characteristics was justified. He thought that Weismann's acceptance of the notion of random occurrence of heritable adaptive variation was "contrary to the laws of logic."[54]

At the end of December 1909 the Twelfth Congress of Russian Naturalists and Physicians convened in Moscow. At the opening meeting of the zoological section, N. Iu. Zograf, lecturer in zoology at Moscow University, read a paper on "the most recent developments in biology." He opened the paper by paying homage to Darwin as the founder of a great tradition in science—particularly in embryology, comparative anatomy, and paleontology—which came to be known as Darwinism. From the very beginning Darwinism attracted wide attention by its reliance on facts and by its sense of reality, which clearly distinguished it from "the hazy and speculative theories that sprang up during the second quarter of the nineteenth century."[55] In time, the empirical view of science prevented evolutionary biologists from extending their research to previously unexplored domains of living nature. The most singular contribution of August Weismann was in carrying evolutionary research far beyond the established limits. His work marked the beginning not only of a break with narrow empiricism in evolutionary research but also of a broad experimental attack on the riddle of variation and heredity.

Russian scholars quickly recognized Weismann's two major contributions. First, he received credit for playing a major role in laying the foundations for "nuclear biology," a new branch of knowledge that helped pave the way for the rediscovery of Mendel's theory, the cornerstone of modern genetics. Second, he injected a Darwinian strain into the "nuclear" theory of heredity, which made him an early harbinger of the synthetic theory of evolution.

The famous Spencer-Weismann debate in 1893 brought Lamarckism into open conflict with the biological view that stressed autogenesis and the stability of hereditary factors. In an article published in *Contemporary Review,* Weismann reiterated the two major positions in his theory: the claim that the germ plasm was the corporeal basis of heredity; and the assertion that acquired characteristics were not heritable. Within a few months the same journal carried Spencer's response attacking both the idea of the internal causes of variation and the postulate of the non-inheritance of acquired characteristics. In Russia the response to the debate was heated and penetrating. Some critics leaned more toward Spencer and some more toward Weismann; a large majority appeared to have been eager to point out the possibility of a reconciliation of the classical evolutionary theory, grounded in the ideas of Lamarck and Darwin, and modern efforts to anchor the study of evolution in a "nuclear" approach to heredity. Most critics agreed with Spencer that the inheritance of acquired characteristics was a strong factor of evolution, but under the spell of Weismann's arguments they were reluctant to abandon the idea of the internal causes of variation altogether. A typical commentator saw the feud as an internal Darwinian affair.

The debate started by Weismann and Spencer proceeded in several directions. At the one extreme was Timiriazev, who viewed Weismann's theory and Darwinism as irreconcilable opposites. At the other extreme, V. M. Shimkevich and M. S. Ganin thought that Weismann had opened a new domain of critical and promising research, even though his arguments appeared as logical constructions rather than as empirically supported scientific statements. Ganin made a strong impression on the scientific community with a systematic survey of developments in "nuclear biology," as he called the current effort to reach the depths of the cellular—or molecular—base of heredity. The future of the theory of evolution, he said, belonged to the experimental study of the particulate agents of heredity and to the study of cell division. He thought that the problem of the material carriers of heredity had not yet been clearly formulated, let alone resolved. In his opinion, the most promising were the theories that assumed that all cells were carriers of heredity. He preferred the generality of Darwin's theory of pangenesis to more recent theories that assumed the existence of specific structural units of the germ plasm, such as Nägeli's "idioplasm." Ganin sided with the experts who saw the hereditary substance in both the "nucleus of the cell" and the "protoplasm."[56]

Ganin's survey of scientific literature showed that the study of the cellular basis of heredity had grown into a scientific enterprise of massive proportions—an enterprise that actually started in 1875 when Oscar Hertwig conducted research on fertilization among sea urchins and received a particularly strong impetus from Nägeli's introduction of the idea of idioplasm. The Weismann-Spencer debate led to far-reaching consequences. It helped the pressure to amalgamate Darwin's evolutionary theory with experimental studies of the material causes of heredity become a strong factor in Russian biology. It influenced the Russian community of biologists to engage in research that led to the revival of Mendelian laws in 1900 and to the emergence of modern genetics. The Russian biologists now agreed generally that Darwin's gemmules were the solid beginning of an active concern with the physical aspects of the transmission of heritable characteristics that led to Nägeli's idioplasm, Weismann's biophores, and de Vries's pangens.

Weismann appealed to many biologists not only because he advanced a particulate approach to the experimental study of heredity, but also because he argued that biology should not be totally dependent on the methods of physics and chemistry. In this effort, he succeeded in raising his research above the limited vision and narrow methodological perspectives of mechanical models. According to M. M. Filippov, however, Weismann's main contribution was in preparing the ground for a future "molecular biology."[57] Weismann drew criticism for advancing cytogenetic generalizations unsupported by experimental data; he was praised as a man who laid the groundwork for experimental biology. His support of natural selection made him a particularly attractive figure in the circles caught in the Darwinian whirlwind of evolutionary thought.[58]

Vladimir Vagner advanced the thesis that neo-Lamarckism (with Spencer closely allied with it) and neo-Darwinism (with Weismann as its chief architect) were "two branches of the same tree, two orientations built on Darwin's theory." It should be remembered also that both Weismann and Spencer were firmly identified as Darwinists. The major difference between the two orientations was that each emphasized a distinct aspect of Darwin's legacy: neo-Lamarckism regarded direct adaptation to the environment as a primary factor of evolution; neo-Darwinism explored the suggestive ideas offered by Darwin's view of pangenesis. In Vagner's view, the neo-Lamarckians built upon a principle that Darwin considered of primary significance; the neo-Darwinists, by contrast, built on a principle that Darwin treated as "secondary" or "tertiary" in importance.[59] Only by abandoning their more extravagant claims could the two schools become productive partners in the common search for new evolutionary light. Vagner resented the neo-Lamarckian flirting with vitalism and the neo-Darwinist, particularly Weismann's, propensity for overelaborated structures of abstract thought.

Neovitalism

The neovitalists came from two different intellectual backgrounds: from philosophy and from science. Henri Bergson was the best-known and most influential neovitalist with a home base in philosophy; Hans Driesch was the most distinguished neovitalist to come from and to operate within the realm of science. The philosopher-neovitalists depended primarily on metaphysical rhetoric; the scientist-neovitalists made maximum use of the most modern biological language. The rise of philosopher-neovitalists received a strong boost from the revival of metaphysics at the end of the nineteenth century; the scientist-neovitalists benefited from the growing opposition in the scientific community to the absolute reign of mechanistic principles in science.

The two groups were united in a strong belief that life is a reality sui generis, irreducible to physicochemical analysis and only partially conducive to laboratory experimentation. Both groups, each in its own way, saw the future of biology in surrendering its key domains to metaphysics. When at the beginning of the twentieth century Henri Bergson talked about élan vital and Hans Driesch about entelechy, they gave names to a substratum of life, an all-permeating and integrating force of life that could not be fathomed with the tools of science alone. In defining the ontological uniqueness of life, the neovitalists gave preference to teleology over causality, to internal influences over external influences, and to holism over atomism. Philosopher-neovitalists thought that the dilemma of life could be resolved by making biology a subsidiary of metaphysics; scientist-neovitalists thought that the best solution lay in emancipating biology from the fetters of the mechanistic paradigm— in creating a new biology qua science. Like a strong wing of neo-Lamarckians, neovitalists emphasized the psychological foundations of

organic evolution. They saw the future of biology in a union of evolutionary theory and psychology.

Neovitalism, in both major varieties, was essentially a metaphysical reaction to mechanistic philosophy. The philosopher-neovitalists
allowed little room for Newtonian explanations and models in biology.
Eduard Hartmann, one of the more uncompromising philosophers of
neovitalism, stated that the new study of life depended on an integration
of causal and teleological explanations of life processes. Causal explanation, as he envisaged it, emphasized strict determinism; teleological
explanation was indeterministic and was based on the notion of the
freedom of the will.[60] In the proposed synthesis of the two types of explanation, Hartmann attached primary importance to teleology and
underlined the preeminence of idealistic metaphysics. He showed no reluctance in treating vitalistic philosophy as superscience. His main arguments concentrated on the impotence of "deterministic" and "materialistic" science—and of science in general—in explaining the very
essence of life. Consistent and relentless attacks on the mechanistic bent
of traditional biology represented the most distinguishing feature of this
group. Friedrich Lange noted that no philosopher of his day talked
more on behalf of and contributed less to science than Hartmann.

The scientist-neovitalists showed more tolerance toward mechanistic views in biology, but they too assigned them a secondary position in
the overall system of organized research. In their effort to create a new
biology, they depended heavily on metaphysical metaphors, which they
tried to elevate to the level of science. They encountered serious opposition from contemporary developments in experimental biology which
were no less "deterministic" and "materialistic" than Newtonian biology. While the philosopher-neovitalists emphasized the impotence of
science to answer the crucial questions of life, the scientist-neovitalists
stressed the need for making selected metaphysical notions a key part of
scientific methodology. The differences between the two groups were
more in the mode of expression than in substantive arguments.

The neovitalists did not exercise strong influence on the development
of evolutionary biology, because they were far more concerned with
spelling out the weaknesses of Darwin's Newtonianism than with providing a coherent and balanced theory of their own. As George Gaylord
Simpson has pointed out, the modern vitalists did not search for a scientific explanation of evolution: their main task was to show that the experimental method was inapplicable to the study of evolution.[61]

In Russia, no less than in the West, the label neovitalism was often used in a derogatory sense. A typical Russian scientist-neovitalist tried to point out the reasons why this label did not refer to his particular orientation. This, however, did not prevent him from relying heavily on standard neovitalist arguments. Only a relatively small number of Russian scientists were actively engaged in defending the principles of neovitalism. Among the champions of these principles, however, there was a nucleus of eminent scientists, including several members of the St. Petersburg Academy of Sciences. Most of these scientists did not write more than one article on the subject. None allowed the identification with neovitalism to influence his engagement in empirical research.

In 1902 the noted physiological psychologist V. M. Bekhterev made an effort to set down the record of the early history of neovitalist thought in Russian biology.[62] He noted S. I. Korzhinskii's address at Tomsk University in 1888; I. P. Borodin's paper "Protoplasm and Vitalism," delivered in 1894 before the members of the St. Petersburg Society of Naturalists; Aleksandr Danilevskii's article "The Living Substance," published in 1896; and A. S. Famintsyn's long essay "Modern Natural Science and Psychology," published in 1898. Not one of these scientists identified himself as a neovitalist, but all recognized the pertinence of neovitalist criticism. All saw the future of biology in a synthesis of mechanism and vitalism, based on the mutual acceptance of the universal validity of the law of the conservation of energy.

Aleksandr Danilevskii, a recognized biochemist working on the molecular structure of protein, gave the most vivid and forceful presentation of neovitalist arguments, modified to match his particular views on the critical dilemmas of contemporary science. He noted the basic philosophical division of science into two major orientations: materialism and vitalism.

> The first of these orientations does not see anything in the living substance except perceptible matter and its forces, as well as regularity in the links between various kinds of material changes. The second, vitalistic, orientation does not overlook the material foundations of life, but it also recognizes the presence of characteristics that do not obey the mechanical principles of the interaction of atomic particles and invite a recognition of forces that are not dependent on matter.[63]

Like many Western supporters of neovitalism, Danilevskii worked primarily on "distilling a healthy kernel" from vitalist thought as a basis for new biology. While not denying the physicochemical attributes of the molecular basis of life, he argued that the real power of the protoplasm lay in its role as a "physiological" phenomenon endowed with a "vital force." Danilevskii made an effort to reconcile mechanism and neovitalism, but he did not hide his belief that the real essence of life began where the current style of physicochemical analysis ended. The explanation of life required recognition of the existence of a "special factor" beyond the comprehension of chemists and physicists. Relying on a mechanistic analogy, Danilevskii thought that just as "cosmic ether" kept the inorganic world in an ordered and predictable balance, so "biogenic ether"—an entity governed by purposefulness rather than by causality—assured the existence and the functioning of the living world. Only the existence of "biogenic ether," or a similar substance, can explain why protein substances, unstable outside of organisms, are very stable and durable within organisms.[64] The biogenic ether protects the protoplasm from disintegration by supplying it uninterruptedly with new molecules. What Danilevskii named "biogenic ether" was neither more nor less than the "vital force" of the old vitalists.[65]

Flirting with neovitalism did not affect Danilevskii's reputation as a biochemist: he kept his experimental research in the biochemistry of protein at a safe distance from his forays into the domain of neovitalist speculation. In a way, his philosophical discourse was aimed not at denying the primacy of matter in scientific inquiry but at redefining it. His discussion of biogenic ether both expanded and narrowed the notion of matter as advanced by Newtonian science: it expanded it by adding "living substance" as a new scientific category; it narrowed it by refusing to accept causality as the universal mode of scientific explanation. Danilevskii thought that present-day science was ready to undertake initial steps in advancing an experimental approach to biogenic ether.[66]

Danilevskii did not mention Darwin's views on evolution, nor did he have to. His description of vitalism as a scientific orientation of the future made it clear that he looked at Darwinism, with its Newtonian stamp, as a science that belonged to the past. The future transformist theory, he thought, would eliminate the view of evolution as a mechanical aggregate of synchronized physicochemical forces and as a dynamic network of cause-effect sequences. In the future, the notion of evolution would cover the transformation not only of the forms of life but also of the "vital substance," "vital energy," or "biogenic ether," the cornerstone of neovitalist thought.

Many Russian biologists found little use for neovitalist ideas. N. A. Kholodkovskii, a prominent St. Petersburg parasitologist and biological theorist, spoke for this majority when he criticized Driesch's effort to

make a revived version of Aristotelian entelechies the core of a tele-
ologically oriented system of biological ideas. He gave Driesch's ideas a
fair airing in Russia, but he also expressed a firm opposition to every
metaphysical intervention with biological research. Teleology, according to Kholodkovskii, is a "philosophical notion," and natural scientists
must reject it as "unnecessary" and "useless." [67]

N. D. Vinogradov noted that neovitalism erred in identifying biological mechanism with philosophical materialism. Philosophical materialism, he contended, was built on the premise that all natural phenomena, organic no less than inorganic, could be explained exclusively as
functions of material objects. Biological mechanism, by contrast, was
guided by the general assumption that all vital processes obeyed the
same laws of mechanics that applied to the inorganic world.[68] Following
the arguments advanced by F. Lange, he rejected philosophical materialism and expressed a general approval of biological mechanism, which
he also identified as an application of the methods of physics and chemistry to the study of living phenomena. Although neovitalism was justified as a critique of the excesses of mechanistic orientation in biology, it
proved actually inadequate as a source of positive suggestions for a new
theoretical and methodological orientation in biology.

In the *Meaning of Life,* V. A. Fausek, professor of zoology at Moscow
University, presented a collection of Western essays on the theoretical
conflict between mechanism and neovitalism in contemporary biology.[69]
By allocating the most space to a selection from *Die Welt als Tat* by
J. Reinke, he showed clearly that his main interest was in publicizing
neovitalist ideas. He, too, recognized neovitalism as a source of criticism of mechanism rather than as a system of acceptable theoretical
principles and methodological designs for a major reorientation in contemporary biology. In none of his empirical studies did he make an
effort to apply neovitalist guidelines and interpretive schemes. He helped
popularize neovitalism as a formulation of challenging questions that
could not be answered within the mechanistic framework and that
promised to help biology reach new heights of scientific achievement.
He did not have much faith in neovitalism alone to achieve this goal.[70]

V. V. Lunkevich, a competent commentator on the major developments in contemporary biology, expressed a typical opinion when he
stated that neovitalism must be welcome only insofar as it pointed out
the gaps in the mechanistic world view, and that biology must depend
not only on physics and chemistry but also on its own theory, expressing the ontological uniqueness of the living substance. He stated also

that "heretic mechanists" supplied particularly sharp criticism of mechanistic epistemology. Lunkevich placed Ernst Mach among the "heretics"
ready to admit that physics, despite its "inordinate power," did not answer all the needs of the scientific study of the living world. Mach's pronouncements achieved two ends: as antimetaphysical statements they
challenged the ontological basis of vitalism, and as epistemological
clarifications they narrowed, but did not eliminate, the idea of "mechanics, physics, and chemistry" as the sole sources of models for biological
studies.[71]

Lunkevich saw no future for neovitalist efforts to substitute teleology
for causality. "Every scientific explanation is and will forever remain a
causal explanation; teleology is allowed only as a transitory or preparatory method, and only when it is unavoidable—when it presents
the best way to approach a problem." What Lunkevich did not mention, however, was that Darwinists, now more than ever before, were
willing to treat "purposiveness"—a more neutral term for teleology—
as an important explanatory principle in biology. To make "purposiveness" more realistic, however, the Darwinists made it a special mechanism of historical dynamics, unrelated to the whims of a supernatural
authority.

M. A. Antonovich, author of *Darwin and Our Time,* claimed that
one of Darwin's greatest contributions was in reducing purposiveness of
the living world to natural causes—by showing that the purposiveness
of the organic universe does not require an assumption of developmental goals predetermined by a higher intelligence.[72] This view received
help from Mach's interpretation of the study of purposiveness as a preliminary step in the study of causality. In an article published in 1892,
N. N. Strakhov acknowledged the pertinence of the vitalist criticism of
the dominance of mechanical principles in contemporary science, but he
was equally ready to admit that the spokesmen for vitalism did not
produce firm and clearly stated arguments.[73]

The *Progress of Science,* a multivolume survey of the achievements of
modern science published in 1912, noted that such leaders of neovitalism as Driesch, Reinke, Hartmann, Pauly, and Bergson were "great
and original thinkers" who gave a clear formulation of the most critical
questions in biological theory. The chapter on the origin of life stated
that "neovitalism is a fresh current, the need for which was felt strongly
during the last decade of the nineteenth century, when biology found
itself in a blind alley, not knowing where to turn for new ideas." It also

noted that, aside from asking pertinent questions, neovitalism was generally a negative development. It was an escape into such abstract notions as "entelechies," "dominants," and "vital principles," which satisfied metaphysicists but provided no help for natural scientists.[74]

It is not difficult to assess the role of neovitalism in the history of Russian evolutionary thought. Although its principles attracted much attention, it did not have strong and consistent representatives. All its supporters in the scientific community continued to do their empirical work in laboratories and other scientific workshops that made no use of its principles. Part of a general criticism of the mechanical orientation of nineteenth-century science, neovitalism represented a unique echo of idealistic metaphysics and its war on the "materialism" of Newtonian thought. The neovitalist criticism of biological materialism climaxed at the time when experimental biology made the particulate basis of heredity its major concern.

A typical Russian spokesman for neovitalism made his reputation not as a master of a distinct set of philosophical principles but as a scholar deeply engaged in experimental research focused on the physicochemical attributes of life. While zoologists provided the chief defenders of Darwinism, botanists, including a relatively large group of plant physiologists, were the bastions of neovitalism. In criticizing the cognitive limitations of mechanical approaches to living nature, Russian neovitalists were actually involved in self-criticism. Spurred by the revival of idealistic metaphysics, on the one hand, and by the rapid growth of a new physics critical of the basic principles of the Newtonian-Laplacian paradigm, on the other, the advocates of neovitalism worked in an atmosphere that encouraged the search for new theoretical orientations and research strategies. They invited metaphysical ideas, not to replace the established research procedures in biology but to complement them. Most biologists, however, resented every effort to inject vitalist "mysticism" into biology. The academic philosophers, by contrast, saw little promise in the efforts to dissociate the new vitalism from metaphysical idealism.

In 1897 the *New Word*, a popular journal devoting much attention to current developments in science, published a Russian translation of Max Verworn's "Vitalism," an effort to clarify the crisis caused by the neovitalist ideas. Verworn's conclusions reaffirmed the views of a solid majority of Russian biologists: "The future of biology is in no way connected with, or dependent on, the streams of neovitalist thought."

The future of biology—he referred primarily to physiology—lay in advancing the techniques for a more efficient study of the physics and chemistry of the corporeal basis of life.[75]

N. K. Kol'tsov, a Darwinist and true pioneer of molecular biology, gave a fairly accurate description of the mechanism-neovitalism controversy:

> Persons who have a natural inclination to pursue research and to conduct experiments cannot remain convinced vitalists for long. It is true that every success in natural science adds to a negation of vitalism. . . . Many unknown factors—factors of supersensory, teleological, irrational, and mystic nature—have been eventually explained and subordinated to causal regularity. In the struggle between vitalism and mechanism the latter has preserved its great importance as a superb working hypothesis: only mechanism gives us faith in the omnipotence of science.[76]

A. N. Severtsov, the most original and productive Darwinian scholar in Russia, considered neovitalism a real threat to both Darwinism and evolutionary studies in general. Speaking at the joint session of the Congress of Russian Naturalists and Physicians and the St. Petersburg Society of Naturalists on January 3, 1910, he unleashed a furious attack on Hans Driesch, who claimed that the study of animal phylogeny had produced nothing but "fictitious constructions," and that the study of evolution did not have a base in empirical facts and would forever continue to be a science of "inferior" quality. Severtsov blamed neovitalists particularly for their growing commitment to nonevolutionary embryology.[77] He found neovitalists guilty of making biology a home for all kinds of metaphysical notions that had no affinity with science.

Darwinism found no need to compromise with neovitalism. After reaching a high point in the work of Hans Driesch and Henri Bergson during the first decade of the twentieth century, neovitalism lost much of its zeal and resilience; it was a victim not of Darwinian "Newtonianism" but of the streams of new biological thought emitted by experimental biology which in due time made the triumph of molecular biology possible.

Russian biologists limited their concern with neovitalism primarily to comments on Western developments. They were divided into two clearly separated groups. One group, represented most eminently by Aleksandr Danilevskii, advocated an expansion of biology beyond the rigid limits imposed by the mechanistic orientation. To achieve this goal, it was willing to seek answers to questions that traditionally constituted an exclusive domain of metaphysics. Clearly, it sought to widen the domain of biology by blurring the border line between science and metaphysics. It criticized the mechanistic orientation in biology for its failure to search for answers to questions pertaining to the origin of life and consciousness and to the inner workings of teleology in organic nature.

The second group, represented by the cytologist S. M. Luk'ianov, demanded that biologists make a concerted effort to refrain from asking questions that did not fall within the competence of science—that they surrender difficult questions to other modes of inquiry. Science, in his opinion, must withdraw from all efforts to explain the cellular basis of mental activity, the cellular origin of life, and the work of purposiveness at cellular and organismic levels.[78] Darwin's effort to explain the place of teleology in the living world, for example, did not produce worthwhile knowledge, for the simple reason that it relied exclusively on the scientific method. Under the spell of the current revival of idealistic metaphysics and its challenge to science, Luk'ianov wanted the scientific community to recognize neovitalism as both a valid critique of the intellectual limitations of science and an extrascientific effort to answer the basic questions related to the origin, organization, and evolution of life. In his effort to deny science the right to study certain aspects of life, Luk'ianov acted as an ideologue rather than as a scientist. In addition to violating the ethos of science, which recognizes no limitation on scientific inquiry, Luk'ianov erred in identifying the limitations of mechanistic orientation in science as limitations of science in general.

Behavioral Evolutionism: A. S. Famintsyn and V. A. Vagner

A. S. Famintsyn, a member of the St. Petersburg Academy of Sciences, and V. A. Vagner, a privatdocent at Moscow University, were the best-known representatives of a growing effort to place the psychological study of the phenomena of life on a solid scientific base. The major goal of this orientation was to avoid the strangulating effects of mechanical models in biological studies without inviting metaphysical interference. Famintsyn worked primarily on the epistemological foundations of the new orientation. Vagner, by contrast, concentrated on giving the new orientation an integrated methodological apparatus. Of the two scientists, only Vagner engaged in empirical research. Both scholars, each in his own way, helped build a scientific tradition that served as a base for the emergence of ethology as a distinct discipline.

A broad-minded and erudite scholar, Famintsyn fought on three main fronts in the cause of liberalism: he participated in the *zemstvo* movement, a grand experiment in local self-government and independent activity; he wrote articles explaining the urgent need for broader guaranties of academic autonomy; and he signed petitions demanding extensive reforms in the structure of the Russian government. In 1879 he served four days in jail—a sentence imposed by local police for "pernicious influence on students." In March 1896 and February 1899 he took an active part in student demonstrations against the policies of the Ministry of Public Education.[79]

The evolution of Famintsyn's views on Darwin's theory went through three distinct phases. During the first phase he clung closely to the basic principles Darwin had built into the foundations of his theory. His public lecture at the University of St. Petersburg in 1874 marked the high point of his allegiance to Darwinism.[80] At this time Famintsyn had no reservations in viewing Darwin's legacy as the main source of ideas for the future development of biology. In *Metabolism and the Transformation of Energy,* a bulky monograph published in 1883, he did not deal directly with the problems of organic evolution. He did, however, endorse one of Darwin's guiding ideas: the unity of the organic world.[81] He undertook a detailed analysis of the basic similarities of plant and animal life, endowed with the same metabolism, chemical composition, cellular structure, excitation mechanisms, and respiration processes.[82]

During the second phase Famintsyn gave prominence to the argument that Darwin's theory did not tackle some of the key questions of organic evolution. While continuing to recognize natural selection as the primary factor in the mechanism of adaptation to the external environment, he now placed a strong emphasis on the close cooperation of experimental psychology and evolutionary theory for the benefit of both. In this effort he was particularly determined to point out the inadequacies of mechanical models and physicochemical methods in the scientific study of vital processes.

During the third phase Famintsyn argued in favor of a systematic and experimental study of symbiosis as an evolutionary mechanism producing more complex forms of life. His theory of symbiogenesis broadened the role of biochemistry in recasting the general theory of evolution. He worked much more on collecting empirical data illustrating the work of symbiogenesis than on advancing a clearly stated and elaborate theory.[83] Particularly interesting and challenging was Famintsyn's notion of the cell as a "symbiotic complex." Vegetal cells, according to him, are products of the evolution of a symbiosis of two simple "organisms": a green organism with one or more chloroplasts, and a colorless, amoeboid organism consisting of a plasm and a nucleus. The basic organelles of the cell did not arise through differentiation of the plasm but were independent symbiotic structures, "cells within a cell." Famintsyn thought that the symbiogenetic theory provided an example of progressive evolution, expressed in the growing complexity of organisms, a problem to which Darwin did not devote much attention.[84] He intended his symbiogenetic theory to complement rather than to negate the Darwinian factors of evolution.

Famintsyn is remembered mainly for the ideas he advanced during the second phase in the evolution of his scientific and philosophical views. He opened this phase with a report to the Eighth Congress of Russian Naturalists and Physicians, held in St. Petersburg in 1890, in which he tried to show that the "mental" factor was a key evolutionary element even among "the lowest forms of life."[85] In 1898 he published a long essay, "Modern Natural Science and Psychology," in which he presented an assortment of philosophical and methodological arguments in favor of a psychological approach to the essential attributes of life and its evolution. Not a model of logically precise presentation of theoretical arguments, the study was an effort to blend vitalism and mechanism by eliminating the excesses of both.

In writing his essay, Famintsyn was inspired by the ideas the German biochemist Gustav Bunge had advanced in a special chapter of his popular textbook in "physiological and pathological chemistry." In this chapter Bunge advocated a new orientation in biology which would assign a key role to the principles of vitalism. The main target of his attack was the reign of physicochemical analysis in the scientific study of life processes. He wrote: "The more we strive to undertake a detailed, well-rounded, and basic study of the phenomena of life, the more we become convinced that the phenomena we thought we could explain on physical and chemical grounds are too complex to be limited to mechanical explanations."[86] Bunge assumed the existence of an "internal sense," reminiscent of Johannes Müller's "specific energy of senses" as the mainspring of vital processes.[87] He treated introspection as the only effective method of penetrating the innermost secrets of life.[88] Psychology held the keys to the understanding of the mainsprings of physiological processes.

In Russia, more than in Germany, Bunge was widely known as a leading defender of neovitalism, an honor he did not cherish. In his article on vitalism, published in the well-known Brockhaus-Efron *Encyclopedic Dictionary* in 1895, Ivan Tarkhanov, an eminent professor of physiology at the Academy of Military Medicine, dealt exclusively with Bunge's ideas.[89] Small wonder, then, that Famintsyn, who gave Bunge fullhearted support, became also known as a champion of neovitalism. Famintsyn resented this label and was anxious to point out his full opposition to making biology a discipline open to metaphysical speculation. He went out of his way to disown every idea of a vital force not governed by the laws of nature.[90]

Obviously encouraged by the profusion of signs indicating a rising attack on the extensive and growing dependence of biology on the methods of physics and chemistry, Famintsyn directed his offensive against orthodox Darwinists who, he thought, believed in the superiority of "mechanical" models in biology.[91] He readily admitted that the laws of physics and chemistry are uniform and universal—that is, that they apply to both inorganic and organic nature. This did not prevent him, however, from claiming that these laws cannot explain the very essence of life. In addition to physicochemical laws, biology must search for laws of nature that are uniquely its own. Famintsyn did not criticize Darwin's theory directly: he criticized it as the most notable expression of the mechanical orientation in biology. While the inorganic world, in Famintsyn's view, can be studied only from an "external" perspective, vital phenomena can and should be studied from both "external" and "internal" sides, and must recognize the "internal" side as preeminent. To approach vital processes from an "internal" perspective amounted to grounding biology in psychological methodology. He took it for granted that a scientific study of behavior can explain the deepest secrets of life.[92] Lamarck, in his view, made an immortal contribution to science by recognizing the active role of habits, inclinations, and ideas in the phylogeny of animals and men.

In an elaborate criticism of the philosophical views of John Stuart Mill, Famintsyn found an opportunity to state his views on the role and the nature of evolution in modern biology:

> Mill and his philosophical followers erred in not allowing for a theory of the evolution of species and for a closely related theory of the evolution of mental life. Little did they know that in the future such a theory would become a point of departure for all biological research. They ignored the phylogeny of sensory organs, as well as the generally accepted view of biologists on the role of these organs as mechanisms for preserving life and ensuring procreation. The philosopher-psychologists have not come around to recog-

nizing the biological interpretation of sensory organs as the most powerful tools in the struggle for existence.[93]

A contemporary writer offered a fair and balanced interpretation of Famintsyn's position between the extremes of vitalism and mechanism:

> Famintsyn is not a metaphysicist: materialism, spiritualism—and even dualism—are to him purely arbitrary categories. He sees them as philosophical constructions without particular significance for science. He rejects all efforts to deduce the entire world from a metaphysical notion of matter or a metaphysical notion of spirit, independent of matter. "The question whether spirit emanates from matter, or matter from spirit, differs little from the question which came first, the chicken or the egg." "Material and spiritual phenomena are different only in methods by which we perceive them. They may appear merely as different sides of the same reality." "Mental life" is not reducible to "the laws of mechanics." Every effort to explain perception and consciousness by the motion of atoms is a sterile effort, and equally sterile is every effort to deduce the origin of life from a "vital force." The mechanistic world view, which reigns supreme at the present time, suffers from a major deficiency: it fully ignores the role of a capital factor in the variation and development of plants and animals. That factor is "mental life" (*psikhika*), understood in the broadest meaning of the term. *Psikhika* constitutes the essential characteristic of every living organism, regardless of its composition. Its development reflects the growing complexity of the organism, and its beginnings are in rudimentary sensations and flickers of consciousness at the lower stages of life. In other words, morphological evolution and psychological evolution are inseparable.[94]

In choosing to emphasize the behavioral basis of organic evolution, Famintsyn was influenced much more by a current fad than by carefully thought out theoretical considerations. At this time, the tendency to reduce all scientific and philosophical knowledge to psychological elements found many enthusiastic advocates in both the Russian and Western worlds of scholarship. In 1887 N. K. Mikhailovskii informed the readers of *Russian Wealth* about the growing interest in the psychological reality of organic nature. To prove his point, he mentioned Mechnikov's decision to attribute a "protozoic psychic activity" to phagocytes, Haeckel's assertion that every atom has a soul, Nägeli's reference to "an intellect of cosmic systems," and the current interest in "cellular psychology."[95] At the same time, sociology was moving rapidly from biological analogies to psychological formulas: Leon Petrazhitskii had established a flourishing "psychological school" in jurisprudence; A. S. Lappo-Danilevskii explored the psychological underpinnings of the "methodology of history"; and Boehm-Bawerk's theory of limited utility, based on a psychological examination of market behavior, found many supporters in Russia.[96]

The Psychological Society in Moscow, founded at the very end of the 1880s, grew rapidly into an important center of intellectual exchange and influence. Among the founders of this society was a small contingent of scholars with positivist leanings who supported the development of experimental psychology as a promising branch of natural science. Timiriazev, Vernadskii, and M. M. Kovalevskii, the leaders of this group, quickly became powerless in shaping the society's interests and philosophical views. Taken over by the philosophy professors, the society quickly became a bastion of antipositivism and idealistic metaphysics.[97] With much ardor, it worked on strengthening the institutional base of the ongoing revival of academic metaphysics. As described by Aleksandr Vvedenskii, the work of the society was responsible for a "resounding victory" of idealism and spiritualism over materialism and positivism.[98] The society had such a high respect for Eduard Hartmann, a leading metaphysical critic of Darwinism, that it elected him an honorary member.

Without giving details, Famintsyn endorsed the current fad of relying on hypnotism as an experimental method of psychological analysis.[99] He favored the psychological legacy of W. Wundt, particularly its antimetaphysical stance and its emphasis on experiment, concrete experience, and comparative method. Famintsyn's thinking contained two major flaws: he assumed that to emphasize the physicochemical study of life processes means an automatic acceptance of the reign of the mechanistic orientation in biology, and he thought that every psychological study was automatically antimechanistic. He overlooked the prospering schools in psychology dedicated to building their discipline on the models borrowed from physics and chemistry.

Famintsyn did not go much beyond marshaling philosophical arguments in favor of a psychological approach to organic evolution.[100] He was much more concerned with showing the futility of rapidly expanding efforts to make "physicochemical analysis" the sole approach to the dynamics of living nature than with setting up an empirical and systematic evolutionary study based on psychology. He collected numerous examples illustrating, but not analyzing, mental activities among plants and animals occupying the lowest rungs on the ladder of evolution. At a time when psychology had begun to develop a firmer basis in scientific methodology, his research interest shifted to the theory of symbiogenesis.[101] Particularly as presented by K. S. Merezhkovskii, this theory considered symbiosis a basic—but not an exclusive—mode of the origin of species.

Despite extensive criticism of its "oversights," Famintsyn stayed close to Darwin's theory: he considered evolution a core notion of biological theory; he attributed an important role to natural selection as a mechanism of organic evolution; and he subscribed to, and elaborated, the general theory of the psychic unity of animal species, including man. In his opinion, "most natural scientists have accepted Darwin's theory" and "considered anti-Darwinian sectarianism" harmful.[102] He thought that Darwin's theory of organic evolution had not yet received incontrovertible empirical confirmation. But he also thought that, in its general principles, this theory was basically correct and had already made substantial contributions to science. His major objection to Darwinism was in its viewing plant and animal organisms as "passive" rather than as "active" agents of organic evolution. According to Famintsyn, if Darwin had only examined the parameters of behavior more seriously he would have hit upon the idea of organisms as active makers of their own evolution.

Vladimir Vagner, a prolific writer in comparative psychology, went far beyond Famintsyn in advancing the cause of behavioral evolutionism. At first, he argued that zoopsychology—the field of his major concern—should be built on factual material obtained with the help of a "subjective method" and an "objective method," an idea he borrowed from George Romanes and Wilhelm Wundt.[103] The subjective method, as he defined it, was anthropocentric: it looked at animal behavior through the prism of human psychology. The objective method followed a reverse procedure: it proceeded from amoeba to man. The latter approach, he thought, called for a scientific study of the physiological underpinnings of behavior.[104] Romanes provided him with enough arguments to defend the idea of instinct as a component of "intelligent" behavior and as a starting point of mental evolution.[105] To both Vagner and Romanes, instinct always involves mental operations; reflex action, by contrast, is a "non-mental" and "neuro-muscular" adaptation to specific stimuli.[106] The domain of psychology begins not with reflexes but with instincts. Physiology, rather than metaphysical speculation, explains the threshold of consciousness as a foundation of mental life and a vehicle of organic evolution. Vagner gave the clear impression that he fully accepted Darwin's idea of the interaction of "free intelligence" and "instincts" as the foundation of more elaborate complexes of behavior.[107]

In his later work, particularly after 1910, Vagner abandoned his original position on the complementary relations of "subjective" and "objective" methods in zoopsychology.[108] Now he became more directly involved in zoopsychology as a comparative or evolutionary discipline and was ready to reject both anthropocentric subjectivism as an approach "from above" and physiological objectivism as an approach "from below." His comparative psychology saw no usefulness in comparing the behavior of amoeba with that of man, or of man with that of amoeba; it allowed only for a comparative study of the mental life of closely related species.[109] It promised to give psychology what Darwin gave to biology: a firm base in natural-science methodology and a consistent historical perspective. Vagner worked on two fronts: he carried evolution to psychology (his main task), and he carried psychology to evolution. On the latter front he joined Famintsyn as a Russian pioneer in the scientific study of the psychological motors of organic evolution.

Vagner is remembered primarily for his effort to advance a theoretical framework for the comparative and evolutionary study of zoopsychological phenomena. He identified evolutionary zoopsychology as a "paleontology of instincts," concerned with the origin and evolution of instinctive and "rational" behavior, both with historical roots in reflexes. In his empirical research Vagner was guided by the idea that every phase in the evolution of animal behavior represents a distinct adaptation to changes in the struggle for existence. The St. Petersburg Academy of Sciences published his monograph on the changing instinctive base of the nest-building techniques among city swallows (*Chelidon urbica*) in a district of Moscow. He observed and described the changes in nest-building from a "hanging" type (attached to a vertical surface) to a "sitting" type (placed on a horizontal surface) and the role of this change in the struggle for existence of this particular bird. The "sitting" type ran virtually no risk of becoming detached from the surface, and it produced a much higher survival rate of chicks.[110] This was one of the rare concrete studies of the struggle for existence in operation. Vagner described the architectural "improvements" that accompanied the change from "hanging" to "sitting" nests.

Vagner was clearly one of the most original and productive Russian evolutionary biologists of his age.[111] He was the only scholar to make an extensive and sustained effort to apply a psychological method to the study of the evolutionary process. His influence on contemporary thought, however, was rather limited. As a privatdocent he occupied too low a rung on the academic ladder to become the leader of a scientific school. He worked at Moscow University, dominated by Timiriazev and

Menzbir, who provided little encouragement for the type of evolution- ary research he preferred. Nor was he helped by the unsettled nature of his theoretical and methodological views and by the excesses of his out- look on the instinctive basis of human behavior.

Vagner tried to present an ordered and unified picture of the psycho- logically relevant empirical material available in his time. In gathering empirical data, he depended both on existing studies and on the results of his own research. If his effort to crystallize a systematic approach to the psychological aspect of evolution did not produce more impressive results, the reason must be sought in the general underdevelopment of psychology. In his time most of psychology was highly speculative and, particularly in Russia, was dominated by philosophers imbued with a cosmic view steeped in spiritualism and mysticism. The so-called scien- tific psychology, still in the embryonic stage, was too narrow and too detached from the problems of evolution: it dealt mainly with instincts and reflexes as building blocks of consciousness. Full-blown ethology was still to come.

In most of his works—but primarily in *The Descent of Man, The Ex- pression of the Emotions,* and *The Formation of Vegetable Mould*— Darwin made comprehensive and empirically supported statements on animal behavior. Michael T. Ghiselin has observed that "one may trace out the systematic development of a comprehensive system of neu- rophysiological theory, from its roots in the hypotheses of the *Origin of Species,* to its application as a fundamental component in the arguments of the *Descent of Man.*"[112] Although he did not make a deliberate effort to establish an evolutionary discipline of behavior, Darwin produced a veritable treasure of suggestive insights and astute observations on the dynamics of plant and animal psychology.[113] Both Famintsyn and Vagner referred extensively to Darwin's psychological comments, but Vagner was much more systematic and thorough. He also showed a more intimate familiarity with current work in experimental psychol- ogy, often building on Darwin's suggestive ideas. Vagner's studies in zoopsychological theory may rightfully be considered an effort to pre- sent a systematic study of Darwin's leadership in building the concep- tual foundations for a psychological study of organic evolution.[114]

Haeckelism

Haeckelism stands for an unusual brand of Darwinism. It combines a high respect for the entire set of Darwinian principles and a most liberal reliance on these principles in building a philosophical and ideological system of views and maxims that have little in common with the intellectual legacy of the English naturalist. No person had supported Darwin and spread his fame as energetically and enthusiastically as had Ernst Haeckel; nor did any scientist surpass Haeckel in recasting Darwin's theoretical principles to make them cornerstones of a personal philosophy.

Although Haeckel did not follow every detail of Darwin's scientific arguments, he became widely recognized as one of Darwin's most loyal followers. In his speech at Cambridge commemorating the one-hundredth anniversary of Darwin's birth, he expressed his true sentiments when he stated that "the monumental greatness of Charles Darwin, who surpasses every other student of science in the nineteenth century by the loftiness of his monistic conception of nature and the progressive influence of his ideas, is perhaps best seen in the fact that not one of his many successors has succeeded in modifying his theory of descent in any essential point or in discovering an entirely new standpoint in the interpretation of the organic world."[115] Darwin had credited Haeckel for making Germany a bastion of the new evolutionary theory, and in *The Descent of Man* he went out of his way to acknowledge his indebtedness to Haeckel's suggestive ideas.[116]

For several decades Haeckel was one of the better-known foreign popularizers of Darwin's theory in Russia. In 1869 I. I. Mechnikov produced a condensed Russian translation of his principal work, *The General Morphology of Organisms*. In the mid-1870s the newly founded journal *Priroda* (*Nature*) published Russian translations of two of Haeckel's essays dealing explicitly with the theory of evolution. The purpose of the periodical was to inform the rapidly growing reading public about current developments in science. By 1890 Russians could read at least six of Haeckel's major studies in their own language, including *The Natural History of Creation*.

After the publication of *The Descent of Man* in 1871 the criticism of Haeckel's theoretical views came from both Darwinists and anti-Darwinists. All anti-Darwinists were automatically anti-Haeckelians. During the early 1870s Strakhov and Pogodin, the most volatile anti-Darwinists, led a most scurrilous attack on Haeckel's efforts to carry Darwin's ideas to the general public. Darwinists had mixed feelings about Haeckel: they respected his defense of Darwin's theory, but they did not favor the way he went about it. Some paid more attention to his great loyalty to Darwin, others concentrated on his digressions from Darwin's theory and on his flirting with metaphysics.

In a long essay published in the journal *Znanie (Knowledge)* in 1875, V. D. Vol'fson offered one of the most thorough and systematic early Russian inquiries into Haeckel's effort in the popularization of Darwin's scientific ideas. He deliberately avoided any discussion of Haeckel's involvement in building a system of philosophical monism rooted in Darwin's contributions to the triumph of evolutionary thought in biology and in a mechanistic view of the universe. As Vol'fson presented him, Haeckel was the first scientist to make a broad and methodical effort to add "to all branches of biological knowledge," to construct genealogical tables of animal species, and to carry Darwin's laws to their logical conclusions.[117] Haeckel's works, he said, stood out as a monumental expression of the most sublime achievements of contemporary biology. His numerous studies were instructive even when they contained unfounded generalizations and misdirected suggestions. Vol'fson was among the very first Russian commentators who tried to integrate V. O. Kovalevskii's paleontological work into the mainstream of evolutionary thought. He relied on Kovalevskii in rejecting Haeckel's suggestion that the two-toed Anoplotheridae of the Eocene formations represented the original ancestor of ungulates. In general, Haeckel performed a historical role in transforming the "chaos" of pre-Darwinian biology into the systematic and integrated thought of modern biology.

I. F. Tsion was the leading anti-Darwinist who waged a total war on Haeckel. His criticism of Haeckel was actually directed at Darwin: Haeckel did not so much distort Darwin's theoretical ideas as carry them to their logical conclusions. He stated explicitly some of the more radical—and unsupportable—ideas that were implicit in Darwin's theory. Darwin, according to Tsion, presented natural selection as a hypothesis; Haeckel presented it as a universal law of living nature.[118] Tsion's decision not to carry a direct attack on Darwin was a concession to the scientific community, which was heavily pro-Darwinian.

Few Russian Darwinian biologists emulated the sweep and the firmness of Borzenkov's opposition to Haeckel. The influence of Haeckelism, Borzenkov wrote in 1881 and 1884, could best be compared with that of *Naturphilosophie* during the first half of the nineteenth century: both made general statements based on a low quality of observation, and both relied heavily on sheer "fantasy."[119] Haeckel, as Borzenkov saw him, digressed from the measured style and high standards of Darwin's work in two regards. First, he was careless and superficial in handling empirical material without which Darwin's theory could not have become the mainstay of modern biology. He had more flair for the cosmic range of metaphysical speculation than for a scrupulous handling of empirical data. Second, he sought to establish a universal genealogical tree of the animal world without waiting for carefully assembled products of empirical research.[120] As a result, his genealogical links were often based not only on inadequate but also on erroneous data. While Darwin tried to bring biology to the level of scientific realism, Haeckel trudged in the morass of metaphysical speculation. While Darwin consistently acknowledged the weak components of his theory, Haeckel had no reservations about placing the law of natural selection on the same plane of exactitude with Newton's law of gravitation.

In his major works of early vintage—*General Morphology* and *The Natural History of Creation*—Haeckel showed clearly that he was interested not only in achieving a scientific synthesis of evolutionary thought but also in advancing a new synthesis of philosophical thought which he identified as monism. This philosophy rejected idealism, a "dualistic" view that recognized both spirit and matter, but assigned the former a reigning position in the configuration of cosmic forces. It rejected materialism which also recognized the dualism of spirit and matter, but, in contrast to idealism, assigned matter the role of ontological primacy. Monism was built on the idea of the unity and inseparability of matter and spirit. This did not prevent it from reducing all science to the laws of physics and chemistry, from preaching atheism, and from making the struggle for existence a supreme law not only of organic nature but of human society as well.

During the 1870s Haeckel developed also a strong flair for social and political criticism. This became particularly clear in his bitter feud with Rudolf Virchow, a powerful figure among German biologists. At the Congress of German Naturalists and Physicians held in Munich in 1877 Virchow told his listeners that Haeckel's writings about the anthropoid descent of man represented an attack on the moral foundations of human society, and that the theory of evolution should not be taught in public schools. Haeckel retorted with a passionate plea for unrestricted scientific inquiry into the secrets of nature. He commended Darwinian evolution as one of the most advanced scientific ideas of the age, promising to fertilize an entire series of sciences and to strengthen the scientific foundations of a new "moral doctrine," independent of the whimsical power of revelation. The debate attracted much attention, primarily because it treated Darwinism within a political context. Virchow's speech, entitled "The Freedom of Science in the Life of the Modern State," found strong echoes in Russia.[121] It not only supplied conservative writers with anti-Darwinian arguments, but it also provided a new justification for the current effort of the Ministry of Public Education to introduce stricter controls over the teaching of the natural sciences in universities and gymnasiums.

Prior to 1890 Haeckel did not allow philosophical and social criticism to interfere with his productivity on the scientific front. He made a strong impression on the scientific community with his ideas on the embryological evidence in favor of evolution, on the origin of multicellular animals, on the tectological hierarchies that explain the morphological unity of the animal universe, and on the problems of a universal genealogy of animal forms. In addition to efforts to synthesize the rapidly growing knowledge on organic evolution—such as the two-volume *Anthropogenie,* published in 1874—he produced serious monographs on specific taxonomic groups. He also wrote a series of articles for the explicit purpose of widening the base of popular support for the idea of evolution.

During the 1890s Russian commentators were divided in their views on Haeckel's place in evolutionary thought. At the beginning of the decade, the zoologist N. M. Knipovich represented the prevalent tendency in the scientific community to emphasize Haeckel's contributions to science and to ignore his rapidly growing involvement in a militant philosophical movement. Knipovich admitted, however, that some of Haeckel's more daring theoretical ventures did not receive support from empirical data, and that some of them had subsequently proven to be "either untrue or one-sided." [122] According to Knipovich, Haeckel deserved much credit for his effort to illumine many key questions in the theory of evolution and for stimulating a lively discussion in zoological circles. The philosopher N. N. Strakhov held an opposite view. He thought that Borzenkov's criticism of Haeckel's metaphysical distortions of the scientific approach to organic evolution was essentially correct. He thought that this criticism applied to Darwin's theoretical constructions as well. While Borzenkov tried to protect the purity of Darwinian thought from the Haeckelian assault, Strakhov saw nothing but misguided thought in both Darwin's and Haeckel's theoretical structures. Indeed, Strakhov depended on Borzenkov's attack on Haeckel to rekindle and reinforce his own deeply rooted animosity toward Darwin. [123] While Knipovich was unwilling to look into Haeckel's philosophical elaborations, Strakhov could do no better than accuse Haeckel of repeating the errors of Darwinian thought.

During the second half of the 1890s, Haeckel became much more aggressive and belligerent in his philosophical and ideological pronouncements. First, he wrote and lectured ecstatically about *Pithecanthropus erectus,* the newly discovered "link" between man and ape. One of his key lectures on the subject was translated into Russian in 1899. Second, in 1896 he was the leading figure in establishing the Foundation of Monists in Germany, an organization responsible for arranging congresses of his followers and for publishing papers presented at these gatherings. Third, he placed the primary emphasis on expanding and publicizing monist philosophy. This philosophy was essentially an effort to integrate Darwin's theory of evolution, as he interpreted it, into a unified scientific world view built upon mechanistic principles and firmly opposed to teleological intrusions of any kind. *The Riddle of the Universe,* the most notable product of this effort, was published in 1899. It quickly sold over 400,000 copies, a sure sign of the "immense impression" it had made on contemporaries.[124]

All these developments produced strong echoes in Russia. From now on, Haeckel attracted increasing attention for his philosophy no less than for his science. The encyclopedia articles made certain to record his philosophical engagement. The contributor to a new encyclopedia stated that Haeckel's philosophical ideas in no way differed from those of Democritus and Goldbach. Haeckel was presented as a man who "rejected the world of transcendental reality and defended the mechanistic world outlook from all kinds of teleological explanations." He treated consciousness as a function of the nervous system, rejected the freedom of will and the idea of immortality, and placed morality on biological and sociological foundations.[125] The Russian readers were duly informed about the founding of the Kepler Society in Germany with the primary purpose of combatting Haeckelian animosity toward organized religion.

With the exception of Miklukho-Maklai, well-known for his ethnographic explorations in New Guinea, Haeckel did not have personal friends in Russia. Miklukho-Maklai was Haeckel's student at Jena University during the late 1860s and participated in Haeckel's biological fieldwork on the Canary Islands. In a letter to Darwin, reporting on his research activities, Haeckel made direct reference to Miklukho-Maklai, one of the more important early funnels for the flow of Darwinian ideas to Russia.[126] Miklukho-Maklai died in 1888, leaving Haeckel without a close link to Russia. This was the reason why he went unnoticed when he attended the international congress of geologists (and paleontologists) held in St. Petersburg in 1897, after which he made a trip through Russia which took him to the Volga region, the Caucasus, the Crimea, and the city of Odessa.[127] His signature on a letter in which a group of foreign visitors thanked their Russian hosts for their hospitality was the only concrete record of his visit to Russia.

During the 1890s—and particularly after 1900—Haeckel's works were translated into Russian with increasing rapidity, and his views became topics of heated and controversial discussion. In addition to books, a selected array of his essays—usually reports given at scholarly gatherings—appeared as pamphlets in Russian translations. Among the most popular were *Present-Day Knowledge of the Phylogenetic Development of Man, The Origin of Man, The Struggle for the Evolutionary Idea,* and *World Views of Darwin and Lamarck.* The publishers of these and related pamphlets presented Haeckel as a person who occupied a place of honor among the pioneers in evolutionary theory and as a most outstanding representative of Darwinian orthodoxy.

Most Russian biologists who supported Darwinism concentrated on Haeckel's scientific work, fully ignoring his elaborations of philosophical materialism and monism. Although he was Russia's most consistent defender of the "mechanistic" base of Darwin's theory, and was known as the leading spokesman for "natural science materialism," Timiriazev concentrated on Haeckel's work directly related to Darwin's views on evolution. He made little effort to distill a philosophical message from Haeckel's work. Nor did Menzbir go into Haeckel's monistic elaborations. He differed from Timiriazev, however, in taking a much more critical view of Haeckel's scientific work. His basic criticisms were that Haeckel made little effort to give empirical support to his theoretical claims and that he indiscriminately embellished Darwin's theory with pure guesswork. Menzbir did find Haeckel's more specialized studies to be solid additions to the mainstream of evolutionary biology.

In a comparison of Darwin and Haeckel—of their style of work, philosophical inclinations, and temperaments—Menzbir found that they stood at opposite poles. He hailed Darwin as an inductive scholar ready to back up his generalizations with a rich assortment of empirical data. In Darwin's methodology he saw an elaboration and application of the principles Bacon had set forth at the beginning of the seventeenth century. Haeckel depended on deductive reasoning—on logical constructions not easily related to empirical substance. Menzbir identified him with Oken and Schelling, the architects of *Naturphilosophie,* in style of thought if not in metaphysical loyalty. Darwin, according to Menzbir, drew sharp lines between "theoretical" and "hypothetical" parts of his biological thought; Haeckel, by contrast, proceeded directly from general assumptions to a construction of nature in the spirit of *Naturphilosophie*. Haeckel's entire *General Morphology* is nothing but "a chain of unproven general propositions." [128] Even when he produced firmly documented ideas, his mode of thinking was metaphysical rather than scientific. Menzbir represented the defenders of Darwinian science who expressed an ambivalent attitude toward Haeckel. While giving Haeckel great credit for a valiant and dedicated defense of the evolutionary principle built on natural selection, he attacked him for his major digressions from Darwinian thought.

The ideological overtones of his monistic philosophy, steeped in materialism, made Haeckel very unpopular in some circles. In 1903 the popular journal *God's World* published in Russian translation a German article on contemporary philosophy which considered Haeckel's *Riddle of the Universe* a national disgrace for the land of Kant, Goethe, and Schopenhauer. Written by Külpe, a well-known psychologist and historian of philosophy, the article found Haeckel's treatment of the "soul" as a function of the "brain" particularly objectionable because it placed a primary emphasis on the neurophysiological approach to mental phenomena. [129] O. D. Khvol'son, the Russian author of a widely used textbook in physics, wrote a book in the German language for the purpose of refuting Haeckel's arguments. [130] He focused his criticism on Haeckel's gross misrepresentation of the laws of physics. Because it carried its criticism too far, this book made little impression on Russian contemporaries.

Despite criticism, the translation of Haeckel's works increased at a rapid pace. At the end of the tsarist era, Russians could read all the major and most of the minor works in their own language. While some read Haeckel to satisfy their scientific curiosity, others read him to quench their ideological thirst. All learned that Haeckelism was a unique prism refracting the rays of Darwinian light. Haeckel commanded two powerful sources of popular appeal: he wrote in simple prose but with the authority of a recognized scientist, and he carried out a persistent and vigorous attack on social ills. [131]

Favorable comments on Haeckel's ideas came from many sources. In 1909 V. I. Lenin stated that Haeckel's *Riddle of the Universe* had two major features: it revealed the partisan [ideological] character of philosophy, and it showed the key questions related to the struggle between

materialism and idealism. The book went through many reprintings because its "materialism"—Lenin was annoyed that Haeckel preferred the
term "monism"—appealed to the "masses" of new readers. According
to Lenin, "this little book became a weapon in the class struggle."[132]
Haeckel made Darwin's theory an organic part of a philosophically articulated ideology. Lenin chose to overlook Haeckel's reputation as an
outspoken enemy of socialism.

In that same year, A. Genkel' informed the readers of *Education*
(*Obrazovanie*), a popular journal with liberal leanings, about the co-
lossal proportions of Haeckel's engagement in science as a body of
knowledge, a philosophical strategy, and an ideology. He asserted that
the theory of organic evolution was the most sublime achievement of
nineteenth-century science and that Lamarck, Darwin, and Haeckel
stood out as the main architects of this triumph.[133] Genkel' tried to
show that Haeckel would have been immortalized even if he had limited
his activities to pure scientific scholarship. Haeckel, he wrote, occupied
a particularly important position in the history of the German commitment to science: he served as a link between the national flair for synthetic thought, dominant during the first half of the nineteenth century,
and the strong national preoccupation with analytical approaches,
dominant during the second half of the nineteenth century.

N. K. Kol'tsov, a noted pioneer in experimental biology, placed
Haeckel among the giants of Darwinian science and among the most
successful popularizers of the evolutionary view. Haeckel showed that
embryology outstripped morphology as a rich source of evolutionary
links in the world of plants and animals. His biogenetic law strengthened the historical orientation in the Darwinian legacy. This law, according to Kol'tsov, contributed to making comparative embryology
one of the most attractive and challenging branches of zoology. Thanks
to Haeckel, phylogeny bcame a key factor in systematizing zoological
and botanical knowledge. Only when Haeckel moved from creating a
new system of science to creating a new system of the world did he enter
the slippery area of philosophical speculation that brought him sharp
criticism from many quarters. As if apologizing for Haeckel, Kol'tsov
stated that the monistic philosophy of the German naturalist was not
intended to be scientific in the first place; it concentrated on answering
questions that reached beyond the competence of science. After all,
"Haeckel's monism, like any religion, addresses itself not only to the intellect but also to the sentiment."[134] Even though he found a comfort-

able home in philosophy, Haeckel earned plaudits as one of the leading contributors to the popularization of science in general and of Darwinism in particular.

A representative group of leading Russian Darwinists showed a generally favorable attitude toward Haeckel's biogenetic law, according to which ontogeny is a brief recapitulation of phylogeny. An old idea in biology, the first suggestion of this law came from K. F. Kielmeyer and É. Geoffroy Saint-Hilaire. Fritz Müller made it an integral part of Darwinian thought, and Haeckel popularized it and gave it an explicit formulation. V. M. Shimkevich thought that Aleksandr Kovalevskii's analysis of the role of "retrogression"—or "simplification"—in the development of lancelets and ascidians provided "an impressive argument in favor of the Haeckelian biogenetic law." [135]

The botanist V. L. Komarov was another representative of the Russian scientific community ready to acknowledge Haeckel's significant role in the advancement and popularization of Darwin's ideas. He commended Haeckel for reinforcing the theory of evolution with "rich material from embryology and comparative anatomy that had escaped Darwin's attention." [136] Haeckel's elaborate theoretical structures and philosophical involvements stayed outside Komarov's concern with evolutionary thought. Komarov did not hesitate to consider Wallace, Huxley, and Haeckel the leading Darwinian scholars.

A. N. Severtsov, Russia's most distinguished expert in evolutionary morphology, belonged to the group of Russian biologists who made certain that Haeckel received due recognition for his original contributions to science. The biogenetic law, in Severtsov's view, was Haeckel's indelible contribution to the triumph of the evolutionary view in biology. Although it required major revisions and reinforcements, this law, he said, must be counted among the leading explanatory principles of organic evolution. Equally great, according to Severtsov, was Haeckel's pioneering work on the genealogical tree of contemporary animals, an undertaking that required sustained and elaborate work in phylogenetic analysis. This work laid the groundwork for the advancement of evolutionary ideas beyond the general theoretical outlines Darwin had formulated in the *Origin of Species*. Generally confirmed by succeeding generations of paleontologists and morphologists, Haeckel's genealogical tree, according to Severtsov, laid a solid foundation for the phylogenetic study of organic nature. Haeckel's influence, however, was not all blessing. It led zoologists to place much more emphasis on "the evolution of individual groups of animals and their organs" than on "the morphological laws of evolution."[137]

No Russian scholar had undertaken a systematic and comprehensive elaboration of Haeckel's unique and exciting mixture of science and philosophy. Russia did produce many scientific and ideological commentators who gave Haeckel's thought an inordinately wide circulation in the country. Without Haeckel's powerful and catalytic influence Russian Darwinism would not have been the same.

CHAPTER SIX

Orthodoxy

The twentieth century brought forth ominous threats to the intricate fortress of Darwinian thought. Mounting challenges came from science, philosophy, and theology. In science, they came primarily from new developments in biological thought, mainly from the general principles of genetics, the most radical branch in experimental biology. Although it had influential spokesmen and was in command of powerful arguments, the opposition lacked inner unity sufficient to present a formidable threat to the reign of Darwinism. The fragmentation of opposing forces worked in favor of Darwinian purists, deeply engaged in a massive effort to preserve and consolidate the orthodox views.

The purists divided the more recent advances in experimental biology into two groups: those that could readily be integrated, as collateral material, into Darwinian thought; and those that must be rejected because of their incompatibility with Darwin's theory. They accepted modern genetics, for example, only insofar as it enhanced Darwin's orientation by helping explain certain phenomena of heredity, a topic to which the English naturalist did not give sufficient attention. They viewed acceptable contributions of modern biology as elaborations of ideas Darwin had suggested in the first place. While agreeing among themselves in principle, the leading guardians of Darwinian orthodoxy differed in their choice and interpretation of new ideas that could be integrated into Darwinian thought.

Three biologists were widely heralded as the most dedicated defenders of Darwinian orthodoxy. Kliment Timiriazev became well known through his efforts to present Darwinism as a stronghold of Newtonian science and through his sweeping and emotionally charged crusade against the developments in contemporary biology that stood in open conflict with Darwin's ideas. Mikhail Menzbir conducted a gentle but broadly based defense of Darwin's theory and presented a popular overview of developments on the Darwinian front. A leading ornithologist, he buttressed Darwinism with a strong ecological interest. Aleksei Severtsov was so much preoccupied with elaborating, refining, and generalizing Darwin's theoretical heritage that he showed little inclination to seek help from "unorthodox" developments in experimental biology. As the creator of a theoretical system of evolutionary morphology, he became well known, far beyond the boundaries of Russia.

Different in style of work, scholarly specialty, temperament, and philosophical involvement, Timiriazev, Menzbir, and Severtsov were united in a massive effort to interpret the rapidly expanding world of evolutionary biology from Darwinian positions. They dedicated their lives to Darwinian thought, protecting it from the inexorable forces of erosion and shaping it into an ordered body of scientific knowledge and into a grand strategy for the study of living nature. Their ideas and actions make up the main substance of this chapter.

K. A. Timiriazev: Darwin's Russian Bulldog

Timiriazev enjoyed a great reputation as a pioneer in the study of photosynthesis, a champion of biological evolutionism, a popularizer of scientific knowledge, and an outspoken defender of academic freedom. With a flair for dramatic expression, he made the popularization and the defense of Darwin's theory integral components of the ongoing struggle for university autonomy, free science, and democratic politics. During most of his scholarly career, which began in the mid-1860s, he made the defense of Darwin's theoretical views the focal point of his writing and public lectures. No Russian scholar of his generation surpassed the scope and the intensity of his concern with Darwin's scientific contributions. He was the editor of an eight-volume collection of Darwin's works in Russian translation, published in 1907–9. Together with M. A. Menzbir, he produced one of the better Russian translations of the *Origin of Species.*

Timiriazev's view of Darwin was clear and simple. Darwin's theory, he thought, represented welcome opposition to every teleology and to every philosophy that relied on transcendental causation. He saw it as the unchallengeable foundation for a unitary view of nature and fully agreed with Boltzmann's statement that Darwin's theory was the only salvation for philosophy. Darwin's theory, in Timiriazev's view, enhanced the ties between biology, on the one hand, and physics and chemistry, on the other. Darwinian rationalism, he argued, was a most powerful and distinctive cluster of ideas shaping the modern world view. Timiriazev did not hesitate to place the tag of vitalism on all the experimental studies of inheritance which he happened to regard as anti-Darwinian.

In his dedicated defense and extensive popularization of Darwin's scientific ideas, Timiriazev was inspired and guided by philosophical judgments of two great Western scientists: Hermann Helmholtz and Ludwig Boltzmann. In 1869, in the opening speech before the Congress of German Naturalists and Physicians held in Innsbruck, Helmholtz praised Darwin for showing that the blind laws of nature rather than the interference of a supreme intelligence account for the development of organisms.[1] In a paper read before a gathering of the Vienna Academy of Sciences in 1886, Boltzmann noted that the nineteenth century will be remembered as a century of natural science, and that, as a century of natural science, it will be remembered for two great achievements: the full victory of the mechanistic view of nature and the triumph of Darwin's theory.[2] Since biology benefited from both triumphs, it was, in a way, a symbolic expression of the century's most sublime achievements in science. In a sense, Darwinism was a high mark of the century because it covered the two basic attributes of the living world—its evolution and its physicochemical foundations. When Timiriazev protected Darwinism, he in effect protected an evolutionary theory built on Newtonian foundations.[3]

At the beginning of the twentieth century Timiriazev was convinced that Darwin's scientific authority was stronger than ever before. Darwin's contribution, he argued, was not just another step in a long series of evolutionist developments in biological thought; on the contrary, Darwin opened a new era in the history of biology by dealing a fatal blow to the idea of antitransformism, the dominant view prior to 1859. At the time of the publication of the *Origin of Species* biological thought was dominated by the antievolutionism of Cuvier, Agassiz, and Owen, as well as by Lyell's efforts to discredit Lamarck's transformist ideas. The effort of É. Geoffroy Saint-Hilaire to fit biology into an evolutionary mold encountered such formidable enemies that it could not become part of established thought.[4] Darwinism won the day—and wrought a revolution in biology—not because it had a long list of ancestors but because it succeeded in identifying "evolution" with "progress," and because it rested on the principle of adaptation, the key to a realistic understanding of the "perfection, purposiveness, and harmony" of living nature. Timiriazev compared Darwin with Vico: Vico tried to transform history into natural history; Darwin succeeded in making natural history true history.[5]

In another essay, devoted to the basic functions of plant physiology, Timiriazev gave a clear picture of his full allegiance to mechanistic biology:

> To study the phenomena of plant life successfully, physiology does not need arbitrary suppositions that the sciences dealing with inorganic matter have rejected a long time ago. It needs no help from outmoded beliefs in the existence of a special organic substance. It needs descriptions and general laws applied to inorganic bodies. It does not need a special, elusive, and self-propelled force that ranks above the law of causality and above number and measure. The basic laws of physics meet its needs no less than those of the inorganic matter. Finally, it does not need to assume the existence of a universal metaphysical principle of purposeful development—the vitalist's last resort. As Darwin has shown, it is satisfied with the real historical process that guides the organic world to inevitable perfection and harmony. We have a full right to assert that, in its effort to explain the phenomena of life, plant physiology must study three categories of causes—chemical, physical, and historical. . . . Each category is dominated by a distinct general law that helps shape our world view, and each is associated with a leading figure in modern science. The three laws are: the law of the conservation of matter, the law of the conservation of energy, and the law of the transformation and unity of life. The three leading figures are Lavoisier, Helmholtz, and Darwin.[6]

Timiriazev's unequivocal emphasis on the methodological superiority of physicochemical analysis of vital processes was a specific expression of an overarching belief in the unity of nature. As a plant physiologist, with photosynthesis as his central interest, he was a natural recruit for the idea of cosmic unity—and of the unity of science. The life of plants depends on photosynthesis, a complex process whereby chlorophyll harnesses solar energy to produce carbohydrates and other organic compounds necessary for the subsistence and growth of green plants. Photosynthesis, as Timiriazev saw it, represents the clearest possible example of the working of the law of the conservation and transformation of energy. A physical explanation of chemical substances holds the key to the understanding of the processes of life without resorting to an imaginary "vital force." In addition to reaffirming the physicochemical unity of nature, the law of the conservation of energy reinforces the mechanistic foundations of natural science. In his discussion of organic evolution as a specific expression of cosmic unity, Timiriazev depended primarily on a combination of philosophical discourse and poetic effusion.

In his defense and elaboration of Darwinian thought Timiriazev fought on several fronts: he repeatedly returned to Lamarckism for the purpose of showing the fundamental nature of its incompatibility with Darwin's theory, but he also fought the revived vitalism and the threats to Darwinism by modern orientations in experimental biology. He defended Darwinism by waging an uncompromising and bitter war on its enemies, both real and imagined.

Asking himself the rhetorical question, "What requirements must an evolutionary theory meet in order to be scientific?" he made his opposition to Lamarckism unequivocal:

> [An acceptable theory] must (1) indicate the role of an evolutionary, or a historical, process in the transformation of species; (2) point out the natural factors that transform this process into progress, that is, into improvement in the adaptation of the organization of living forms to the conditions of existence; and (3) explain the apparent contradiction between the notion of the unity of organic nature as a whole and the presence of discontinuities within and between living groups of all orders. Darwinism is the only theory that meets these requirements. No other effort—not even Lamarckism—has met these standards in a satisfactory fashion.[7]

Lamarck's idea of the direct influence of the environment on the transformation of living forms, according to Timiriazev, did not receive support from "convincing facts." In fact, modern research has fully rejected the notion of the primary role of use and disuse of organs in the adaptation of living forms to the environment—and in the acquisition of new heritable characteristics, the chief mechanism of organic evolution.

On another occasion Timiriazev was much more conciliatory. He wrote:

> If Darwin spoke harshly about Lamarck, it was only in relation to his unfortunate effort to explain the evolutionary process by relying on the behavior of plants and animals, a criticism that has been justified by subsequent developments in science. From the very beginning, Darwin recognized the dependence of organic forms on environment—that is, he recognized the part of Lamarck's theory that has survived as an important contribution to

modern biology. Darwin's appreciation of this idea increased as time passed. Only a merging of this side of Lamarckism with Darwinism can solve the key problem of biology. . . . Many contemporary German neo-Lamarckians suffer from a misunderstanding of the interdependence of the two theories.[8]

This citation has a double edge. In the first place, it expresses a view of the complementarity of Lamarckism and Darwinism, widespread among Russian biologists. This does not mean, however, that this complementarity was a topic of comprehensive and systematic discussion. Lamarckism existed primarily as a valuable, but little debated, corollary of Darwinism. In the second place, Timiriazev's clearly expressed distaste for the ideas of the German supporters of psycho-Lamarckism represented a fair expression of the general sentiment in Russia. Only isolated Russian biologists concerned themselves with the psychological aspect of evolution, and all firmly opposed the strong metaphysical bias of the more extreme German psycho-Lamarckians, led by A. Pauly and R. Francé.

In the resurgence of vitalism Timiriazev saw not only an unwelcome attack on the spirit of Newtonian science but also a deceitful effort to subordinate science to idealistic metaphysics. The rebirth of vitalism was more than an attack on the reigning orientation in contemporary science; it was an effort to limit the intellectual authority of science.[9] Timiriazev viewed science as a unique approach to reality that would never reach the point of diminishing returns. The neovitalists, by contrast, saw biology, heavily committed to physicochemical analysis, as a science that had reached a point of diminishing returns and that could be saved only by a heavy reliance on metaphysics. In attacking the neovitalists, Timiriazev concentrated on their pessimistic view of the future of science. His scathing attack on the botanist I. P. Borodin, who welcomed the resurgence of vitalism, helped limit the spread of neovitalist ideas in the Russian scientific community.[10] Timiriazev's optimism about the future of science was an expression of unlimited faith in physicochemical techniques in the study of life processes.

After 1900 Timiriazev made "the Mendelians and mutationists" the main targets of his war on "anti-Darwinism." The more the criticism of Darwin's theory grew in depth and in fervor, the more he clung to the unmodifid principles of Darwin's theory and the more uncompromising he became in his war on real and imagined enemies. In his criticism of the mutation theory he pursued two lines of attack: on the one hand, he argued that mutations are aberrations rather than regular components of the natural course of evolution; on the other hand, he contended that Darwin did not ignore mutations and that de Vries and Korzhinskii learned about them from the first edition of the *Origin of Species*.[11]

Mendel's followers troubled Timiriazev most because, in his opinion, they advanced the most pernicious arguments against Darwin's theory. In his criticism of Mendel's laws of heredity he asserted that evolutionary research needed most of all physiological experiments as a substitute for "a statistical record of observations."[12] The claim of Mendel's followers that their genetic laws were universal did not meet with Timiriazev's approval. In contrast to popular opinion, Mendel contended that crossing does not produce either a "fusion" of alternative characters or a disappearance of individual characters. Without adducing empirical evidence to support his claim, Timiriazev argued that a fusion of characters was also possible, and that therefore Mendel's assertion did not constitute a universal law of nature.[13] He quickly added that Mendel, "a wise man and an experienced scholar," did not intend to attribute universality to his discoveries.[14] Timiriazev gave Mendel credit for conducting "a careful statistical study" of heredity which led to an empirical generalization in full agreement with Darwin's theory. He considered the fast-rising genetics only a minor partner in the family of biological sciences.

Obviously, Timiriazev tried to make it clear that he opposed Mendelians, not Mendel. In his opinion, the "anti-Darwinism" of the new genetics was the main reason for Mendel's fast-rising popularity. In Germany, he wrote, anti-Darwinism was both a by-product of the growing influence of "clericalism" and an expression of nationalism, a unique sign of anti-English sentiment.[15] He was particularly angered by an English writer who in 1907 elevated Mendel to the ultimate heights of scientific achievement—a favorable comparison with Newton.[16]

Whether he criticized mutationism or Mendelism, neo-Darwinism or neovitalism, Timiriazev made Darwinism part of a materialistic ontology of science, of a positivist orientation in epistemology, and of a firm devotion to scientific knowledge as a liberating social force. An exponent of positivist philosophy, he counted Auguste Comte among Darwin's leading forerunners.[17] In protecting Darwinism Timiriazev also protected a view of science as the primary source of moral progress in the modern world. He undertook the task of popularizing the contents of Marcellin Berthelot's *Science et morale*, particularly the claim that from the seventeenth century on, science has been the only true contributor to "the improvement of the material and moral conditions of social life."[18]

To defend Darwinism, Timiriazev thought, was to be allied with a political ideology that saw the future of Russia in democratic institutions. He saw Darwinism as a bastion of rationalism and as the most powerful tool against metaphysical vestiges in modern scientific thought. In particular, it was the basic weapon in the war against neovitalism, a metaphysical orientation steeped in spiritualism and allied with a strong antiscientific movement.

Timiriazev's scorching attacks on the "enemies" of Darwinism did not evoke serious rebuttals. Powerful groups in and outside the academic world shared his total devotion to Darwin and his thought. He was a charismatic speaker, and everything he said at scholarly assemblies produced strong echoes in educated society. His claim that the triumph of Darwin's theory was a key factor in the rapid growth of natural science in Russia evoked approving responses of massive proportions. Much of his writing consisted of popular accounts of science as a cultural value of the first order. Darwinism, as he presented it, was the real force behind the developments that led to the victory of rationalism, historical outlook, and secular thought in modern Russian society. He benefited from his clear identification with the liberal professoriate, representing the dominant and most influential ideology in the scholarly community. Working in his favor was also the continued high respect for Darwin among the radical intelligentsia, including the most determined critics of the struggle for existence as a prime mover of organic evolution.

Timiriazev fought his enemies either by totally rejecting the scientific merit of their theoretical claims or by treating their work as inconsequential challenges to Darwin's scholarly achievement. His attacks were sweeping and acrimonious; his willingness to compromise was of the most evasive kind. For this reason his contemporaries remembered him primarily as a bitter enemy of all modern developments in biology that did not, in his opinion, follow Darwinian thought.

In 1894 M. M. Filippov, editor of *Science Review*, published a lengthy article criticizing Timiriazev's defense of Darwinism. He said that Timiriazev oversimplified the raging conflict in the scientific community by identifying all critics of Darwinian orthodoxy as anti-Darwinists.[19] Filippov was essentially correct. To be admitted to the ranks of Darwinists, according to Timiriazev, a biologist must recognize the evolutionary significance of Malthus's law, the indirect influence of the environment on the evolution of plants and animals, natural selection as a universal factor of organic transformation, and the evolutionary process as a pure product of external natural causes, unmarred by internal teleological considerations. Filippov, by contrast, claimed that Darwinism needed a broad reformulation that would recognize both the direct and the indirect influence of the environment, would treat the struggle for existence as one of many evolutionary factors (he placed particular emphasis on symbiotic coexistence of different life forms as a source of new species), and would make the idea of purposiveness an integral part of the evolutionary model. He directed particularly harsh remarks at Timiriazev's repeated efforts to view all criticism of natural selection as a flirtation with vitalism.[20] Timiriazev, he thought, did injustice to Darwin by transforming his ideas into a dogma and by trying to preserve some of Darwin's less defensible suggestions.

Filippov asserted categorically that he was not part of an anti-Darwinian movement, and his published essays attest to the truthfulness of his claim. He wrote that without the scientific legacy of the founder of evolutionary biology, the study of life would not have reached the heights it occupied at the time of his writing. Biology, he added, could advance only by building upon, rather than by negating, Darwin's contributions.[21] The future of biology was neither in N. Ia. Danilevskii's unfounded anti-Darwinism nor in Timiriazev's dogmatic Darwinism. In assessing the state of evolutionary thought in Russia, Filippov made one major error: he ignored the Russian echoes of the neo-Darwinian efforts to combine Darwinian natural selection with the primacy of internal causes of variation. He was closer to Lamarckism than to any other strand in contemporary evolutionary biology.[22] As a free-lance writer, mainly on scientific themes, he did not pose a serious threat to Timiriazev, who decided not to be involved in yet another bitter debate. A brief but biting footnote to an article in *Russian Thought* was the extent of Timiriazev's rebuttal.[23]

Timiriazev was not always ready to recognize the work of those who contributed to the popular literature on Darwinism as a positive force. For example, he took no note of M. A. Antonovich's *Darwin and His Theory,* published in 1896, most of which was previously serialized in the journal *Russian Thought.* Nor did he refer to Antonovich's "Theory of the Origin of Species," published in *Sovremennik* in 1864, one of the clearest and most laudatory early essays on Darwin's theory to be published in Russia. Antonovich's book covered important topics from the history of Darwinism which were not previously discussed in Russia in a systematic fashion. It had two strong features: it traced the growth of Darwinian thought from the appearance of the *Voyage of the Beagle* to the series of studies that appeared after *The Descent of Man;* and it provided special discussions of the reception of Darwinian ideas in England, Germany, France, the United States, and Russia—with every country accorded individual treatment.

In one important respect, Timiriazev's adherence to the theoretical substratum of Darwin's biology was somewhat ambivalent. He accepted the struggle for existence but did not give it a precise definition. He "solved" the dilemma simply by avoiding a direct confrontation with this unpleasant and tortuous task. By his own admission, he made a conscious effort to avoid any reference to the struggle for existence. This reluctance had an ideological rather than a scientific explanation: he echoed the opinion of the leading ideologues among the radical intelligentsia, who tried to reconcile Darwinism with social theories that regarded cooperation, rather than competition or conflict, as the prime mechanism of progressive evolution.

In his search for an acceptable notion of natural selection, Timiriazev received assistance from Thomas Huxley's essay "The Struggle for Existence," which he had helped translate into Russian.[24] In this article, Huxley, in an obvious effort to erase the path leading from Darwin's theory to Social Darwinism, described the difference between the cosmic process of life in general and the ethical process which has dominated human society after its ascent to the higher rungs of evolution. As human society passes from a "natural" to an "ethical" state, natural selection also passes from conflict or struggle to cooperation. To talk about the struggle for survival is to talk about the "natural" phase in the evolution of human society. "The unfortunate expression 'struggle for survival' has nothing in common with the doctrine of morality, because human morality is built by the social structure—'by society'—rather than by the biological endowment."[25] Or: "The theory of the struggle for survival was abandoned at the threshold of history; the entire rational and cultural activity of man is only a struggle against the struggle for existence."[26] Neither Huxley nor Timiriazev worked hard to explain the origin and social dynamics of the ethical process; both made scientific concessions to alleviate the ideological pressure coming from many sources. This was one of the very few times that Timiriazev showed a willingness to modify the Darwinian legacy to satisfy the extrascientific pressures of the day.

While shifting the emphasis from conflict and struggle to cooperation, specifically in reference to human society, Timiriazev showed no
inclination to abandon the Darwinian notion of natural selection as the
basic mechanism of organic evolution. He continued to resist the arguments of critics who emphasized the practical impossibility of giving the
notion of natural selection adequate empirical support. Admitting that
the search for empirical verification of natural selection faced grave difficulties, he contended that solid steps in that direction had already been
taken. As a specific case of natural selection in operation, he cited
W. F. R. Weldon's biometric study of the differential adaptation of a
small shore-crab (*Carcinus maenas*), a denizen of Plymouth Bay, to
waters with a low level of sediment and to waters with a high level of
sediment.[27] Timiriazev found it necessary to refute de Vries's claim that
artificial selection did not play as important a role in the origin of new
varieties of plants and animals as Darwin was predisposed to think. In
his rebuttal of de Vries's view, Timiriazev leaned heavily on Luther Burbank's successes in effecting rapid changes in "almost all characteristics" of domestic plants on which experiments were made. Timiriazev
particularly welcomed Burbank's open admission that in all his experiments he operated strictly from Darwin's theoretical positions.

Timiriazev's major contribution was in defending Darwin's theory at
the time when it was much in need of a strong defense. He helped make
Darwinism a high point in the growth of modern secular thought in
Russia, and more clearly than any other Russian scientist, showed the
unsurpassed proportions of Darwin's contributions to reunifying man
with nature and to expanding the range of subjects open to scientific
scrutiny. He contributed to making Russia a rare country that did not
encourage the emergence of Social Darwinism as an ideological factor
of serious consequence. Inasmuch as he tied his defense of Darwinism—
and of the evolutionary point of view in general—to a relentless war on
the key developments in contemporary experimental biology, his contributions to the advancement of modern biological thought in Russia
were clearly negative.

M. A. Menzbir:
A Quiet Defender of Orthodoxy

In 1882, three years after graduating from Moscow University, Mikhail Aleksandrovich Menzbir (1855–1935) wrote two studies which
opened two distinct avenues for his future scholarship. The *Ornithologi-*

cal Geography of European Russia, immediately accepted as a masterwork in zoogeography, started him on a distinguished career in the
study of the ecological features and taxonomy of Russian birds. Going
against the established practice in zoogeography, he divided the vast territory he studied into distinct regions by relying on faunistic rather than
on geographical features. This study kept Menzbir close to the current
work in natural history, which acquired massive proportions particularly after the founding of naturalist societies at all the leading universities. It brought quick and impressive rewards: in the same year he was
appointed editor of the serial publications of the Moscow Society of
Naturalists, a job he held until his death; two years later he was appointed docent at Moscow University, the institution with which he was
affiliated most of his life. The second work was an essay commemorating the death of Charles Darwin, the first in a long series of publications
Menzbir devoted to the scientific contributions of the founder of the
evolutionary theory. Published in the popular journal *Russian Thought,*
this article presented the main components of Darwin's theory, showing,
in particular, the usefulness of the evolutionary orientation in ornithology. Unlike most of his Russian contemporaries, he made no objection
to Darwin's reliance on Malthusian progressions.[28]

In the scope and fervor of his defense of Darwin's scientific legacy,
Menzbir was Timiriazev's equal. The two men, however, did not fit the
same pattern of scholarly engagement. They brought different backgrounds to their involvement in Darwinian studies. Timiriazev, a plant
physiologist, brought an awareness of the importance of physiology to
the study of evolution, as well as a conviction that physicochemical
analysis should be the primary method of physiology. Menzbir, by contrast, came from a background that combined zoogeography and systematics, a common combination at this time. His ornithological studies
appealed to bird lovers, ecologists, and taxonomists alike; most of them
stayed at a low level of theoretical considerations.

Both Timiriazev and Menzbir addressed most of their writings on
evolution to two audiences: professional biologists and interested lay-
men. The modes of their writing were quite different. Timiriazev was
particularly eager to show the intricacies of Darwin's construction of the
conceptual base of his theory of evolution. Menzbir was much more interested in showing how evolution, as Darwin interpreted it, worked in
real life. Timiriazev used empirical data to illustrate the intricacies
of the general theory of evolution; Menzbir used evolutionary theory
to trace transformation processes in specific subdivisions of the ani-

mal universe. There were other differences as well. Unlike Timiriazev, Menzbir was calm, careful, and systematic in reacting to "unorthodox" views in evolutionary theory. In his concessions to the new discoveries in experimental biology he was more gracious than Timiriazev, and his arguments did not transgress the limits of academic propriety. In one important respect the two scientists acted as one: both contended that Darwin's theory of evolution needed only clarification and refinement to be fully established as the primary factor of organic evolution. Despite the measured tone of his rebuttals, Menzbir stuck firmly to the idea that all new developments in biological theory must be judged by their contributions to the universal appeal and revolutionary significance of Darwin's science.

There was another pronounced difference between Timiriazev and Menzbir. Timiriazev was very close to Darwin: he cited him profusely, he stuck closely not only to the substance but also to the style of Darwin's scientific argumentation, and he approached Darwin's theory as a symmetrical, explicit, and integrated system of thought. Menzbir was much less prone to make direct references to—particularly to cite—Darwin: in some essays on evolution, written in the Darwinian spirit, he did not make a single direct reference to Darwin. His Darwinism was selective, asymmetrical, and rather personal. He ignored most of Darwin's suggestions for future exploratory work and his views on more sensitive problems of methodology. Yet in his dedication and basic theoretical commitments Menzbir was one of Darwin's most loyal and enthusiastic followers.

Menzbir's loyalty to Darwin's theory made him a staunch defender of the common origin of man and animal. In 1889 R. Virchow, in a speech at the Congress of German Anthropologists held in Vienna, made a plea for dissociating anthropology from evolutionary ideas and from Darwinian theories. He justified his action on the ground that all efforts to produce a scientific explanation of the origin of man had fallen by the wayside.[29] He argued that no lower animals could be identified as ancestors of man; nor did he think that science would ever be in a position to answer the question of man's animal origin. Four years later Menzbir published an article analyzing and refuting Virchow's argumentation point by point, mainly by relying on a string of authorities in evolutionary biology. Menzbir considered the animal origin of man a foregone conclusion; his major aim was to defend the power and resourcefulness of Darwinian theory and to uphold the supremacy of "scientific materialism." In the process, however, he gave Russian readers an opportunity

to learn about the basic issues that came in the wake of Darwin's suggestions on the ancestry of man.

P. P. Sushkin, a leading ecological evolutionist, noted in 1916 that in
his lectures and published studies Menzbir relied on Darwinian models
and analytical designs to give his own ideas inner consistency and overall unity.[30] Although Darwin's ideas, in Menzbir's view, were not unchallengeable dogmas that discouraged criticism, they were impervious
to sweeping criticism and flippant negation. Held together by firm principles, they needed only additional supporting facts, more precision,
and wider vistas of research. Only the theory of sexual selection required a more radical reconstruction.[31] Menzbir relied on the *Origin of
Species* not only for help in general biological theory but for empirical
information as well. To students in his zoology classes he presented
Darwin both as the creator of a grand theory of evolution and as a true
master of the inductive method. Darwin, he said, wanted his followers
to treat the theory of evolution as an open system of ideas, inviting criticism and obeying the rules of the inductive method. Menzbir wrote:

> Darwin's legacy stresses not a blind allegiance to his authority but a re-
> liance on the inductive method of studying nature, discovering its laws, and
> unraveling its secrets. In none of his volumes do we encounter passages writ-
> ten in a dogmatic tone. Moderation and modesty give weight and credence to
> the proofs he uses in support of his theses.[32]

Darwin's close adherence to the rules of the inductive method evoked
most favorable—and very frequent—comments in Russia, among opponents no less than among followers. Particularly appealing was Darwin's
abundantly displayed willingness to point out his own conclusions that
required additional empirical testing and support. In *The Variation of
Animals and Plants Under Domestication* Darwin made concrete suggestions on how to proceed in empirical testing of natural selection:

> This hypothesis may be tested—and this seems to me the only fair and
> legitimate manner of considering the whole question—by trying whether it
> explains several large and independent classes of facts; such as the geological
> succession of organic beings, their distribution in past and present time, and
> their mutual affinities and homologies. If the principle of natural selection
> does explain these and other large bodies of facts, it ought to be received.[33]

While praising the power and rich perspectives of Darwin's inductionism, Menzbir was only too ready to point out and to criticize the
speculative constructions built into the theories of some scholars who
considered themselves true Darwinists. Ernst Haeckel, as Menzbir saw

him, tried to dress Darwinian ideas in the abstractions of "the recently expired *Naturphilosophie.*"[34] Right or wrong, Menzbir was convinced that individual German biologists were deeply steeped in the metaphysical habits of the masters of *Naturphilosophie,* even though this orientation had ceased to exist as an organized movement. Haeckel copied the logical forms, not the ontological foundations, of *Naturphilosophie.* The complete system of Haeckel's phylogenetic propositions built into the gastraea theory showed a full disregard of accumulated empirical evidence. Haeckel made no reference to the geographical distribution of plants and animals, and he had little respect for paleontology. He spent more time in creating nature than in studying it.[35] As a systematizer of morphological knowledge, however, he made contributions of lasting value. In general, he was guilty, according to Menzbir, of not following Darwin's advice that the future triumph of the idea of evolution would come from the cooperative work of many branches of biology.

Contemporary developments in biology attracted Menzbir's attention only insofar as they referred to the key positions of Darwin's theory. In Theodor Eimer's theory of orthogenesis—built on the idea of organic evolution as a process moving in a definite direction and clearly affiliated with Lamarckian thought—he saw an unwelcome and dangerous invitation to teleological interference. In elaborating his orthogenetic views, Eimer refused to recognize random variation as a source of organic evolution. By rejecting the random occurrences of variation, he eliminated the need for natural selection. Nor did Menzbir favor Petr Kropotkin's theory of mutual aid, which categorically rejected Darwin's theory of the struggle for existence and was closely tied to Lamarck's idea of the direct influence of the environment on the evolutionary process.[36] Menzbir was not impressed with the "successful" efforts of Kammerer, Standfuss, and E. Fischer to present the direct influence of the environment as a leading factor of evolution. Laboratory conditions, he said, accelerate the evolutionary process and make it unrepresentative of the work of nature. He was willing to accept the direct influence of the environment only in cases where it did not interfere with the work of natural selection.

In 1891 the popular journal *Russian Thought* invited Menzbir to contribute an article on "the present state of biology." Responding to the invitation, he wrote that all contemporary biologists were clearly divided into evolutionists (Darwinists) and antievolutionists (anti-Darwinists), that each group had made its position clear and irrevo-

cable, and that the differences between the two groups were irreconcilable, which made the ongoing strife a totally useless exercise. At this time
Menzbir was firm in his belief that the anti-Darwinists were "soundly
defeated." He saw the triumph of Darwinism not only in its answering
specific questions related to organic transformation but also in its giving
biology a general method.[37] The task of modern biology was not to
question the validity of Darwinian principles but to carry them to previously unexplored domains of nature. Arguing from a Newtonian position, Menzbir stated that to demand new proofs for Darwin's theory
was "as unnecessary, and as harmful, as to enter into a polemic with
Laplace."[38] At this time, it should be noted, Laplace's views on the absolute and universal validity of causal explanations stood for the quintes-
sence of Newtonian science.

In 1893 Menzbir no longer clung to the view of a sharp and irreconcilable split in biology between Darwinists and anti-Darwinists. Now he
recognized the clear existence of an intermediate group of biologists
who were neither anti-Darwinists nor orthodox Darwinists. This group
recognized the generality of the Darwinian struggle for existence, but it
also advanced an anti-Darwinian theory of heredity that placed exclusive emphasis on internal causation. Menzbir took the trouble to explain the details of Weismann's neo-Darwinian theory for the readers of
Russian Thought. While welcoming Weismann's strong support for
Darwinian natural selection, he argued that his interpretation of the
mechanism of heredity required extensive refashioning and refinement.
Weismann's theory of heredity, as Menzbir viewed it, suffered from conceptual overelaboration that made it far removed from both the experimental and natural-historical base of biology. In Weismann's "biophores," "determinants," and "ids," he saw unreal categories created
by philosophical playfulness detached from the empirical underpinnings
of modern science.[39]

Despite these reservations, Menzbir did not hesitate to voice a favorable opinion about the potential usefulness of Weismann's theory of heredity. Weismann, he wrote, must be given credit for helping create a
tradition in Russia that encouraged efforts to build a theoretical bridge
between Darwin's legacy and experimental biology, particularly the
branch concerned with heredity and variation.[40] Although it was a product of a particular tradition in German philosophy rather than of solid
empirical research, Weismann's theory was destined to serve for a long
time as a source of fertile ideas.[41]

Hugo de Vries's mutation theory, which viewed saltation rather than "slight modification" as the real source of variation in the forms of life, acquired widespread reputation as a major assault on Darwin's theory. At first Menzbir looked at the new theory with a great deal of suspicion. He was unwilling to attach more importance to "sudden leaps" than to "slight modifications" in morphological features. Nor could he see how the extremely rare saltatory digressions from the norm of heredity could explain the universality and unceasing movement of evolution. He was not even sure whether de Vries's "mutation" and Korzhinskii's "hetero-genesis" were one and the same thing, as they were generally assumed to be. The struggle for existence, he said, played an important role in de Vries's theory but did not figure at all in Korzhinskii's thinking.[42] His first reaction was to count de Vries among the anti-Darwinists, whose ranks made significant gains at the turn of the century. He softened his criticism as soon as he learned that de Vries's allegiance to the idea of evolution, and to Darwin's natural selection, was not in doubt. In his judgment:

> It is not generally known that de Vries is an evolutionist; if he were not an evolutionist he would not have spent long years in search of an answer to the question of the origin of species. Never questioning the transformability of species, he worked carefully and impartially on experiments aimed at decod-ing "the secret of secrets," as the origin of species was previously termed.[43]

After having completed a critical survey of de Vries's mutation the-ory, Menzbir drew a conclusion that characterized his rather uncertain attitude toward the challenges of the new experimental biology to the Darwinian orthodoxy:

> De Vries's theory, still requiring elaboration and still unsettled in its foun-dations, cannot replace Darwin's theory, which is well balanced and general, and can answer an infinite series of questions. Nor can it cause a crisis in the development of Darwinism, which has brought together all branches of biol-ogy. But my last word about de Vries's research shall not be negative. There, where so much work has been done to answer the most important biological questions, even if that work has contributed only a few grains of truth to science, there can be no negative judgment. Even if we cannot agree with the conclusions of the creator of the mutation theory, we sincerely hope that he will continue his experimental work. Without doubt, this work will produce many new ideas, which may or may not sustain the mutation theory.[44]

Menzbir was less kind to the Russian botanist S. I. Korzhinskii, whose theory of heterogenesis, made public in 1899, attracted much at-tention in the West. Like de Vries, Korzhinskii recognized saltatory changes, not caused by external influences, as the primary source of morphological variation in plants and animals. In one respect, however, the two scientists were far apart: while de Vries recognized natural selection and the struggle for existence as factors of evolution and was full of praise for Darwin's contributions to the triumph of the evolutionary view in biology, Korzhinskii had nothing kind to say about Darwin's conclusions and was ready to reject evolution as a useful biological notion. Menzbir lamented Korzhinskii's premature death, which prevented him from completing the work on his new theory. He claimed that Korzhinskii ignored the actual processes linking heterogenetic change to the emergence of new species.[45] Menzbir reminded his readers that Darwin was fully aware of saltatory changes and their role in the origin of new species. He was convinced, however, that their exceedingly rare occurrence makes it impossible to treat them as a major factor in the general evolutionary process.

Menzbir was untypically slow in commenting on the rediscovery of Mendelian laws of heredity in 1900. The third edition of his popular university textbook in zoology and comparative anatomy, published in 1906, surveyed the basic principles of the mutation theory, but it fully ignored Mendel and current elaborations of Mendelian genetics. He waited more than twenty years after the rediscovery of Mendel's laws of heredity to comment on the relations of genetics to Darwinism. At that time he acknowledged Mendel's contribution to "the field of hybridization" and to the study of the "mechanism of heredity," but he was sure to add that all this did not explain "the evolution of the organic world" and that Mendel did not make such a claim in the first place.[46]

In Menzbir's view the future of Darwinism, and of science in general, was in the expansion and refinement of the mechanistic view of the universe. He made little effort to examine the growing attack on Newtonianism by the swelling ranks of contemporary scientists and philosophers. He wrote in 1901 in the widely circulated *Russian Thought:*

> The nineteenth century was dominated by the idea that the phenomena of life are open to mechanical explanation, a conviction that met no opposition. The superiority and the power of mechanical explanation are built on the obvious fact that this mode of scientific thinking offers simple and comprehensive interpretations of phenomena, uses the established methods of the exact sciences, and penetrates the depths of the most complex and mysterious phenomena of life. The mechanical approach offers a powerful analytical tool that can handle even the most complex living phenomena and their minutely segmented components. . . . The so-called philosophical school continues to claim that, despite their brilliant discoveries, the scientists continue to be unable to explain force, matter, and life. This is not true;

every scientist has an excellent understanding of life. Regardless of how we view organisms, we know that a plant lives as long as chlorophyll gives it energy by dissolving carbon dioxide and as long as it produces starch. We know that an animal lives as long as its nervous system receives impressions from the outside world and reacts to them. We also know that both plants and animals live only so long as they exist in both space and time. Inasmuch as our study of vital phenomena depends on the methods employed in the study of all other forms of motion, it produces the same results as long as we do not separate the notion of the organic from the notion of the inorganic. We know the difference between these two categories of phenomena. But we also know that all forms of motion in nature—from molecular motion to the motion of celestial bodies—give us an idea of uniformity and harmony superimposed upon the innate diversity of natural phenomena. The nineteenth century built a bridge between inorganic substance and protein compounds, between protein as a cosmic body and protein as a self-destroying body, and between plants and animals, the latter endowed with strikingly complex behavioral mechanisms. It made all phenomena parts of an indissoluble chain. While coming quietly to a close, it presented all those who showed interest in science with a picture of the grand unity of nature, revealed in the gradual development of the universe.[47]

Menzbir was not very comfortable in discussing philosophical issues. He thought that idealism and materialism, each as a distinct philosophical strategy, had contributed to the accelerated progress of modern science. Idealism helped elaborate and refine the deductive method and the growing role of mathematics in the growth of scientific knowledge.[48] Materialism made a major contribution to the creation of a unified picture of the universe: it made "matter and motion" the common denominators of inorganic and organic worlds. Materialism had shown that "nature is one"—that every natural phenomenon is part of a causal link, and that there is no predetermined purposiveness in nature, even though there is harmony. While making tactical concessions to the idealistic strain in science, Menzbir appeared to have been leaning toward materialism—toward the idea of the sovereignty of "matter-in-motion." In another paper he stated explicitly that the triumph of "mechanical explanation" in biology was one of the greatest achievements of the nineteenth century.[49]

Menzbir admitted that Darwin's theory had not yet achieved full victory. He believed, however, that Darwinism would not only survive all the challenges that came from experimental biology, various branches of neo-Lamarckism, and organismic metaphysics, but would acquire new strength and authority. The new biological theories, he said, can survive and become parts of modern scientific thinking only insofar as they contribute new information and modes of inquiry to the grand edifice of Darwinian science. The new theories enriched Darwinism, and they also benefited from it. "It is difficult for us to imagine how Darwin would have reacted to the data put forth by his enemies. But we do know that Darwin was aware of the ideas that went into the making of de Vries's mutation theory and Mendel's law of heredity." [50]

Despite accumulated challenges to its principles, Darwinism, as Menzbir viewed it, represented one of the greatest intellectual challenges of modern time. It inspired all those who cherished the spirit of free inquiry and who showed no fear of transgressing the limits of thought imposed by tradition. "Darwinism became a horrifying specter for all those who were steeped in prejudice and egotism." Darwin's theory, Menzbir argued, was too viable and too timely to be crushed by its enemies. The bitter attacks on it during the 1880s—the Danilevskii-Strakhov era—succeeded, in his opinion, neither in beclouding its rising star nor in undermining its scientific principles; they merely showed the deep pessimism of a society in crisis—a society devastated by intellectual "darkness," "moral crisis," and "loss of faith in itself." Menzbir made it clear that "pessimism" and "moral crisis" had taken deep roots in a large part of the civilized world. In his view, the bitter attacks on Darwin's theory were the purest symptoms of the intellectual and ethical disorientation of modern society. [51]

Darwin, as Menzbir saw him, wrought a revolution in biology and radically transformed the modern world outlook. His works are rich in theoretical thought and research perspectives and are well stocked with precious empirical data and cogent insights. Menzbir was determined to present Darwin as a model scientist—a personification of the highest standards of contemporary scientific methodology. Guided by "the spirit of free inquiry," Darwin tolerated only the ideas open to challenge. He freed science from metaphysical and theological doctrines and modes of thinking, as well as from intellectual subservience to authority of any kind. By purifying the inductive method, he contributed to the modern emphasis on experimental studies and to a methodological recognition of the unity of inorganic and organic nature. Menzbir joined the commentators on the general development of modern scientific ideas who called Darwin "the Newton of biology." [52]

With Darwinism uppermost in his mind, Menzbir worked on many fronts of biological scholarship. He carried on empirical and synthetic research in ornithology and wrote special essays showing the effectiveness of Darwin's theory as a research tool. In a series of biographical sketches of the pioneers of modern evolutionary biology he illumined the key ramifications of Darwin's theory and the most potent challenges germinated by more recent developments in evolutionary thought. Not avoiding a confrontation with the philosophical aspects of Darwinism, Menzbir worked with remarkable consistency on clarifying and strengthening the Newtonian base of evolutionary biology. He participated in public debates on the issues of evolution and served as a busy functionary of the Moscow Society of Naturalists. The writer of a successful university textbook on zoology and comparative anatomy, he also translated several biological-evolutionary classics from the English language, including Alfred Russel Wallace's *Darwinism* and *Natural Selection*. He was partly responsible for one of the better Russian translations of the *Origin of Species*.

In 1900 Menzbir published a series of essays under the general title "The Leading Representatives of Darwinism." Individual essays presented the evolutionary views of Wallace, George Romanes, Haeckel, and Weismann. A few years earlier he had published a similar essay on Thomas Huxley. His aim was clearly to show the flow of evolutionary ideas from various disciplinary sources to a central pool of knowledge and to point out both the commendable contributions to and the lamentable digressions from the basic principles of Darwin's theory. The study of "digressions" produced a significant by-product: it led Menzbir to point out the domains of evolutionary thought, such as the theory of heredity, which needed help from the non-Darwinian effort in biological research. In the work of Haeckel and Weismann, Menzbir detected a strong influence of the cognitive habits and metaphysical propensities of *Naturphilosophie,* which separated the idea of evolution from the reality of empirical data. He did not approve of Wallace's claim that mental differences between man and animal are differences of kind rather than of degree;[53] and he thought that Romanes's "theory of physiological selection," intended to replace Darwin's theory of natural selection, consisted of incongruous parts that prevented it from acquiring empirical support.[54] Despite the shortcomings of certain aspects of their theories, Menzbir admitted that these scholars provided the main wheels for the triumph of Darwinian ideas and for carrying the theory of natural selection to new areas for research. Haeckel's empirical work represented the high point in the development of evolutionary morphology; Wallace was the most consistent and most thorough defender of Darwin's general theory; Weismann showed the path for the experimental study of heredity; and Romanes carried Darwin's theory to the vast area of "mental evolution."

A. N. Severtsov: Morphology and Evolution

Severtsov belonged to the same group of evolutionary biologists as Timiriazev and Menzbir: he was a thorough Darwinist, not interested in accommodating himself to anti-Darwinism, neo-Darwinism, or quasi-Darwinian developments in evolutionary thought. Unlike Timiriazev and Menzbir, he dedicated his life much more to a search for new parameters and logical elaborations of Darwin's theory than to a popularization and defense of Darwinism. He firmly accepted Darwin's views on the struggle for existence, natural selection, heredity, and variation as the basic factors of the evolutionary process. His research interests, however, did not embrace these problems, the key categories of Darwinian thought. The causes of evolution did not attract his attention: he concentrated on the paths and modes of evolution. In this enterprise his task was not to create a new general theory of organic evolution but to give Darwin's theory more depth and broader compass. He was both an astute defender of Darwin's biological legacy and one of the more original and profound contributors to the elaboration of Darwinian evolution.[55]

As a scientist, Severtsov displayed two general characteristics. First, in his work on the general theory of evolution he depended heavily on the empirical studies conducted by himself and his numerous disciples. His studies radiate an air of fresh exploration both on empirical and on theoretical levels. Second, his approach is interdisciplinary. Although he concentrated on elaborating and codifying the high theory of evolutionary morphology, he drew extensive help from comparative anatomy, embryology, and paleontology. If he ignored physiology and its contributions to the general theory of evolution, it is not because he did not appreciate its strength but because it was outside his competence. He also thought that modern physiology developed in a direction that did not particularly encourage a direct concern with the essential problems of evolutionary biology.

Severtsov's father—Nikolai Alekseevich Severtsov—was a well-known biologist, combining a strong interest in natural history with biological theory. He was particularly noted for his faunistic and floristic surveys of the Tien-Shan Mountains and the Pamirs. In his widely read *Reminiscences,* L. F. Panteleev reported that in 1861 Severtsov gave a series of public lectures on the *Origin of Species.*[56] During the 1870s his studies showed clearly a firm affiliation with Darwinism. "Darwin's theory," he wrote at this time, "is a convenient and solid philosophical umbrella or a common denominator for all the observable general phenomena of organic nature."[57] In 1876 he visited Darwin in Down. In a special analysis of the emergence of new species among the shrikes of the central Tien-Shan, he placed a strong emphasis on natural selection and felt that he had helped confirm Darwin's theory.[58] His contribution to Darwinism, he noted, came not from books but from direct study of Central Asian nature. *The Vertical and Horizontal Distribution of Turkestan Animals,* the first systematic and comprehensive survey of the vertebrates of the region, is rich in analysis inspired by Darwin's theory.[59] His description of the fifteen newly discovered species of mammals and forty-nine species of birds included rich data on the effects of geographical distribution on evolutionary change. Severtsov showed a particularly keen interest in ecology and systematics as evolutionary disciplines. In his strong emphasis on the direct role of environment in the transformation of species he was a follower of Lamarck.

The family atmosphere helped Aleksei Severtsov acquire a profound appreciation of Darwin's contributions to evolutionary biology and a keen interest in the type of scientific research that combined empirical inquiry with a search for regularities in the work of nature. In 1890 he graduated from Moscow University at the time when this institution was firmly established as the national center for Darwinian studies, thanks primarily to the work of Timiriazev and Menzbir. Four years later he read a paper on the primary and secondary metamerism of the head of vertebrates, at the Ninth Congress of Russian Naturalists, held in Moscow. In 1896–97 he was in western Europe working in laboratories, conducting empirical research, writing papers on sharks and electrical skates, participating in seminars on current developments in cytology and histology, and observing the Western scientific community at work.[60] He benefited most from his association with Kupffer and Boehm at Munich University and with Anton Dohrn at the Naples Marine Zoological Station.

After a long teaching engagement at Dorpat and Kiev universities, Severtsov returned to his alma mater in 1911 as professor of comparative anatomy. His transfer coincided with Timiriazev's and Menzbir's resignation from Moscow University in protest against the efforts of the Ministry of Public Education to suppress the student movement and to curtail academic autonomy. A professor at Moscow University until 1930, he offered courses in general zoology, vertebrate zoology, comparative anatomy, and the theory of evolution.[61] Clearly demarcated and individualized, all courses were united by an overarching interest in the theoretical aspect of biological evolution. In 1920 he was elected to full membership in the Russian Academy of Sciences in Petrograd. Contrary to the time-honored custom, he was allowed to remain in Moscow and to continue his teaching activity at Moscow University.[62]

Severtsov's works published during the Soviet period, particularly his *Morphological Laws of Evolution*, gained him an international reputation as a leading scholar in evolutionary morphology.[63] These works offered a grand synthesis and elaboration of the theoretical ideas presented in three essays he had published before the October Revolution: "Evolution and Embryology" (1910), *Studies in the Theory of Evolution* (1912), and "Modern Problems of Evolutionary Theory" (1914). These early works represented an impressive effort to make morphology the core system of theoretical categories in evolutionary biology. Unlike Timiriazev, he added new ramifications to Darwinian thought while, at the same time, living in peace with rapid developments that opened new and challenging vistas in the experimental study of evolution, particularly in the domain of heredity and variation.

Severtsov's empirical work showed a rich diversity of topics closely related to the theory of evolution and a high degree of technical competence. During the 1890s he concentrated on the study of the metamerism of the vertebrate head, that is, on the emergence and evolution of the head among the vertebrates as the center of sense organs, brain, and skull. During the first decade of the twentieth century he advanced a new theory of the origin of the pentadactylous extremity in the multiradial fin of primordial fishlike forms. Subsequently he worked on the origin and evolution of lower vertebrates and drew a genealogical tree of these animals by adding reconstructed missing links to presently existing and paleontologically documented species. The purpose of his empirical research was to test specific theoretical ideas or to illumine general issues that invited controversial interpretations.

General theoretical concerns, all related to organic evolution, occupied a central position in Severtsov's involvement in science. The biological tradition, he said, had advanced three major phylogenetic approaches to evolution: paleontological, comparative-anatomic, and comparative-embryological. Readily acknowledging the strengths and weaknesses of all these approaches, his research experience convinced him that the time had come to accord morphology a central position among the evolutionary disciplines in biology. Morphology, he thought, should not only occupy the strategic position among the biological sciences but should also undertake the task of producing a general synthesis of evolutionary thought. He claimed that the time had come for morphology to go beyond its traditional concern with genealogical trees—and with empirical generalizations—and to become engaged in a search for "the morphological laws of evolution."[64] The contemporary study of phylogeny, he felt, formed only a preliminary phase in the general search for morphological laws of evolution.

Severtsov worked most successfully in two areas of evolutionary theory: the theory of phylembryogenesis, an elaborate approach to the relationship of phylogeny to ontogeny as a mechanism of evolution, and the general theory of morphology, concerned with universal regularities or laws of organic evolution.[65] Since in both areas evolution figured as the main object of inquiry, he preferred to label his approach as a historical orientation in biology. In both the theory of phylembryogenesis and the general theory of evolutionary morphology he made an effort to produce a new integration of generalized knowledge.

The theory of phylembryogenesis is "the theory of evolution seen through changes in the course of embryonic development." It is a search for answers to such fundamental biological questions as the correlation between "individual" and "historical" development of characters—between ontogeny and phylogeny. In a way, this theory is an elaboration of Darwin's effort to transform the "law of parallelism," referring to similar characters in different lineages, into the "law of correlation," referring to similar characters in ontogenetic and phylogenetic developments. The theory of phylembryogenesis is in essence a systematic coordination of ontogenetic and phylogenetic developments of "morphological" characteristics. It has laid the groundwork for evolutionary morphology.[66] Severtsov contended that "ontogeny is a function of phylogeny," that is, that without an understanding of phylogeny we cannot understand the laws of individual development—strictly speaking, the laws of life.[67] But he also contended that changes in ontogeny are the real source of changes in phylogeny; ontogeny not only repeats phylogeny but also makes it a historical process, subject to constant change.

The theory of phylembryogenesis created a wide base for a systematic inquiry into a complex set of challenging problems. The work on four such problems had proven particularly attractive: regularities in the evolutionary aspect of ontogeny; bonds between the ontogeny of animals and the conditions of their existence, including the adaptation of individual organs to functional changes; comparative data on the ontogenetic development of closely related species and genera; and ties between theoretical inquiry and practical problems, particularly those related to animal breeding.[68] Russian biologists were particularly attracted to the general theory and ecological dynamics of phylogenetically significant ontogeny. A new discipline—ecological embryology—grew in the tradition of Severtsov's biological thought.

Severtsov's extensive empirical research, particularly among lower vertebrates, led him to recognize three basic modes of phylembryogenetic changes: anaboly (extension of the last stages of morphogenesis), deviation (changes during the middle stages), and archallaxis (changes during the initial stages). Severtsov was particularly anxious to point out the relationships of individual modes of phylembryogenesis to the tempo and general character of evolution. Anaboly, as he saw it, refers to changes that are exceedingly slow but that lead to highly diversified phylogenetic alterations. Archallaxis, by contrast, depicts discontinuous leaps and relatively fast changes. Deviation occupies a position between anaboly and archallaxis. The notion of three modes of evolution led Severtsov to revise Haeckel's biogenetic law. He accepted the idea of the ontogenetic recapitulation of phylogeny, but he linked it only to anaboly.

In general, Severtsov's theory of phylembryogenesis is an elaboration of two general ideas. First, phylogenetically significant change can take place in every phase of ontogenetic development.[69] Or, as Bernard Rensch has stated in reference to the work of the Severtsov school: "The results obtained so far indicate that phylogenetic alterations may appear first in very different stages of the ontogenetic development."[70] Second, during the growth of an organism, phylogenetic alteration takes on different forms and intensities: the earlier the stage of ontogenetic development, the broader the phylogenetically significant changes. Early stages produce alteration in the characteristics of families, classes, or other larger taxonomic groups and affect entire complexes of organs. Phylembryogenesis, in Severtsov's view, is the primary fountain of information on the evolution of the forms of life.

Haeckel studied the relationship of ontogeny to phylogeny only insofar as it helped him in his lifelong project to construct a universal genealogical tree of the animal world. For this reason he was interested in ontogeny as a true recapitulation of phylogeny. Severtsov had an additional interest: he treated the study of this relationship as a preliminary step in a broader effort to formulate the general morphological laws of evolution.[71] He criticized Haeckel for his inclination to limit evolutionary biology to the study of phylogeny, but he also rejected the neovitalist claim—most explicitly stated in Rádl's *History of Biological Theory*—that "Darwinian morphology" could never become a historical discipline, for the simple reason that it had no access to historically significant data.

Severtsov's second major activity concentrated on the formulation and elaboration of evolutionary morphology as a system of integrated theoretical principles. He wrote in 1912:

> In recent times, there has emerged a new orientation in the study of the evolutionary process. Independently of each other, paleontologists (paleo-biologists), comparative anatomists, and embryologists have adopted this orientation. Previously, all these specialists limited their study exclusively to the evolution of individual groups of animals, and to construction of phylogenetic trees, that is, to preparation of as complete a [genealogical] record of the animal kingdom as possible. At the present time biology has advanced so much that it has become possible to undertake another task that had eluded old zoologists—the study of phylogenetic regularities in the evolutionary process.[72]

Early Darwinists laid the foundations for a phylogenetic classification of larger taxa, established the phylogeny of a series of organs from head and brain to gill and fin, and unveiled some regularities in the evolutionary process, typified by Haeckel's biogenetic law.[73] They had spent much time in gathering empirical material related to evolution; the time had now come to concentrate on the search for morphological laws, the quintessential expression of the regularities of organic evolution. Severtsov observed, however, that notable efforts in that direction had already been made, as was shown by such general principles as Osborn's "adaptive radiation," Anton Dohrn's "change of function," N. Kleinenberg's "substitution of organs," L. Plate's "extension and intensification of functions," and D. M. Fedotov's "physiological substitution."[74] These principles, all pointing in the right direction, required further elaboration, particularly inasmuch as they were interrelated. Above everything else, morphologists were required to search for previously undetected regularities in the evolutionary process.[75] According to Ernst Mayr, Severtsov contributed to an explanation of the "intensification of function" as a distinct regularity in morphological evolution. Mayr referred specifically to Severtsov's explanation of the conversion of the five-toed foot of primitive ungulates into a two-toed or one-toed foot.[76]

Severtsov worked on two fronts: he looked for morphological regularities as revealed by direct empirical research, and on a purely theoretical level he endeavored to formulate and codify the morphological laws of evolution. On the empirical level, as a matter of personal taste and predilection, he worked mainly on those embryonic changes that are preserved in adult organisms as adaptation to the environment.[77] In his empirical search for general modes of morphogenesis he relied heavily on ecological material. Much of his general theoretical work concentrated on giving morphology a firmly postulated and comprehensive evolutionary orientation.[78]

In elaborating the general principles of evolutionary morphology, Severtsov was noted particularly for his effort to make the notion of progress an integral part of evolutionary studies.[79] He drew a line between morphophysiological progress in the organization of animals and biological progress in the survival potential of individual species. The first refers to the increasing and decreasing morphological complexity of living forms; the second is based on three criteria: increase in the vitality and rate of growth of a species; the expanding size of the area occupied by a species; and the splitting up of a species into subordinate taxonomic groups, the so-called adaptive radiation.

In his later work Severtsov developed a scheme of four modes of "morphological progress." Aromorphosis, the first mode, designates "the most general adaptive change in the organization and functions of animals, which normally increases the vital energy of animals and the diversity of their forms (differentiation)." This mode of change is important biologically inasmuch as it marks "a useful adaptation of animals to changes in the environment and inasmuch as it produces stable characteristics."[80] In brief, aromorphosis stands for the evolutionary process as postulated by Darwin's theory. I. I. Shmal'gauzen, the most authoritative interpreter of Severtsov's scientific legacy, viewed aromorphosis as "a more substantial step in the process of transformation, signifying the rise of an organization to a higher level on the [evolutionary scale]."[81] B. S. Matveev, another disciple, did not hesitate to identify aromorphosis as saltatory change.[82] Idioadaptation, the second mode, usually follows aromorphosis. It denotes adaptation to specific conditions of life, which, however, does not add to the complexity of organization.[83] This mode of evolution explains the mosaic character of present-day flora, made up of forms that, by their morphological character, belong to various geological eras. Cenogenesis, the third mode, stands for embryonic adaptations not reflected in the structure of grown organisms.[84] These adaptations are progressive inasmuch as they help increase the proportion of animals reaching the age of maturity. Regression, the fourth mode, denotes morphological simplification as an adaptation to specific conditions of existence. It is best illustrated by animals that change from a roaming to a stationary way of life or from active to parasitic feeding.[85] "The process of partial or general regression," according to Severtsov, "should in no case be identified as a phenomenon of degeneration. In all cases we have analyzed, the process of regression is useful to a group of animals and enhances their adaptability to the given conditions of existence: morphological regress becomes a condition of biological progress." The adaptability of organisms may be strengthened by the increased efficiency of existing organs; but it may also be strengthened by the disappearance or atrophy of individual organs. He recognized two kinds of regressive changes in organs: changes in which the functions of eliminated or atrophied organs are not taken over by other organs, and changes in which they are transferred to other organs. Severtsov contended that no mode of adaptive evolution occupies a dominant position—that "animals can exist and flourish for an unlimited time regardless of the evolutionary path they have followed." [86]

In essence, evolution, as Severtsov interpreted it, is a result of the adaptation of organisms to the environment. Adaptive change is primary or correlative. Primary change is the main course of evolution; correlative change is a by-product of primary change and is totally dependent on it for survival—if primary change is erased, correlative change becomes atrophied. Since it is not hereditary, correlative change may or may not recur with regularity. [87] While recognizing the Lamarckian use and disuse of organs as an evolutionary factor, Severtsov gave it only a correlative significance. He rejected the Lamarckian use and disuse of organs as a source of heritable characteristics. [88]

In paying homage to the Russian scientists who contributed to the advancement of the general theory of organic evolution, Severtsov selected P. P. Sushkin for special praise. An ecologically oriented evolutionary biologist, Sushkin studied the role of major climatic and geomorphological changes in the evolution of vertebrates. [89] He produced suggestive ideas on the combined effects of these changes in giving a unique direction to the evolution of mammals. Severtsov thought that Sushkin's work opened wide possibilities for a systematic and comprehensive study of the evolutionary role of environmental factors. He felt much closer to the environmentalism of the Lamarckian tradition than to the Mendelian legacy, which regarded environment as an evolutionary factor of secondary importance.

The scientific study of the evolution of animals, as Severtsov saw it, must take into account two separate factors: the organization—the degree and the complexity of the structure—of animals at the time a change takes place in the habitat, and the general direction of change in the habitat. The first factor determines the change in a given organism, for it is well known that a specific change in the environment may be differently reflected in various animal and plant forms. The second factor determines whether the particular change in a given organism will lead to progress or regress, divergence or convergence. "These are regularities," he said, "of very general character that worked in the past, that work at the present time, and that most probably will work in the future."[90]

Severtsov accepted the struggle for existence as an empirically verified fact. He also accepted the idea that the influence of the environment on the process of evolution is "indirect," rather than "direct," as Buffon and Lamarck had emphasized.[91] He dealt, however, much more with the morphological aspects of variation than with the prime causes of transformation processes.[92] Despite powerful challenges from genetics, Severtsov stuck steadfastly to the Darwinian view of environment as a primary—even though an indirect—factor shaping and directing the process of evolution.

The question of evolution as a reversible process did not escape Severtsov's attention. L. Dollo, the well-known Belgian paleontologist, suggested that evolution is an irreversible process, by which he meant specifically that an organ cannot repeat evolutionary phases it had left behind in its long historical journey. Some biologists generalized this suggestion into a universal law of the irreversibility of evolution. Severtsov thought that Dollo's suggestion could be accepted only with reservations. When we state, for example, that a fully atrophied organ cannot be restored to a normal state, we are making an empirical statement that we do not know of any case of a restoration of such organs. In some cases, however, the reversibility of the evolutionary process can be readily documented. In the evolution of a part of the skeleton of some species of deepwater fish, to give one example, flexible cartilage was replaced by inflexible bone which, in turn, was replaced by flexible cartilage. Severtsov was ready to admit that the cases of the reversibility of evolution were exceedingly rare.[93]

Severtsov made an explicit statement on the nature and the strength of Darwin's theory and, by implication, on the reason for his belief that the basic postulates of this theory were unchallengeable. He wrote in his classic *Studies in the Theory of Evolution* (1912):

> For our purposes, it suffices to note the following features of Darwinian evolutionism that separate it from all preceding evolutionary constructions and make it a solid basis for further research in the theory of evolution. According to Darwin, the ultimate cause of phylogenetic change in the organization of animals is always the change in the conditions of existence, that is, in the surrounding environment, taken in a broad meaning of the term. For

the most part, change in the conditions of existence exercises an indirect influence on the organism, that is, it determines the direction of the accumulation of individual variations acquired by natural selection. Without a change in conditions, there can be no change in species. . . . From a theoretical point of view, it is equally important that Darwin does not rely on an unverifiable and unprovable internal evolutionary principle, not related to changes in the environment. In other words, Darwin's theory deals with facts open to direct verification by experiment and observation. In his opinion, phylogenetic changes, as products of natural selection, always appear as useful adaptations of organisms to their environment. They are purposive alterations in the structure and functioning of a given animal. It is possible that Darwin was in full agreement with the principle of purposiveness in the living world, but, unlike Karl von Baer, he treated it, not as a basic—and unfathomable—quality of organisms, but as a problem open to examination. His entire theory serves as a brilliant explanation of the phenomenon of organic purposiveness, which at first sight appears so mysterious.[94]

In von Baer's view, purposiveness antedates organic evolution—evolution is a creation of predetermined purposiveness. Darwin, by contrast, viewed purposiveness as a creation of evolution; the higher the place of a species on the scale of evolution, the more complex the inner mechanics of its purposive action. In Darwin's view, as Severtsov interpreted it, purposiveness is an empirical phenomenon subject to natural causation; it is an evolutionary variable, not a "constant" of organic life.[95]

Despite his deliberate effort to avoid sociological involvement, philosophical construction, and popular elaboration, Severtsov reached a wide-ranging reading public. The magic of his influence was not only in the innovative character of his theory but also in the lucidity, appealing precision, and natural flow of his writing style. Particularly appealing was his attitude toward Darwinism as an open theory, inviting new research on substantive, methodological, and conceptual levels. His criticism of anti-Darwinian orientations was mild but not perfunctory. Unlike Timiriazev, he did not subject his "enemies" to unmitigated scolding.

Severtsov's close adherence to the uniformitarian roots of Darwin's thought prevented him from taking serious note of theories advocating multiple lines of evolution and of the emergence of new causal factors in the transformation of species, such as, for example, the accumulated effects of human intelligence. As an uncompromising critic of autogenesis, he could not help but look with suspicion at some of the most dynamic and promising branches of experimental biology. Suspicion, however, did not prevent him from anticipating a convergence of Darwinism and the new genetics.[96]

Severtsov is remembered for his effort to draw a clear line between Darwinism as a science and Darwinism as a world view, or as an ideology. Darwinism as science is a system of theoretical thought inviting challenge and open to change. Darwinism as a world view is a system of beliefs and sentiments that discourages any criticism of its basic premises and is intolerant of change in basic evolutionary theory. The representatives of resurgent vitalism, particularly strong in Germany but not without influential spokesmen in Russia, criticized the scientific foundations of Darwinism from the vantage point of a world view steeped in idealistic metaphysics. Hans Driesch, in his view, rejected Darwinism on ideological grounds, but his effort was unsuccessful on scientific grounds. Severtsov stated:

> It is clear that the antagonism between neovitalism, on the one hand, and the general evolutionary theory and Darwinism, on the other hand, is based on historical and psychological arguments rather than on logic. Despite the scientific veneer and mathematical formulas of Driesch's theory, neovitalism is not a scientific theory but a philosophical world view shared by a group of modern biologists. Driesch is linked to the modern idealistic orientation in metaphysical thought to the same degree that the *Naturphilosophie* of Oken and Carus was linked to the philosophy of Fichte and Schelling.[97]

Severtsov was careful to point out the basic differences between neovitalist and Darwinian interpretations of the work of purposiveness in the living world. To a typical neovitalist, purposiveness is an "internal teleological principle" that cannot be analyzed into component parts and into empirically ascertainable elements. It is inaccessible to scientific scrutiny. To Darwinists, as Severtsov represented them, purposiveness is a complex process that can be divided into functional components and is open to empirical ascertainment and scientific inquiry. Whereas to neovitalists purposiveness is a metaphysical category, totally inaccessible to science at the contemporary level of its methodological competence, to Darwinists it is an empirical category and a legitimate object of scientific study.

We know, wrote Severtsov, that every great scientific discovery shapes the world view of the given time. Galileo and Newton helped create the world view of their time, and more recently Darwin had played a similar role. The world views of Galileo and Newton had gone out of style, and it is possible that the same fate awaits Darwin. Even if this happens, it will not reduce the value of Darwin's contributions to science: Darwinism will continue to serve as a "working theory" for zoologists and botanists insofar as they concern themselves with the relations between the organism and the environment. "Although we attach great value to Darwin's theory, we have no right to be dogmatic and to be satisfied with the achievements of Darwin and his immediate followers: scientific theories are strong inasmuch as they open possibilities for new discoveries and for asking new questions."[98] Severtsov showed exemplary loyalty to the cardinal principles of Darwin's theory, but he never abandoned the idea that Darwinism needed much elaboration and refinement.

Compared with Timiriazev, Severtsov showed more subtlety and more precision in separating Darwinism as a world view from Darwinism as a turning point in modern science. He readily admitted that science and the world outlook are overlapping cognitive and sociological categories. The modern world outlook contains intellectual ingredients drawn from science; science, in turn, harbors residual beliefs drawn from the world outlook. He merely noted that, in validating relevant knowledge, science and ideology use different standards. A scientist depends on the rigor of logic, theoretical consistency, and empirical verification in accepting new knowledge and in integrating this knowledge into larger systems of secular wisdom. An ideologue, by contrast, validates knowledge by establishing its compatibility with preferred values and philosophical prejudgments. A scientist depends on the weight of "proofs"; an ideologue depends on the "harmony" between the facts of science and his scale of values.[99] A vitalist in biology errs in judging scientific theory from the standpoint of idealistic metaphysics, a unique world outlook. Ernst Haeckel, on the other hand, was guided by the criteria of materialistic metaphysics when he translated Darwin's theory into a rigid monistic system. Severtsov was ready to recognize Haeckel's specific effort to make Darwin's theory the cornerstone of a scientific theory of organic evolution. He was not ready to recognize Haeckel's specific effort to make Darwin's theory the cornerstone of a closed philosophical system. He knew that Darwinism was caught in a double crisis: the crisis generated by new advances in science (particularly in genetics), and the crisis caused by ideological conflicts and uncertainties.

Timiriazev placed Darwinian study within a narrow framework: he endorsed all the cardinal principles that went into the making of Darwinian evolutionism, but he did very little to elaborate, refine, and ramify these principles. He was a doctrinaire Darwinist in a very strict sense. Severtsov was different by virtue of his temperament, intellectual bent, and theoretical achievement. Like Timiriazev, he made no effort to disrupt the symmetry and logical harmony of the theoretical principles that gave Darwin's theory its distinctness and inner unity. Unlike Timiriazev, however, he carried Darwin's principles to previously unexplored, or little explored, domains of evolutionary biology, and he gave these principles a more precise meaning and authority. He had a livelier predilection for exploratory work in the realm of general theory and a clearer picture of the countless research vistas Darwin's work had bequeathed to future biologists.

In the process of giving morphology an evolutionary orientation, Severtsov gave the theory of evolution both new support and new research challenges, and in the end produced a very active school of biological thought. In one important respect, however, Severtsov did not keep pace with contemporary developments in biology: he did not go beyond general pronouncements about the promising perspectives of a closer cooperation of evolutionary morphology with experimental biology in the common search for a general theory of organic transformation. Mark Adams has given a judicious appraisal of the evolution of Severtsov's attitude toward genetics:

> Severtsov's attitude toward genetics was less a product of his own reorientation toward experimentalism than a response to the changing orientation of geneticists toward evolutionary theory. He had bemoaned the antagonism of such geneticists as Bateson, Johannsen, and Filipchenko toward Darwinism, but came to applaud genetics when it began to disprove neo-Lamarckian views, confirm natural selection, and favor evolutionary theory as he knew it. However, genetics had still not really provided a theory of the causes of evolution in Severtsov's terms because genetics, and even population genetics, had still not made any progress toward explaining the major macroevolutionary phenomena.[100]

Severtsov was firmly convinced that the pioneers of genetics—as represented by Bateson and Johannsen—were at first much more interested in advancing a "static" approach to heredity than in joining evolutionary studies. Gradually, the leading geneticists accepted the idea of casting their experimental studies within evolutionary frameworks, thus opening the doors for fruitful cooperation with Darwinian scholars. The new biology widened the base for the study of organic transformation and for work on an integrated evolutionary theory.

In his attitude toward the new branches of experimental biology, Severtsov showed much uncertainty and considerable wavering. For example, he recognized mutations as a key factor of organic evolution, but, at the same time, he reaffirmed his position as an unbending foe of autogenesis. It is most probable that he thought of mutations as environmentally induced leaps in the evolutionary process, subject to natural selection. Despite all the reticence and uncertainty, he deserved credit for alerting his students to the great achievements and promise of experimental biology. The time was fast approaching, he said at the end of his distinguished career, for the students of ecology, genetics, and developmental mechanics (W. Roux's *Entwicklungsmechanik*) to join forces in the search for new perspectives in evolutionary thought.[101] He wrote at this time: "We are fully justified in claiming that the vast and valuable material collected by geneticists and students of mutations will be used by the champions of natural selection."[102] I. I. Shmal'gauzen not only followed his teacher's counsel but also became one of the most accomplished contributors to the grand theory of evolutionary synthesis. Severtsov's students, particularly Shmal'gauzen, helped abate the traditional conflict between genetics and Darwin's theory of organic evolution by making elaborate recommendations for an integrated approach to evolution. They accomplished this not by turning against their teacher but by adding to the breadth and depth of his theoretical insights.

Deeply aware of the need for bringing together Darwinian and genetic evolutionary strategies, Severtsov was convinced that neither Darwin's theory nor genetics was advanced enough to justify immediate work on such a project. He dedicated his life to preparing the Darwinian side of the evolutionary equation for a grand synthesis by giving it more depth and precision and by helping raise it to higher levels of scientific abstraction.

Severtsov had a logical and disciplined mind. In the intricate conceptual structure of his evolutionary morphology, only the notion of "regression" appeared to be insufficiently clarified and imprecisely stated, particularly as a unique mechanism of "biological progress." His extensive use of the adjective "morphophysiological" did more justice to morphology than to physiology. He did not make the historical orientation of the theory of phylembryogenesis and the structural orientation of morphology parts of a unified strategy in evolutionary research. Concentrating on the "how" rather than on the "why" of organic evolution, he showed much more interest in bringing genetics closer to Darwinism than in bringing Darwinism closer to genetics.[103] Severtsov's achievements were, nonetheless, of major proportions. Particularly noteworthy was his generally successful effort to raise evolutionary study to higher levels of conceptualization and to make morphology a central integrating link in the modern approaches to organic change. He succeeded in elevating the theory of biological evolution above the recurring and ever threatening intrusions of neovitalism and speculative branches of neo-Lamarckism. In all his studies, Severtsov adhered closely to the laws of evolution stated in the *Origin of Species*.

In reflecting on the history of evolutionary biology, Severtsov summed up his views on the history of Darwinian thought. Darwin, unlike Lamarck, was successful because the scientific community was ready to absorb his ideas, which "attracted talented and brilliant representatives of many sciences: the comparative anatomists Huxley and Weismann, the embryologists Balfour and A. O. Kovalevskii, and the paleontologists Cope, V. O. Kovalevskii, and Gaudry." [104] In his opinion, the first ten to fifteen years after the publication of the *Origin of Species* produced such an abundance of great results that contemporaries became convinced that the salient points of the evolutionary theory were firmly established and that the future generations of biologists would be expected only "to fill in details." This did not happen, however. The deeper the biologists went into Darwin's theory, the more they became convinced that it needed much more than details giving added strength to the established principles. According to Severtsov, it also needed additional principles applicable to the newly carved out domains of inquiry. The new evolutionary biology grew in two directions: horizontally, by widening the base of evolutionary studies, and vertically, by endeavoring to reach higher levels of scientific abstraction. Severtsov distinguished himself primarily by his comprehensive effort to make morphology, as a core discipline of organic evolution, an elaborate network of scientific laws and principles.

Severtsov's heavy reliance on high theory was only one mode of defending and advancing Darwinian orthodoxy. N. V. Tsinger, a professor at the Novo-Aleksandriia Institute of Agriculture and Forestry, relied on a different mode. He devoted no time to building up an elaborate theoretical structure true to Darwin's basic principles. Working on a purely empirical level and taking Darwin's principle of natural selection for granted, he undertook to show how it worked in concrete situations. Particularly noted was his study of the origin of *Camelina linicola,* an annual false-flax plant producing small yellow flowers, which grows in fields under flax cultivation, a condition indicating that it is a relatively new species. After a long study, Tsinger concluded that this plant has evolved from a xerophytic to a hydrophytic species, and that it no longer could revert to xerophytic existence.

Tsinger's study attracted particularly wide attention because it dealt with a unique case of natural selection. Tsinger concluded that one of the xerophytic *Camelina* plants of the family Cruciferae was the ancestor of *C. linicola*. The question was why the other *Camelina* species continued to live outside the cultivated flax fields. The answer was in the common technique of selecting flax seed for planting. The farmers usually relied on special sieves to isolate the large seed of flax from the smaller seed of undesirable wild plants. Of all seeds of wild plants, only those of the ancestor of *C. linicola* happened to be large enough to remain with flax seed during the sieving process; in due time they produced a new species that took on many characteristics of cultivated flax and became fully adapted to the wet habitat of flax fields. *C. linicola*, the new species, survived because it emerged victorious in its struggle for survival—which happened to be a struggle against human effort to eliminate it from its new habitat. This selection was not artificial, because "artificial selection"—at least the way Darwin knew it—refers to human effort to produce new varieties of domestic plants and animals. Nor was it natural: "natural selection"—at least as Darwin saw it—does not provide so much room for human interference. In one important respect, however, it was natural: it was clearly a product of the struggle for existence.[105] Tsinger's study provided one of the rare empirical and direct descriptions of the work of "natural selection" as a mechanism of evolution. What Severtsov showed on a high theoretical level, Tsinger achieved on an empirical level: both produced evidence showing that the struggle for existence and natural selection are dynamic and flexible processes with infinite ramifications.

Darwinism as a World View

A strong group of scientists depended on philosophical metaphors in expressing their Darwinian affiliation. These scholars were not professionally conversant with the specific ramifications of evolutionary principles built into Darwinian thought. Nor were they necessarily involved in evolutionary research. They merely looked at the work of Darwin as a triumph of modern intellect and as a most sublime achievement of science as a body of knowledge, a method of thinking, and a world outlook. In this department Darwin's contributions found a great and eloquent supporter in the famous neurophysiologist Ivan Pavlov. In "Experimental Psychology and Animal Psychopathology," a paper delivered at the Fourteenth International Congress of Medicine, held in Madrid in 1903, Pavlov professed his strict adherence to the Darwinian view of the scientific approach to the adaptation of organisms to the dynamics of environment. This approach had no use for the philosophical notion of teleology, a lure to vitalism.[106] In a speech delivered in London in 1906, Pavlov identified Thomas Huxley as an energetic supporter of the theory of evolution, the "greatest idea" in biology.[107] In a later speech he noted the profound influence of the *Origin of Species* on the intellectual world in general and on the scientific community in particular.[108]

Pavlov's article "Natural Science and the Brain," delivered in December 1909 at the Twelfth Congress of Russian Naturalists and Physicians, appeared in *In Memory of Darwin* (1910), a collection of essays honoring the one-hundredth anniversary of Darwin's birth and the fiftieth anniversary of the publication of the *Origin of Species*. The article made no specific reference either to Darwin or to the theory of evolution. But that did not make much difference: important was the fact that a widely read and discussed book made Pavlov's sentimental and philosophical allegiance to Darwinism firmly documented and generally known. Loyal to Darwinian principles, he claimed that only an experimental study of animal adaptation to the external environment can produce scientific knowledge on the elementary modes of behavior, which Pavlov identified as "higher nervous activity."

In 1913, in a paper presented at the Moscow Scientific Society, Pavlov made his assessment of Darwin's place in modern science clear and forthright:

> In all justice, Charles Darwin must be given the major credit for stimulating and inspiring the modern comparative study of the higher [psychic] manifestations of animal life. In the second half of the last century, as every educated person knows, his brilliant explanation of [organic] development fertilized the entire intellectual sector of human activity, particularly the biological branch of natural science. The hypothesis of the animal origin of man has led to an absorbing interest in the study of the higher manifestations of animal life. The post-Darwinian period faces the task of finding the best way to answer this question and to organize a study of the problem.[109]

In 1916 Pavlov contributed a paper to a symposium honoring K. A. Timiriazev, the most ardent and consistent Russian Darwinist. Pavlov ended the paper by identifying Timiriazev as a "tireless champion of a real scientific analysis in a region of biology" in which it was easy to stray along "false paths."[110] By "real scientific analysis" Pavlov could have meant only a defense of Darwin's evolutionary views, the main theme of Timiriazev's involvement in biological scholarship. Pavlov and Timiriazev were united in a firm belief that neovitalism was a metaphysical aberration in biology and that physiology, depending exclusively on the experimental method, was the backbone of biology.

Although Pavlov's expressions of a favorable attitude toward Darwin's contributions did not usually go beyond parenthetical statements, they placed the eminent neurophysiologist solidly in the camp of Darwinism, bolstering its defenses and carrying its influence to a new domain of scholarly activity. At this time the Russian anti-Darwinists found it much easier to attack the ghost of Darwin than to challenge Pavlov's firm hold on the scientific community. At a time when Darwinism was besieged from all sides, Pavlov's admiring attitude toward Darwin's philosophy contributed immensely to the strength of a vital tradition in national science.

Despite his frequent testimonials, Pavlov did little to place his own research within an evolutionary framework. He did not accept Karl von Nägeli's challenging idea that the real study of organic evolution belonged to a physiology built on the model of Newtonian mechanics. Severtsov saw organisms as morphological unities and as systems of adaptive mechanisms responding to changes in the habitat. Neither Pavlov nor any other Russian scientist undertook a systematic and comprehensive study of the physiological unity of organisms as adaptive or evolutionary mechanisms. Without mentioning Pavlov by name, Severtsov pointed out the difference between the evolutionary bent of morphology and the nonevolutionary interest of physiology:

> For a given species it is irrelevant whether one of its organs has performed the same function for both ancestors and descendants. What is relevant is whether the performance of this function has become more efficient in the struggle for existence. It is exactly here that our view differs from that of physiologists: they study the functions of organs per se, we study these functions as means that help individual species in their struggle for survival. The changing organs are only the tools that help the descendants of a given living form create biologically significant active or passive adaptation mechanisms.[111]

During and immediately after World War I, Pavlov showed slight interest in linking the theory of conditioned reflexes to the Lamarckian principle of the inheritance of acquired characteristics.[112] Subsequently, he stated publicly that he was not a partisan of the theory of the heredity of conditioned reflexes.[113] W. H. Gantt, Pavlov's American student, acknowledged his teacher's Lamarckian leanings, but he noted that Pavlov's "Lamarckism" consisted only of a few passing remarks. Working on the experimental foundations of the theory of unconditioned reflexes, Pavlov gave the problem of the inheritance of acquired characteristics only minor significance. Gantt thought that Darwin's and Pavlov's theories were mutually complementary: while Darwin treated adaptation of individual species to the habitat over long periods of time, Pavlov dealt with adaptation—by means of conditioned reflexes—within the life-spans of individual organisms.[114] In Severtsov's terminology, whereas Darwin's approach was historical, Pavlov opted for an essentially ahistorical approach.

Among the contributors to *In Memory of Darwin* the physicist N. A. Umov, a well-known professor at Moscow University, enjoyed an enviable reputation as one of the earliest Russian commentators on quantum and relativity theories. An expert in the theory of earth magnetism, the diffusion of aqueous solutions, and optical qualities of opaque media, he expressed clearly articulated philosophical views, which in their total effect can be labeled scientific humanism. Like several other Russian scientists, including I. I. Mechnikov and V. I. Vernadskii, he equated the evolution of human intelligence with the steady expansion of a force of cosmic proportions.[115] He was the moving force in the founding of the Ledentsov Society, the first serious effort in Russia to set up a private foundation providing financial support for research in the natural sciences and technology. Umov's attitude was best expressed in the following statement:

> The great thinkers Helmholtz and Darwin have removed the barriers from the roads leading to the understanding of human origin and life. The first disproved the existence of a special [vital] force in living nature . . . ; the second adduced incontrovertible evidence in support of genetic bonds between man and animal. New scientific studies have carried the idea of evolution to the study of inorganic nature as well. In consequence, science has shown that without exception all natural phenomena, regardless of their inner organization, are rational, that is, open to human reason for examination. Although there are secrets which cannot be decoded at this time, [we must remember that] everything is part of nature—that there are no secrets outside nature.[116]

In an earlier paper, clearly inspired by Darwin's evolutionary idea, Umov postulated "the third law of thermodynamics," which worked in the opposite direction from the law of entropy. The processes of life, he said, lead to an increase in both the quantity and the harmony of the organic world. In the "selective adaptation" of organisms he saw the real propelling force of this law. In the living universe, "selective adaptation is a weapon in the struggle against both disharmony and entropy"—it is Maxwell's demon selecting favorite molecules.[117] Umov did not pursue his intriguing ideas; he merely wanted to show how the various branches of modern physics combined with the grand idea of organic evolution to create a world view permeated by scientific optimism. He rejected vitalism for its negation of the possibility of creating a unified picture of nature.

Pavlov and Umov shared a cosmic outlook dominated by an organic blend of Darwinian evolutionism and the Newtonian mechanistic view of nature. P. P. Lazarev, a new physicist with old ideas, went so far as to claim that Newtonian physics was not only a source of models for a scientific study of organic evolution but also a fountain of information corroborating the basic principles of Darwin's theory. In an earlier article published in 1915 in *Priroda* he cited Helmholtz to show that Newtonian physics, which he thought was the high point in the evolution of science, gave full support to Darwin's notion of the purposive adaptation of organisms to their environment. He also credited Boltzmann with having adduced "physicochemical" arguments in favor of the struggle for existence as a prime mover of evolution.[118]

More than any other Russian scholar, Maksim Kovalevskii carried Darwin's ideas to the domain of sociology. In his studies of the history and social organization of the *obshchina* and in his analysis of the Caucasian legal customs he expressed himself in favor of an evolutionary approach to the universal history of human society, based on the use of the comparative method as the most reliable tool of sociological inquiry. In 1890, in a monograph on the origin of the family and private property, he noted the indebtedness of modern sociology and social history to Darwin's biological contributions. Darwin's "great law of evolution," he observed, exercised a growing influence on the study of social phenomena and had become a guiding force in "the philosophy of history and in the sciences of religion, law, and morality."[119]

In a later study Kovalevskii offered a detailed assessment of the influence of Darwinian biological thought on the development of modern social theory. He made four points. First, Darwin's major contribution was in influencing the emergence of an empirical approach to social phenomena and in adopting causal analysis as a method of sociological explanation. Second, there was no modern sociological orientation free of the influence of Darwin, his precursors, and his followers.[120] From the first published essays of Herbert Spencer to the contemporary French school, which emphasized both struggle and solidarity as the nexus of social existence, all sociology, according to Kovalevskii, built upon biological foundations, and all sociological laws appeared as variations on the laws that, in Darwin's opinion, explained the origin of species.[121] Third, natural selection required major modifications to become a sociologically useful notion. Durkheim's "organic solidarity," Petr Kropotkin's "mutual aid," Lester Ward's "psychological method," and G. Tarde's *"opposition universelle"* indicated the preferred sociological responses to Darwin's ideas. Fourth, by planning for a society devoid of class struggle, Marx transformed sociology from a budding science into a utopia. No doubt Kovalevskii was reading his theoretical impulses and social vision into the sociological empire of Darwinism. He translated Darwin's science into his own world view. Nonetheless, he was generally considered the leading Russian sociologist of his time, and it came as no surprise in 1914 when the St. Petersburg Academy of Sciences elected him to full membership.

Russian historians, concerned with the methodological and theoretical problems of their discipline, made a habit of carrying Darwin's suggestive ideas into their own academic fields. P. G. Vinogradov, whose expertise in the social history of medieval England earned him an academic position at Oxford University, believed that the notion of evolution, as applied to both nature and society, stood out as the greatest and potentially most fertile product of the modern scientific mind. In his view, "natural science began to exercise a real influence on the study of history only after it absorbed the ideas of transformation and development and adopted the historical method."[122] The main contribution of Darwinism consisted not only of making evolution a central problem in biology but also of pointing out its significance for the social sciences. While Darwin concentrated on biological change, his followers laid the foundations for a comprehensive "philosophical explanation of the theory of evolution."[123] The future of human society, as Vinogradov saw it, depended primarily on knowledge of the universal laws of nature.

The Ministry of Public Education kept particularly vigilant control over the selection of professors who taught philosophy on the university level. During the 1890s it became clear that only persons identified with idealistic metaphysics were allowed to hold the title of professor of philosophy. For this reason there was not a single philosophy professor in Russian universities willing to match the forthright assertion of John Dewey, a Columbia University professor of philosophy, that the real strength of Darwinian influence on philosophy was in conquering "the phenomena of life for the principle of transition," and in freeing "the new logic for application to mind and morals and life." Darwin, Dewey continued, "emancipated, once for all, genetic and experimental ideas as an organon of asking questions and looking for explanations." [124] Nor did Russia produce a single university professor of philosophy ready to agree with Harald Höffding, professor of philosophy at the University of Copenhagen, who wrote that Darwin, whose contributions he valued as much as those of Copernicus and Newton, deserved full credit for pointing out the verifiable base of the notion of the origin of species. [125]

Outside the universities, the philosophical situation was very different. Idealistic metaphysicists were here, too, but so were the writers favoring a close alliance of philosophy and science. One of the more accomplished and influential of these writers was Mikhail Mikhailovich Filippov, editor of the journal *Science Review* and recipient of a doctorate in natural philosophy from Heidelberg University in 1892, with a dissertation on a specific aspect of differential equations. In 1895–98 Filippov published *Philosophy of Reality*, a two-volume synthesis of philosophy and science. He viewed philosophy as a systematic study of the basic principles and ideas of contemporary science and as an effort to create an "integrated scientific world view." [126] He operated on the assumption that his approach was the only acceptable path to sound philosophical discourse and that philosophers must rely primarily on scientific knowledge. The *Philosophy of Reality* presents the ideal of evolution as the most basic and penetrating link between individual sciences. Without evolution, the unity of scientific disciplines could not be based on concrete knowledge. Such topics as cosmogony, evolutionary paleontology, organic evolution, and psychological and sociological evolution form the heart of the study. The huge volumes stand out as a critical synthesis of modern scientific knowledge.

The study has some major flaws; for example, it needs a more precise structure of individual chapters, its selection of scientists who contributed to the elucidation of the idea of evolution tends to be fortuitous, and it does not give a clear picture of the relative significance of individual subsidiary topics. Despite these omissions, the study was a major success, primarily because it helped counterbalance the mysticism of academic philosophy. It introduced Russia to the fresh rays of modern philosophy grounded in science and opposed to metaphysical obscurantism. The *Philosophy of Reality* depicted the triumph of the evolutionary principle in modern scientific thought, introduced Darwin's work as a fundamental contribution to modern philosophy and to a new world outlook, and presented the main trends in the development of evolutionary thought after the publication of the *Origin of Species*. Written simply and lucidly, it gave the general reader a comprehensive idea of evolution as the cornerstone of modern secular thought.

Filippov belonged to the group of evolutionists who favored Lamarckism no less than Darwinism. He saw only a limited use for the struggle for existence and natural selection in the evolutionary process, but he was ready to admit that the triumph of the evolutionary principle owed more to Darwin than to any other scientist. It was Darwin who made the universality of transformism both a scientific and a philosophical conception and who opened many new avenues for a future elaboration of the evolutionary idea. Above all, Filippov showed that negative criticism of the struggle for existence and natural selection should in no way interfere with the recognition of Darwin as one of the greatest scientists of all time. Despite the striking "weaknesses" of individual principles built into his theory, Darwin won the day because he allowed no room for "forces and impulses acting independently of the known laws of nature," which made his theory acceptable to "a majority of natural scientists." [127]

CHAPTER SEVEN

Rejection

The defense of orthodoxy during the waning decades of the tsarist era was only one side of the Darwinian equation. The other side was the rapidly growing criticism of the scientific substance and philosophical implications of Darwin's general evolutionary views. The most comprehensive and systematic criticism came from theologians, philosophers, and scientists as separate communities of scholars. Despite their distinctive features, these communities did not act as clearly separated and autonomous groups. The theologians, for example, depended mainly on philosophical and scientific arguments and metaphors. The philosophers concentrated on the fundamental intellectual issues of the epoch and on the essential realities of the human universe, but they did not hesitate to solicit help from selected domains of science or to tackle the moral dilemmas which theologians considered their rightful domain. The anti-Darwinists in the scientific community seldom operated on scientific grounds alone. As the need arose and pressure increased, they made extensive use of philosophical arguments and theological allusions. Despite the extensive crisscrossing of anti-Darwinian argumentation, each community represented a distinct integration of critical views.

Theological Criticism

Russian Darwinists did not mount an open attack on religious beliefs; such an effort would not have passed the tight censorship controls in the first place. In Germany Ernst Haeckel could write that Darwinism provided "the heavy artillery" for "the monistic war against Christianity" and get away with it;[1] in Russia the censors would have considered such a statement too seditious to appear in print. This, however, did not prevent the theologians from issuing constant reminders that Darwinism was indeed a leading modern foe of Christianity. Gradually, these reminders led to the emergence of a strong anti-Darwinian theological literature, occasionally displaying carefully thought-out arguments and logically arranged theoretical insights and intuitive suggestions. After 1890 the community of scholars included a clearly defined group of theologians devoted primarily to an analysis of ideas that dominated contemporary scientific thought in general and Darwinism in particular. Unlike the West, Russia did not produce a single theological scholar willing and ready to search for a compromise, even of the most tenuous and elementary nature, between Darwinism and Christian thought.

There were two main categories of theological criticism of Darwinism. The first category concentrated on the incompatibility of the evolutionary theory, grounded in natural causality, with the moral teachings of Christianity. It sought to ally theology with moral philosophy as elaborated in Immanuel Kant's *Critique of Practical Reason*. In *The Descent of Man* Darwin erred, according to theological critics, in trying to apply the scientific method to the study of the roots of morality and in overlooking the essentially metaphysical nature of the origin and dynamics of ethical precepts. This kind of criticism consisted mainly of old arguments expressed in modern philosophical and scientific metaphors. The exponents of this orientation operated on the assumption that the Darwinian view of the anthropoid origin of man represented a slanderous attack on the divine origin of the moral code. They challenged the competence of science to explain the origin of morality.

The second category, a relatively new phenomenon in Russian theological literature, consisted of efforts to invalidate the scientific foundations of Darwin's theory. This kind of criticism is particularly relevant to a study of the history of Russian Darwinism and deserves closer scrutiny. Theological critics were not specialists in any particular branch of science; Russia did not have scholars typified by Erich Wasmann, the German Jesuit scholar who worked in entomology and ethology as a competent scientist and who, in his criticism of Darwinism, acted as both a scientist and a theologian. Russian critics, however, possessed sufficient knowledge and skill to handle broad aspects of scientific theory and to discuss the basic issues of scientific methodology.

This type of criticism operated on two levels. On one level, theological experts tied their anti-Darwinian writing to a general criticism of Newtonian science; on the second, and more specific, level, they criticized Darwinian evolutionism in the light of post-Darwinian developments in biological theory. They welcomed the scientific revolution at the turn of the century because, in their opinion, it uprooted the nineteenth-century belief in the intellectual hegemony of science. While "old science," in the opinion of a typical theologian, had lost its credentials to serve as an authority on the structure of the universe and on the place of man in nature, new science had not yet appeared. Nor would science ever again rise so high as in the nineteenth century. All this was interpreted as an invitation to theology to fill the vacuum created by the crumbling Newtonianism. As one theological writer put it, the time had come for a broader interest in theology as the true source of absolute and universal descriptions of the work of the universe. The collapse of Newtonianism, as theologians saw it, made it possible to put more faith in "providence" than in "random change" and "natural selection." [2]

Among theoretical contributions to anti-Darwinian literature, the most imposing were general treatises covering the pivotal controversies. [3] This genre was more popular before the advent of the mutation theory, the rediscovery of Mendelian genetics, and the formation of various neo-Lamarckian schools, when the picture was much simpler and criticism of Darwinism could easily be presented as a criticism of the evolutionary idea in general. One of the most comprehensive works in this category was *The Origin and Primordial State of Mankind* (1894) by S. S. Glagolev, prefaced with a statement that it was the duty of Orthodox Christian theology to refute the theory of the anthropoid origin of man. The best way to achieve this goal, according to Glagolev, was to fight every nuance of thought that went into Darwin's theoretical elaborations. The book showed the author's considerable familiarity with current developments in biology which challenged Darwinian thought. It featured such prominent members of the scientific community as von Baer, Kölliker and Nägeli, widely known for their astute and stubborn criticism of Darwinian thought.

Glagolev was particularly eager to give his refutation of Darwinism an air of objectivity and generalizing restraint. [4] He achieved this goal by presenting his counterarguments as ideas much in need of additional empirical support. His main satisfaction lay, however, in reducing Darwin's theoretical views to the level of dubious hypotheses. He stated: "I do not claim that my arguments against the evolutionary theory are incontrovertible. But my survey shows that the origin of species has continued to be a question without an answer. There is not a single species for which the question of origin has been answered. Evolution is merely a hypothesis; only the future will tell how, and to what degree, it is true."[5]

Soon after 1900 theological criticism of Darwin's theory became more astute and penetrating. It served as the featured component of a spreading effort to build the foundations for "theoretical theology," the preoccupation of a growing group of professors at theological academies and seminaries. Theological critics took serious note of current developments in biological theory and in the flood of arguments released by the outburst of idealistic thought in philosophy. The new religious commentators concentrated on limited clusters of theoretical problems and their deeper metaphysical meanings. In comparison with their predecessors, they were more sophisticated in the use of philosophical notions, better versed in the intricacies of biological theory, and more determined to show the full collapse of Darwin's theory.

At that time theological journals—among which *Belief and Reason* and *Theological Herald* were the most erudite—often carried articles concentrating on and siding with various "anti-Darwinian" theories. Despite occasional tactical concessions, these articles were part of a unified front against Darwinian evolutionism. Theologians did not lag behind biologists in bringing de Vries's mutation theory and Mendelian genetics to the attention of the Russian reading public. They were alone, however, in bringing to Russia the "Darwinian" literature produced by Western theologians. For example, they alone wrote extensively about Erich Wasmann's *Die Moderne Biologie und die Entwicklungstheorie,* which sought a compromise between theology and Darwinism, an effort they viewed with a great deal of skepticism.[6] Works like Wasmann's helped theological writers marshal arguments against the unacceptable offerings of such specialized and highly technical fields as evolutionary embryology.

Wasmann's work represented a new phase in the evolution of Western religious—particularly Catholic—interpretations of evolutionism.[7] Instead of continuing the church tradition of fighting the evolutionary heresy, Wasmann concentrated on formulating a new theory of evolution that, he hoped, would be fully acceptable to the church authorities. Wasmann recognized two kinds of species: natural species and systematic species. All presently existing forms of life belong to systematic species, all products of a long evolutionary differentiation of natural species, the original products of divine creation.[8] This claim, he said, has two advantages. First, it can be readily verified by the data of comparative morphology, which can trace the surviving common features of natural species distributed among many systematic species. Second, it provides the only sound philosophical ground for uniting Christian and scientific views of living nature—creation and evolution. In this strategy Wasmann saw the best way to disarm the "atheists" who took it for granted that creation and evolution were mutually exclusive. With regard to "the origin of man," however, he was much less willing to compromise; he simply relied on J. Reinke's statement that science can shed no light on this matter.[9]

Wasmann's and similar efforts did not attract serious attention in western Europe, in either the theological or the scientific community. They contributed, however, to the creation of a more tolerant attitude toward Darwin's scientific theory, to a broader diffusion of evolutionary thought, and to a modernization of theological discourse. Russian theologians, by contrast, did not produce a single effort of this kind. They did not show much sympathy for Wasmann's endeavors to use evolution as a meeting ground for theology and biology. Nor did they lose much by not following Wasmann, who attracted only a little more attention in the West than in Russia. He did not appeal to the scientific community, because of his widely heralded affinity with theology.[10] Nor did his effort to reconcile creationism and evolutionism appeal to theologians.

In a lengthy discourse on Wasmann's Christian theory of evolution, Glagolev was careful not to impress Russian theologians with the need for a similar study under the auspices of the Eastern Orthodox Church. He favored Wasmann's determined effort to show that science alone could not answer the question of the descent of man. He was convinced, however, that Wasmann's empirical studies of ants supported "the theory of the evolutionary origin of species."[11] Wasmann "reconciled" evolutionism and creationism by making the scientific explanation of the secrets of life a corollary to theological explanation.

Glagolev, on his part, wanted no compromise with scientific evolutionism. He was an unbending creationist. In his attack on evolutionism, he depended largely on logical constructions built upon selected and isolated scientific premises. Most of all, however, he depended on a simple and elementary religious argument: the theory of evolution was wrong because it was un-Christian and was in open conflict with "the dictates of moral consciousness."[12]

Glagolev's long essay on Mendel's theory was primarily an attack on Darwinism. It stated that Mendel's experimental work caused a full collapse of Darwin's theory and ended the controversies evoked by the *Origin of Species*. Mendelian genetics inaugurated a new era in the history of biology—an era of harmonious relations between religion and the life sciences, between natural causality and supernatural teleology. It showed the advantages of narrowing the vision of scientific inquiry to specific and concrete problems that the experimental method was competent to answer. The success of scientific research was not in the unlimited scope of Darwinian perspectives, but in the narrow limits of Mendelian realism.[13] In elaborating these points, Glagolev noted that Mendel and de Vries had destroyed the two pivotal ideas built into Darwin's theory: the idea of continuity of change (which had no room for leaps in nature), and the idea of making every ontogenetic change part of phylogeny.[14] Mutations, in Glagolev's view, were the best proof that there were natural phenomena that were not conducive to causal explanation. He was convinced that the new orientation in biology would give eugenics a firmer base, would show that there was a passable path between "heredity" and morality, and would make teleology a link between the scientific study of life processes and the theological study of transcendental causation.[15]

Glagolev and most other theological scholars favored the Lamarckian tradition because, they thought, it provided solid arguments against Darwinism. They favored mechano-Lamarckism because it offered strong opposition to the idea of natural selection. They particularly favored psycho-Lamarckism because of its strong anti-mechanistic stand and elaborate dedication to teleology as the supreme law of living nature. To a typical theologian, the recognition of teleology meant the recognition of the supremacy of a superrational force in living nature, irreducible either to the laws of mechanics or to "the element of playfulness in natural processes," such as the randomness of variation.[16] All this did not mean that the church scholars had given unqualified endorsement to neo-Lamarckism. They did not like the strategy of viewing each organism as the locus of a Lamarckian "impulse for perfection"; nor did they endorse the notion of direct adaptation to the environment. Pavel Kapterev cast his argument within a much broader framework: he disagreed with the psycho-Lamarckian argument, particularly as N. Cossmann phrased it, that both causality and teleology are the universal explanatory principles of the phenomena of the living world. He argued that teleology actually functioned in living nature, while causality existed only as a formal principle the scientists had invented.[17] In fact, he considered teleology the real causality. He "saved" causality by emancipating it from Laplacian restrictions. To show the modern spirit of his thought, he gathered logical arguments in favor of a synthesis of Kant's "transcendental teleology" and neo-Lamarckian "immanent purposiveness." [18]

Aside from isolated criticism, the church scholars were happy, however, to make use of the rich neo-Lamarckian reservoir of anti-Darwinian thought. For the psycho-Lamarckians—for A. Pauly and R. Francé— they were the most gracious and appreciative Russian hosts.[19] These writers did not exercise sufficient care in handling the facts of science, but they did make the reader aware of some of the more intricate philosophical questions related to Darwinism. They contributed their share to the great philosophical debate that anticipated and accompanied the post-Newtonian revolution in scientific thought.

The theologians traveled a lonely road. Their articles were buried in theological journals that reached few lay readers and provoked no comment or controversy. Scientists and professional philosophers alike were consistent in avoiding any reference to their writings. Their studies showed, however, that Russian theological scholarship had entered the twentieth century with a new brand of scholars, trained to undertake a delicate and selective synthesis of religious thought and the streams of ideas produced by the ongoing revolution in science. The clerics' war on Darwinism was part of a general criticism of science, occasionally disguised as an effort to bring theological doctrine closer to scientific thought. The new theologians knew that religious criticism could not provide arguments sufficient for an objective appraisal of scientific thought. Their criticism did not affect the course of science in Russia, but it did bring church scholarship closer to the intellectual realities of modern times.

Theological critics helped broaden the base of Darwinian debate. They created an extra channel for the flow of new biological ideas to Russia. They reflected not only the theological interpretation of the Darwinian menace, but also critical views on the reign of mechanism in science and materialism in the philosophy of science, shared by a growing portion of the scientific community. If they declared war on the advocates of the intellectual hegemony of science, so did a surprisingly large representation from the scientific community. They fought the nihilist scientism, but so did many leading scientists, particularly those who sensed the coming triumph of post-Newtonian scientific ideas and attitudes.

Theological scholars viewed Darwinism both as a scientific revolution that overthrew a reigning theory in biology and as an ideological revolution that produced a fundamental redefinition of man's place in the universe. Their attack on Darwinism as a science was primarily a rebuttal of evolutionism as an ideology. Theological criticism did not impede Darwinism either as a scientific or as an ideological revolution. But it did transform the strategy of theological scholarship in two important respects. In the first place, by endorsing the theories of contemporary geneticists, the theologians abandoned the idea of condemning evolutionism in toto. Now they substituted a more guarded and limited theory of evolution for the universal and deterministic aspirations of Darwinian thought. Only with regard to human species did they reject the idea of transformism without the least readiness to compromise. In the second place, they modernized theology by engaging it in discussions of scientific themes and by directing theological criticism to a general search for the explanation of the intellectual effects of the twentieth-century revolution in science.

It would be incorrect to state that the theological inquiry into the theory of organic evolution did not go beyond efforts to adduce anti-Darwinian arguments. On rare occasions theological scholars showed a broad interest in current developments in philosophy and displayed high analytical skills and rich erudition. Pavel Kapterev, for example, published one of the most comprehensive and penetrating studies of the scientific and philosophical foundations of the psychological branch of neo-Lamarckism.[20] Disciplined loyalty to a theological-philosophical outlook did not prevent Kapterev from offering a reasonably fair and careful account of the theoretical intricacies of the representative schools of neo-Lamarckian thought. His elaborate analysis of the fundamental similarity between Schelling's *Naturphilosophie* and contemporary psycho-Lamarckism, as well as his extensive reliance on modern philosophical tools of analysis, did not add new substance to science, but it did enrich Russian literature on the philosophical dilemmas of the critical issues of natural science.

In very rare cases—none after the early 1880s—did theological writers recognize the possibility of effecting a reconciliation of Darwinian and creationist views. V. D. Kudriavtsev-Platonov was one such scholar. In a long essay published in 1883 he argued that it was an error to consider Darwin's theory anticreationist in the full meaning of the term. A full and unconditional opposition to the scriptural theory of the origin of species, according to Kudriavtsev-Platonov, came from the theory of spontaneous generation, which allowed no room for divine intervention. Darwin's theory occupied a middle position between the two extremes: it recognized the creation of the first species as a divine act, but it eliminated the role of God in the evolution of the forms of life from the original species. Kudriavtsev-Platonov thought that creationism and Darwinism were not natural enemies. He contended that it was possible to reconcile the two theories. He expected Darwin's theory, however, to make all the concessions. To be acceptable, the Darwinists would be expected to recognize the divine creation of all species, to deny evolution the power to transcend the limits of species, and to give primacy to internal causation, that is, to deny the causal preeminence of natural selection as a factor of evolution.[21]

Kudriavtsev-Platonov thought also that the time had come to reexamine Darwin's view on the place of purposiveness in the processes of organic nature. He said that nothing was more erroneous than to assume that Darwin allowed no room for purposive activities in the realm of life. In this respect, he hastened to add, Darwin's biological theory must be regarded as a negation of materialism. "Despite his inclination to emphasize the external conditions [of the evolutionary process], Darwin acknowledges the existence of an instinctive, and often a rational, striving of animals to achieve definite goals."[22] Darwin did not conceal his conviction that the higher the biologists looked up the ladder of organic forms the more alienated they became from materialism and the closer they came to the recognition of the psychological propellers of evolution. To prove his point, Kudriavtsev-Platonov cited Kölliker's statement that, in his basic convictions, Darwin was "a teleologist in the full meaning of the term."[23]

Kudriavtsev-Platonov's interpretation of Darwin's attitude toward materialism was simple and direct: by accepting "teleology" Darwin negated the sovereignty of causality as a pivot of scientific explanation, and by reducing the range of causality, he raised the authority of "individual choice" and of freedom. Darwin, however, did not complete the undertaken task: he did not detach "purposiveness" from its natural anchorage, nor did he transform purposiveness into transcendental teleology. Although he "rejected" materialism, he failed to enter the kingdom of idealism. Kudriavtsev-Platonov found Darwin useful in his effort to formulate teleology as a bridge between nature and divine authority.

Philosophical Criticism

A massive involvement of natural scientists in philosophical discourse was one of the most outstanding new characteristics of Russian intellectual culture during the waning decades of the tsarist reign. In 1898 the physicist O. D. Khvol'son wrote that a growing number of Russian scientists followed Helmholtz, Hertz, and Mach by publishing articles in which it was difficult to ascertain where physics stopped and philosophy began.[24] The mineralogist and crystallographer Vladimir Vernadskii assured his colleagues that in every phase of its history science was closely linked with philosophical thought.[25] The Kiev physicist N. N. Shiller propounded on the role of science as a tributary of philosophical thought:

> Science alone cannot provide the material necessary for the formulation of the world outlook. This material must come from all fields of human endeavor: from poetry, art, politics, morality, law, history, and religious teachings. For this reason, a philosopher must perform a broader role than a scientist: he must be not only a literary person, but also, in essence, an artist, a moralist, and a theologian, expressing the broadest meaning of the *homo humanissimus*.[26]

Academic philosophers were divided into two clearly demarcated groups. The first group consisted of professors who wrote extensively on philosophical themes but who did not teach philosophy and were not classified as professional philosophers. The second group was made up of professors of philosophy, the only group classified as professional philosophers. The first group was dominated by individuals engaged in formulating and defending "scientific idealism," a philosophy closely affiliated with science but opposed to philosophical materialism of any description. It opposed every kind of metaphysics and confined most of its discussion to the level of epistemology. The second group was dedicated to a defense and an elaboration of idealistic metaphysics, moored in mysticism and spiritualism. The two academic groups were not involved in a dialogue: generally, each group went its own way in crystallizing a philosophical attitude toward the intellectual dilemmas of the modern era. The proponents of scientific idealism placed the central emphasis on the cross-fertilization of science and philosophy. The idealistic metaphysicists, by contrast, stressed the cross-fertilization of science and religious thought.[27] Their work on the relation of philosophy to science was limited and generally unsystematic.

Many proponents of "scientific idealism" operated within the current streams of neo-Kantianism and neopositivism. The neo-Kantians demanded separate treatments for epistemological and ontological questions, and the neopositivists, strongly represented by the followers of Ernst Mach, identified the philosophy of science as an epistemological discipline, completely disregarding questions of ontological nature. The two currents combined to give Russia what it did not have before: a systematic and sustained discussion of the fundamental philosophical questions of modern science. According to one spokesman, both neo-Kantianism and neopositivism traced their historical roots to the philosophies of Locke and Hume, heavily concerned with the theory of knowledge.[28] The basic argument of both was that modern philosophy must build upon, and closely cooperate with, contemporary science. A typical representative of scientific idealism contended that, in order to become a fertile source of philosophical ideas, science must change some of its old habits and intellectual commitments. As the physicist-philosopher A. I. Bachinskii put it, science must acknowledge the existence of cosmic problems which are beyond the limits of its competence. He was under the spell of the philosophical ideas of Henri Poincaré, a French mathematical physicist with a deep interest in the epistemological crisis of modern science, who claimed that every scientific generalization is no more than a hypothesis and that every scientific question has many answers. Above everything else, science must emancipate itself from the strangulating limitations imposed by the mechanistic view of the universe.

The new idealism went under several names. The jurist-sociologist B. A. Kistiakovskii named it "scientific-philosophical idealism"; Bachinskii preferred to call it "naturalist idealism"; V. I. Vernadskii called it "scientific world view," which he clearly separated from "science." Working outside the academic community and on a more popular level, V. V. Lesevich published a series of articles elaborating a "scientific philosophy" imbued with the spirit of Comtian positivism.

The representatives of this orientation did not agree on the place Darwinism occupied in modern science. Despite the remarkable scope of his interests, Vernadskii made only parenthetical references to Darwin. This was rather strange for a person deeply involved in placing mineralogy into an evolutionary mold. His scanty references, however, never took an anti-Darwinian turn. He was inclined to think that a broader evaluation of the theoretical principles of Darwin's scientific stance should be postponed until the current crisis in physics had been settled.

In his opinion, the time was fast approaching when the revolution in physics would play a key role in refashioning the philosophical, theoretical, and methodological foundations of biology. In his later writings he was ready to suggest the need for a comparative analysis of Darwin's and Einstein's conceptualizations of time.

Bachinskii made it abundantly clear that he did not favor Darwinism. His criticism, however, was limited to one point: he made his disapproval of Darwinism part of a general rejection of mechanistic orientation in science. Working in molecular physics and thermodynamics, he was deeply involved in a critical analysis of the philosophical parameters of the fast-developing crisis in physics—and in science in general. Bachinskii found Darwin's theory a victim of Laplacian determinism and physicochemical reductionism. In his opinion, Darwinism erred in making "the hazy concept of natural selection" an integral part of the general search for a mechanization of the picture of the universe which dominated the thinking of the scientific community during the second half of the nineteenth century.[29] Darwin, he said, made himself a Newtonian by transforming random variation into a specific expression of the universal continuity of organic change. With only a passing interest in Darwinism, Bachinskii cited it as one of many examples of the failure of Newtonian science to meet the needs of modern empirical research. He brought together, but did not elaborate, the criticism of Darwin's theory advanced by the dominant streams of modern philosophy and by the new scientific studies of heredity and variation. Bachinskii was obviously impressed with the discovery of mutations in the evolution of plants and animals and of the quantum nature of radiation, the two natural phenomena that could not be accounted for by the Newtonian—and the Darwinian—exclusive concern with the continuity of motion.

In dismantling the Darwinian edifice, Bachinskii did not hesitate to seek help from the philosophy of Friedrich Nietzsche, which took him outside his major concern with epistemological questions. In *Twilight of the Idols* Nietzsche claimed that if the struggle for existence was a reality it was so only as an exception rather than as the rule. Not the struggle for existence but the "struggle for power"—the "will to power"—is the elemental force of nature and society. Human history is shaped by individuals with strong "will to power," typified by Julius Caesar, Cesare Borgia, and Napoleon, who appear at irregular points in history.[30] Bachinskii's endorsement of Nietzsche's negative view of Darwin's theory did not reflect the most common opinion in Russia.

Leo Tolstoy, a realist and an outsider to the academic community, expressed a more popular view. He noted in 1903 that Darwin had continued to hold his own but that he was under increasing fire from Nietzsche, whose "absurd, half-baked, muddy, and wicked" philosophy was "more in tune with the present-day world outlook."[31] N. K. Mikhailovskii, also outside the academic community, thought that Nietzsche's "evolutionary ethics"—ethics that does not recognize unchanging principles of morality—could have appeared only after the publication of the *Origin of Species*. He added, however, that Nietzsche's ethical views, dominated by an aristocratic bias, had no affinity with scientific thinking.[32] When Nietzsche asserted that "our age is proud of its historical sense," he actually paid homage to Darwin, who played a key role in making historicism the main thrust of nineteenth-century thought.

Kistiakovskii, a more articulate and thoughtful defender of the philosophy of scientific idealism, worked exclusively in the social sciences. Inspired by the ideas of Wilhelm Windelband and Heinrich Rickert, the neo-Kantians of the Baden school, and intimately familiar with the burgeoning sociological theory of Max Weber, he tried to show that there were not only fundamental similarities but also fundamental differences between the natural sciences and the social sciences. The idea of evolutionism, built into the Darwinian tradition, presented fertile leads for the development of modern sociology. But it had serious limitations as well: in its emphasis on the conditions of natural environment and on randomness of variation in plants and animals, the evolutionary theory underplayed the role of conscious action, particularly on the human level. Kistiakovskii was particularly eager to point out the limited usefulness of naturalist models in sociology. His general argument was that the effort to equate social development with organic evolution produced more problems than it could resolve.

According to Kistiakovskii, Darwin must be credited with making evolution a major substantive area of scientific inquiry, but not with making evolution a law of nature. He said, in fact, that "evolution" and "scientific law" are mutually exclusive notions, and that it makes no sense to talk about evolution as a law of nature.[33] Evolution, he said, is always limited in space and time; scientific law, by contrast, must apply to regularities of natural processes unlimited by space and time. If biology and sociology expect to formulate universal laws, unbounded by space and time, they must look for them outside the realm of evolution.

The main thrust of Kistiakovskii's argument was that evolution required much clarification before it could become a useful scientific concept, and that this clarification must come from epistemology rather than from metaphysics. Indeed, he belonged to the group of Russian idealistic philosophers who dealt most systematically with the irrelevance of metaphysical debate for the advancement of scientific theory. His interpretation of evolution was considerably narrowed by his primary interest in sociological considerations.

Metaphysical idealism, the second category of academic philosophy, drew strong support from a variety of closely related groups united by a firm adherence to the intellectual legacy of V. S. Solov'ev. Favored by the government, this orientation received effective backing from the philosophical output of a group of former Marxists, represented by N. A. Berdiaev and S. N. Bulgakov, and from the harbingers of New Christianity, led by Dmitrii Merzhkovskii, strongly inspired by the aristocratic bent of Nietzsche's philosophy. These metaphysical orientations, and their offshoots and close allies, exercised full control over the teaching of philosophy at the leading universities.

The leading university professors of philosophy—from A. V. Vvedenskii and S. N. Trubetskoi to N. O. Losskii and L. M. Lopatin—worked on making metaphysics the most fundamental and comprehensive science. They were involved in integrating various strands of spiritualism and mysticism into the propositions of a grand philosophy in which the limitless flights into the world of transcendental imagination went hand in hand with the exacting demands of formal logic. S. N. Bulgakov, outside the university community, welcomed the dedication of the new metaphysics to the task of examining critically all theoretical postulates of science and of harmonizing scientific knowledge with the guiding values and ideas of modern society.[34] He represented the mainstream of the new idealism when he said that one of the basic functions of metaphysics was to act as a clearing house for scientific knowledge—as a superior judge ruling on the acceptability of current theories. Metaphysics filled the gaps in scientific knowledge by resorting to methods of transcendental reflection, "mystical empiricism" (Losskii), and intuitionism. Despite this ambitious task, representatives of all orientations in idealistic metaphysics either were dilettantes in science or, as typified by Bulgakov, did not have the vaguest idea about the propelling forces and the broader meaning of the ongoing revolution in science.

Russia did not produce a single metaphysicist in the style of Henri Bergson, who combined a sensitivity for philosophical nuance with a broad understanding—and a firsthand study—of the theoretical intricacies of modern biology. The value of Bergson's *Creative Evolution* was in alerting the scientific community to the realities of nature, which must become part of scientific inquiry even though they escaped the formal pathways of Aristotelian logic and the mechanical props of Newtonianism. Russian metaphysicists preferred to believe that the current revolution in science, as a fundamental attack on Newtonianism, was actually an attack on the exaggerated role of science in the modern world outlook. Bergson's *Creative Evolution,* by contrast, is a celebration of the emancipation of science from the rigidity and narrowness of mechanistic views, and of the rapid expansion of scientific vision. Bergson's ideas attracted the attention of Russian biologists as a challenge deserving serious consideration: in 1914 *New Ideas in Biology,* a collection of articles on current developments in biological theory, devoted a special article to a survey of Bergson's suggestive ideas for the future development of biology.[35] By contrast, the scientific community found the Russian metaphysical criticism too irrelevant to deserve attention.

After the publication of William Stern's *Person and Thing* in 1907 and Henri Bergson's *Creative Evolution* in 1908 the ideas of "philosophical evolutionism" attracted the attention of isolated scholars interested in the philosophical aspects of the social sciences. By writing an essay on Stern's views on evolution, S. L. Frank became one of the very few Russian scholars to make an effort to formulate a metaphysical theory of evolution.[36] Preoccupied with the assumed collapse of the mechanistic orientation in biology, he concentrated on energeticism and neovitalism as the harbingers of new biological concerns. Energeticism, he said, destroyed "the fundamental principle" of the mechanistic view, according to which the motion of material particles explained all natural phenomena. Neovitalism reversed the course science had followed since the time of Bacon and Descartes: it introduced a teleological view of nature which made it impossible to reduce life to physical phenomena.[37] Frank noted that energeticism and neovitalism were not isolated phenomena but organic components of a broad view of life.

Clearly under the spell of Stern and Bergson, Frank thought that the evolution of life depended on two natural processes: self-preservation and self-creation. Darwin, he said, erred in tying the struggle for existence only to self-preservation while overlooking the very essential self-creation. Frank said: "Look at the struggle of plants in a dense forest, or the struggle of man in human society: an unbiased observation will show you that the struggle is not limited to the naked facts of existence—it elevates and expands life by giving it new power and creativity. Nietzsche's formula 'Wille zur Macht' gives a more exact characterization of this process than the Darwinian-Schopenhauerian formula 'Wille zum Dasein.'"[38] In the spirit of the central idea of Bergson's Creative Evolution, Frank emphasized the internal nature of the "creative" power that makes evolution independent of the external environment. He went one long step beyond Bergson: he asserted that the ideas of self-creation and self-preservation applied to the physical universe as well, but he made no effort to explain his assertion. By making "creative evolution" an "internally caused" and "universal" natural force, Frank conjured up an evolutionary schema dominated by theistic messages.

Frank was also mindful of the more recent developments in biology, without making specific references to them. In his opinion, the idea of organic evolution included, among other elements, a specific feature of the mutation theory: it allowed for a succession of alternating periods of high and low creativity in the development of living forms. He accepted mutations because he found them much easier to reconcile with creationism than with Darwinian random variations; each mutation may be interpreted as a discrete act of creation, which did not apply to Darwin's "slow and slight successive modifications."

Frank rejected both Darwinism and Lamarckism on the ground that they overemphasized the role of external environment in the development of living forms. He readily admitted, however, that he found Lamarck's theory more acceptable than Darwin's. Darwin viewed the struggle for "self-preservation" as a "passive" adaptation of organisms to their environment. Lamarck, by contrast, viewed it as both a "passive" and an "active" adaptation, the latter form guided by a predetermined direction in the evolution of organisms.[39] Lamarck came much closer to uniting "self-preservation" and "self-creation" processes.

There were university professors who did not teach philosophy but who went a long way toward recognizing metaphysics as a legitimate, indispensable, and superior mode of inquiry. One of these was P. I. Novgorodtsev, a professor of law with a strong flair for philosophical speculation. He suggested a "new evolutionism" as a synthesis of Darwin's theory of natural selection and Bergson's theory of "creative evolution" which placed more emphasis on an innate drive for change than on external influences. In no form, however, did evolution, as he saw it, apply to the development of human society. In opposition to Spencer, who viewed evolution as continuous rearrangement of "primary elements," Bergson and Stern, according to Novgorodtsev, viewed it as a constant creation of new forms and "new phenomena"—as a truly creative process.[40] Novgorodtsev tried to "save" evolution by making it a metaphysical rather than a scientific concept; he treated natural selection as a factor of secondary importance, and he raised "internal" teleology above "external" causality.

Novgorodtsev took it upon himself to show the inadequacies of evolutionism in jurisprudence. Fighting for a revival of "natural law," he claimed that the latter transcended the limits of evolution and historical relativism. He rejected sociological evolutionism in general and had no use for Darwinism as a source of models for legal theory. Assuming a metaphysical position, he argued that the ethical essence of law is given before the evolution or history of legal institutions and practices begins. "Natural law" forms the "internal" core of legal norms. It is independent and above the growth of concrete legal practices, the so-called positive law.[41]

Novgorodtsev's search for an approach to law, linked with "idealistic philosophy" and concerned with transevolutionary legal abstractions, came immediately after the government had begun to suppress the sociological and historical orientations in jurisprudence. In 1899 the government abolished the Moscow Juridical Society as punishment for the Society's disproportionate interest in the social dynamics and evolution of law. This Society ran into trouble with the government because it encouraged a study of law as an instrument of social change, not merely as a tool for preserving the existing social order. Novgorodtsev elevated the role of metaphysics as the theoretical core of jurisprudence and, at the same time, scoffed at the inductionism of the Darwinian evolutionary approach, which in his opinion could not illumine the universal essence of "natural law." The authorities favored metaphysical speculation about the universal aspects of law not directly and concretely related to the Russian realities over critical studies of law in daily operation.

N. A. Berdiaev, who was not a professor, held similar views. As a young Marxist philosopher he was generally impressed with the grandeur of Darwin's scientific contributions. Darwinism, he wrote, was "essentially a progressive theory." In an effort to make it an integral part of its ideology, the bourgeoisie subjected it to extensive changes of a corruptive nature.[42] As his Marxist allegiance began to wane at the turn of the century, Berdiaev came closer to a neo-Kantian stance in philosophy, particularly in ethics. Neo-Kantianism, in turn, was a brief stopover on the way to a metaphysical orientation dominated by mysticism and spiritualism. Evolutionism, as Berdiaev saw it during the final phase of his intellectual odyssey, does not explain the origin and the universality of moral values and duties; it deals strictly with the adaptability of ethical ideals to specific historical conditions. While ethics deals with absolute values, evolutionism is a mechanism of historical relativism. Ethical norms—moral values—are gradually realized in the life of mankind: social development makes them part of actual or empirical reality. "Ethical norms, like the laws of logic, cannot evolve: they are unchangeable. What changes is the degree of our proximity to them."[43] The origin of law is in universal moral values, which are realized in, but are not created by, social reality. Society does not determine ethical norms: it determines only the degree to which these values are realized at a given point in history.

In his discussion of ethics and evolution, Berdiaev actually elaborated on—and expressed in standard metaphysical terminology—a statement the novelist Leo Tolstoy had made somewhat earlier. In an article published in 1895 Tolstoy relied on religious arguments to refute two ideas built into Darwin's theory. First, he scoffed at the assertion that the struggle for existence is a major factor contributing to moral progress. In his view, the idea of the struggle for existence represents a denial of the moral fabric of human society. Second, he refused to accept Huxley's effort to interpret moral progress as a result of social progress: rooted in religious sentiment, moral principles, as he saw them, are independent of social dynamics.[44] Religion, he wrote, defines man's relation to the infinite universe; morality is the conduct of daily activities in the spirit of this relation. With a single move, he placed the world of moral principles outside both the struggle for existence and social evolution. Moral rules regulate human society but are not created by it. Nor are they restricted by the laws of evolution, natural or social. Unlike Berdiaev, Tolstoy did not engage in a metaphysical exercise; nor was he particularly interested in defending the standard assumptions of idealistic metaphysics.

Typical university professors of philosophy and their external allies—represented by Bulgakov and Berdiaev—emphasized the intellectual superiority of idealistic metaphysics over science. Without exception, they avoided making favorable statements about Darwin's contributions to science and to the modern world outlook. The metaphysical criticism of Darwin's theory followed several distinct courses.

The champions of metaphysical antievolutionism directed their main guns at the mechanistic orientation of Darwin's theory. All claimed that Darwin was wrong in making the study of "external causality" the key to true knowledge and in basing his theory on the notion of the unity of the living world. He "erred" in compressing "the universal development of organisms" into a single mechanical model in which discrete, minute, and random variation is transformed into evolutionary continuity. He disregarded purposive or teleological processes in living nature. By ignoring internal causality of organic evolution he treated organisms strictly as passive participants in the transformation process. By fastening his general theoretical orientation to mechanistic philosophy, Darwin made his theory of evolution an organic part of "natural science materialism" and an ideological pillar of social democracy.

Many representatives of idealistic metaphysics argued that Darwinism suffered from the common ailment of "natural science materialism": it asked questions science was not equipped to answer. In their view, a successful study of organic evolution must be preceded by an explanation of the origin of life—the question that science has no tools to handle. Because of this deficiency, Darwin was forced to depend on incomplete and unconfirmed information. He transformed untested hypotheses into the laws of nature. He was found guilty of overextending the logical formalism of the inductive method to make up for the absence of empirical data. Darwin was found guilty on yet another count: he presented a theory that was scarcely more than lightly disguised Lamarckism, whose basic premises Cuvier had subjected to "devastating criticism." Guided by this kind of criticism, the philosopher B. N. Chicherin wrote:

> Since Darwinism has dominated modern minds to such an extent that it has become a reigning theory in biology, it would seem too daring to challenge its scientific foundations. A careful study, it seems, cannot lead to any other conclusion. Darwin has assembled vast factual material on which he claims to have based his theory. But this is only a mirage. His theory does not contain an iota of scientific truth. He selects only those facts that justify various *possibilities* supporting an entire chain of false logical constructions.[45]

Darwin's theory, in Chicherin's view, has no factual basis. The struggle for existence has no empirical foundations. The idea of the evolution of organic forms as an infinite process is a monstrous result of leaps in logic.[46] While it may work in geology, the assumption that the same causal factors have operated in all geological eras does not apply to biology. Chicherin rejected Bobretskii's claim—in *Foundations of Zoology*—that, thanks to Darwin, "the theory of descent occupies the same place in biology that the law of the conservation of energy occupies in mechanics and physics."[47] Bobretskii erred, he said, in equating an established law of nature with an unfounded hypothesis. Whereas Helmholtz, one of the fathers of the law of the conservation of energy, depended on the rigor of mathematical proof, Darwinists could rely only on "fictitious genealogies." Darwin ignored the presence of omnipotent spirits in nature, which compelled him to bypass a deeper inquiry into the universal causes of the evolutionary process. This criticism rested on the assumption that natural causation provides insufficient mechanisms for the understanding of the universal laws of natural processes.

Chicherin rejected Darwin's theory, but he did not reject the idea of evolution in general. He based his view of evolution on two general assumptions. First, he introduced a modified version of creationism: God created, not individual species, but the physical and chemical components which the long process of evolution molded into distinct forms of life. By modifying them, he accepted both creationism and evolutionism. No Russian theologian was ready to take Chicherin's suggestion as the starting point for a serious discourse. Second, in upholding a teleological view of organic nature, Chicherin stuck steadfastly to the idea of evolution as a finite process. Teleology, he claimed, was incompatible with the Darwinian idea of evolution as an unending process. In his view, the evolution of each species is a gradual realization of a design, and as such, it has a beginning and an end.

As a defender and articulator of idealistic metaphysics, Chicherin was the lone representative of a unique orientation, radically different from the Solov'ev tradition. Solov'ev considered the theoretical contributions of science too prosaic to be included in the intellectual mainstream of the age. He viewed metaphysics as a source of ideas revealing the poverty of—and going far beyond—scientific thought. Chicherin, in contrast, did not disparage science. He wanted to add new strength to science by making metaphysics a source of scientific wisdom. As he envisaged it, science is built on three methods: mathematical, experimental, and metaphysical.[48] In his lengthy and tedious explanations, however, he was interested much more in defending the intellectual supremacy of metaphysics than in adding new strength to science. When he wrote about "science"—for example, he offered a metaphysical recasting of atomic theory, Mendeleev's periodic law of elements, and Cuvier's catastrophism—he leaned heavily on scientific ideas prevalent during the first half of the nineteenth century. The scientific community met Chicherin's forays into "science" with complete silence. No evolutionary biologist referred to his anti-Darwinian arguments.

Individual defenders of metaphysical views criticized Darwinism as the main source of "sociological materialism," which reduced the dynamics of human society to material—economic and technological—influences. Bulgakov believed firmly that Marx's historical materialism was a sociological transposition of Darwin's struggle for existence as a biological notion.[49] There were also critics who thought that the Darwinian notion of the struggle for existence was a biological translation of a sociological principle. This charge, however, did not owe its origin to metaphysical speculation alone. It came from new developments in sociology—from Max Weber no less than from Georg Simmel—as well as from the swelling ranks of Marxist revisionists. There was one difference, however. The new metaphysicists were not content merely to point out the difficulties built into evolutionary sociology: they wanted to show the futility of sociological theory in general, particularly when compared with the limitless potential of metaphysical analysis.

A typical defender of idealistic metaphysics was not sure that the evolutionary idea, Darwinian or not, belonged among the keys that unlocked cosmic secrets in the first place. Accustomed to making announcements about absolute truths—truths unaffected by the whims of history and the ravages of time—he was strongly predisposed to place evolution among the sources of uncertain and, at best, ephemeral knowledge. A defender of the new idealism was infuriated with Timiriazev's claim that the study of evolution provides the safest path to the understanding of "the harmony and beauty of nature."[50] Rejecting evolution on scientific grounds, he refused to transfer it to the domain of metaphysics.

Academic metaphysicists—unlike theologians, who formed a much larger community of scholars—did not have a single person dealing exclusively, or primarily, with scientific problems. In general, philosophers of all persuasions showed relatively little interest in Darwinism. They were more attracted to the revolution in physics which opened the theoretical foundations of all major sciences to a thorough philosophical reexamination. No Russian philosopher at this time had enough technical competence to undertake a serious analysis of Darwinian or post-Darwinian biological theories of variation and heredity.

The professors of philosophy—unlike theological critics of natural science "materialism"—were not isolated from the scientific community. The creation of the Psychological Society in Moscow, dominated by university professors, provided a general forum for the cooperation of scientists and philosophers united by a strict adherence to metaphysical idealism.[51] During the 1890s the Moscow Mathematical Society and the St. Petersburg Society of Naturalists boasted small enclaves of scientist-idealists, all dedicated to the proposition of making science an intellectual subsidiary of spiritualistic metaphysics.

Anti-Darwinism in the Scientific Community

The scientific community was not only the bastion of Darwinism but also the main source of anti-Darwinian criticism. In most cases, this criticism was more subtle than that generated by theologians and academic philosophers, but it was also more penetrating and more diversified. In some notable cases it was as unyielding as the anti-Darwinian tirades of the more combative philosophers and theologians. The anti-Darwinism generated by the scientific community made extensive use of metaphysical and ethical arguments. The biologists who criticized individual propositions of Darwin's theory solely on scientific grounds refrained from joining the ranks of anti-Darwinists. Although R. E. Regel' and Iu. A. Filipchenko, for example, were clearly unwilling to treat natural selection as a prime mover of organic evolution, they showed no hesitation in expressing their great admiration for Darwin as the true pioneer of evolutionary orientation in modern biology.[52]

The criticism of Darwinism generated by the scientific community but placed into ideological channels added a distinctive parameter to modern antievolutionism. A. A. Tikhomirov provided a classic example of this kind of criticism. A professor of zoology at Moscow University, Tikhomirov was noted for his anatomical and embryological studies of the silkworm and for his contributions to silk technology, which earned him a faculty position. His original work on artificial parthenogenesis attracted international attention. Gradually, he abandoned experimental work altogether, moving to a very active engagement in the defense of autocratic ideology from mounting attacks staged, he thought, by "science materialism" in general and Darwinism in particular. His lectures at the university were anti-Darwinian harangues, dominated more by moral obsessions and religious fervor than by presentation of current developments in science and philosophy.[53] Many times he appeared at government-sponsored education centers for industrial workers to present scornful and emotionally charged descriptions of the evils of Darwinism. And every time he stressed the pernicious influence of Darwinism on "Russian values" embodied in the autocratic system. His lectures and published papers were aggregates of standard scientific arguments, disconnected philosophical categorizations, and moral exhortations. His line of argumentation differed from theological criticism only in that he displayed more passion and outright bitterness. "There is renewed hope," he wrote in *The Fate of Darwinism* (1907), "that after temporary blindness, biology will return to a view of human nature that is harmonious with the striving for an absolute ideal, about which an internal voice gives us daily reminders."[54] In soliciting favorable responses from his listeners, he asserted categorically that Russia received Darwin's ideas more uncritically and with more "blind servitude" than any other country.

Tikhomirov belonged to a group of the most conservative advocates of university reforms. At a time when liberal professors fought against the university charter of 1884, which imposed serious limitations on academic autonomy, he fought against the same charter but for completely different reasons: he thought that this document did not go far enough in eliminating the last vestiges of academic autonomy guaranteed by the 1863 charter. Despite the intent of government authorities, the subsequent ordinances, as he interpreted them, combined to give the charter a more liberal interpretation.[55] The government appreciated Tikhomirov's help in consolidating its control over university affairs. His reward came in 1911. In that year all leading Darwinists at Moscow University resigned from their teaching positions. They joined a large group of professors from the same university who resigned in protest against the new oppressive measures instituted by Lev Aristidovich Kasso, minister of public education. Exactly at this time Tikhomirov received a substantial reward from the government for his patriotic war against Darwinism: he was appointed superintendant of the Moscow school district, one of the highest positions in the central administration of educational institutions at all levels.

Tikhomirov was among the very few Russian naturalists who criticized Darwin by weaving "scientific arguments" into a fabric of beliefs held together by religious maxims, and who transformed both "science" and "religion" in such a way as to make them parts of the ideology of the Black Hundreds, an ultraconservative political organization that made deep inroads into the academic community during the waning years of tsarist rule.

The great novelist Leo Tolstoy died in 1910. In 1911 Tikhomirov published a pamphlet unleashing a savage attack on Darwin and Tolstoy, the chief creators of "anti-Christian delusions" in "science and art."[56] "The biological theory currently known as Darwinism," he wrote, "is undoubtedly an anti-Christian theory, and its founder must be regarded as an enemy of Christianity." "We must remember," he added, "that the difference between Darwin and Haeckel, and his followers, is only that Darwin has tried to refute the truthfulness of the world view based on the teachings of Christ without making a public statement about it, while Haeckel has loudly demonstrated his hostility."[57] Darwin's major sin, in Tikhomirov's view, was in expressing a sympathetic attitude toward Haeckel's anti-Christian works. He reminded his readers that Darwin stated publicly that Haeckel's *History of Natural Creation* made it possible for him to undertake the writing of *The Descent of Man*.[58] In the bitter and unscrupulous attack on him, Darwin found himself in good company. Tikhomirov viewed Tolstoy as another major source of anti-Christian statements. In *War and Peace,* as Tikhomirov read it, Tolstoy preached "pantheism," which stood in direct opposition to Christianity. In *Resurrection* he made "a blasphemous attack" on everything "dear and holy" to true Christians.

S. S. Glagolev, the leading theological scholar involved in a relentless war on Darwinism, noted in 1911 that only two Russian scholars were anti-Darwinists: N. Ia. Danilevskii and A. A. Tikhomirov. By "scholars" Glagolev obviously meant "scientists." If he meant that Danilevskii and Tikhomirov were the only Russian scientists who carried out a sustained, systematic, and comprehensive war on Darwin's ideas, he was very close to the truth. Even these two scholars took time to pay homage to Darwin as a brilliant naturalist and a man of intellectual and moral integrity.

Criticism of Darwinism had difficulty in establishing a strong foothold in the Russian scientific community. This can best be explained by the continuing strength of Darwinism in the leading universities, by the categorical and enthusiastic support given to Darwin's evolutionary thought by such eminent leaders of the scientific community as Ivan Pavlov, I. I. Mechnikov, K. A. Timiriazev, and N. A. Umov, and by the unwillingness of the critics of Darwinism to undertake a sustained anti-mechanistic campaign or to transform the St. Petersburg Academy of Sciences into a bastion of anti-Darwinism.

As might have been expected, there were biologists who were regarded as anti-Darwinists by some groups and as Darwinists by others. Korzhinskii, Borodin, and Famintsyn were among the better-known scholars who did not escape such a conflicting identification. These three eminent botanists had two traits in common. First, while rejecting the "mechanistic" affiliation of Darwin's theory, they expressed a favorable attitude toward neovitalism, not as a set of acceptable answers to the key problems of organic evolution but as a source of challenging questions on the future development of biological theory. Second, they recognized the preeminence of Darwin's role in making the history of the forms of life the central concern of biology. Of the three scholars, Famintsyn was much more explicit in making his allegiance to Darwin a matter of public record. Borodin was not explicit at all—the main reason his contemporaries remembered him primarily as an anti-Darwinist. He attracted wide attention mainly by his "Protoplasm and Vitalism," a scorching attack on the exaggerated role of physicochemical analysis in modern biology and of mechanical models in the explanation of life processes. One of his biographers went so far as to assert that at no time was Borodin ready to side with the vitalist orientation in modern biology.[59]

Korzhinskii, according to Timiriazev, started his scholarly career as a Darwinist and ended as an anti-Darwinist. Perfunctory in paying homage to Darwin as a founder of modern biology, he was profuse and unbending in his anti-Darwinian exhortations. He added new fuel to the anti-Darwinian fire, which for a while appeared to have been running out of control. On the basis of a mass of published records he concluded that organic evolution takes place not through "slight modifications" in the existing forms of life but through heterogenesis—sudden and saltatory digressions from the existing types. He also claimed that "the struggle for existence and natural selection limit the number of emergent forms and prevent further variation, but in no case do they lead to the emergence of new forms." Adaptation, he said, may be a result of the struggle for existence, but it is not a synonym for progress.[60] He was careful to add that in the world of *homo sapiens* "progress does not depend on victory in the struggle for existence but is guided by an internal principle, a striving toward the ideals of truth, goodness, and beauty, which are deeply rooted in man's soul, and, perhaps, represent only a special expression of the impulse for progress that characterizes all life." Korzhinskii made no effort to elaborate on his philosophical proclamation.[61] Raoul Francé, the well-known Munich botanist and psycho-Lamarckian, was so impressed with the new formulation of heterogenesis that he called Korzhinskii the Columbus of mutationism.[62]

Criticism of Darwin's theory also came from scientists who did not work in biology. It came from all corners of the scientific community, usually in small doses. In mathematics, to give one example, most criticism came from the so-called arithmological school, founded by a group of conservative professors affiliated with Moscow University. Arithmology was based on a theory of discontinuous or discrete functions, a mathematical apparatus for a new orientation in science that went beyond the exclusive Newtonian concern with the continuity of natural processes. In rejecting the universality of "continuity," the principal notion behind the Newtonian clockwork picture of the universe, arithmology rejected the monopolistic position of the analytical orientation in mathematics—of infinitesimal calculus. The arithmological theory of discrete functions "reestablished" free will as the supreme law of nature. Arithmology was a mathematically expressed philosophical attack on causality as a prime instrument of scientific explanation. It was admittedly an effort to establish harmony between science and religious faith.[63] Its basic aim was to separate science from materialism.

N. V. Bugaev, the founder of arithmology, directed particularly sharp criticism at efforts to make the study of "continuous phenomena" the main task of biology, psychology, and sociology. In his opinion, "the theories of Lamarck and Darwin are nothing else but efforts to build biology on the idea of the continuity of change in the living world—the same idea that had reigned supreme in geometry, mechanics, and physics."[64] Darwin, arithmologists argued, found the strongest support among scientists who defended the supremacy of analytical mathematics, a tool of Newtonian science and Laplacian determinism. Darwin's major error was in making evolutionary biology a subsidiary of Newtonian science. Darwin, they wrote, "transformed" the discontinuity of random variation in plants and animals into the continuity of Newton's mechanical motion.[65] In the massive evidence presented in the *Origin of Species* they saw an exercise in "sham empiricism" and an endorsement of rigid causality and absolute determinism, which, in turn, they viewed as the fulcrum of materialism.[66]

The physicist-philosopher A. I. Bachinskii praised arithmology for its rebellion against the monopoly of continuous functions in mathematics and against Darwin's mechanistic bias. A translator of Henri Poincaré's *La science et l'hypothèse* into Russian, he identified Newtonianism as a world view expressed in infinitesimal calculus, the main tool of mathematical analysis that reduces discrete cosmic forces to continuous functions and simple dependencies. In Darwinism, as an integral component of Newtonianism, he saw part of a reigning scientific philosophy dominated by the mechanistic picture of the universe. He did not miss the opportunity to identify the discontinuities of Hugo de Vries's mutationism as a triumph of the arithmological attack on the analytical orientation in mathematics.[67] In his crusade against the "analytical world view," Bachinskii did not spare the writer Leo Tolstoy, who, in a passage in *War and Peace*, viewed differential calculus—the main instrument of analytical orientation—as the most effective method the students of human history had at their disposal.[68] Differential calculus, it should be noted, represents a mathematical response to the Newtonian principle of the continuity of motion. It searches for continuities even where they do not exist.

The arithmologists made it clear that they did not advocate a total rejection of analytical orientation—the mathematics of continuous functions. All they wanted was a recognition of both continuous and discontinuous functions, the latter ready to utilize the enormous potential of probability theory. They wanted a full recognition of both analytical and arithmological orientations in mathematics. They presented their emphasis on discontinuous functions as an original Russian contribution to modern science. By a considerable stretch of the imagination, they made Butlerov's pioneering work in structural chemistry and Mendeleev's periodic law of elements early victories for the arithmological point of view. In their isolation, they failed to note that the theory of discontinuous functions in mathematics had many pioneers. They had no rivals, however, in their effort to fasten arithmology to a political ideology that granted autonomy only to the monarch and the state.[69] The road from castigating Newtonianism and Darwinism to praising monarchical autocracy was bumpy and not perfectly logical.

The arithmologists and their admirers viewed the new mathematics not only as a powerful addition to science but also as a notable development in the broad field of philosophical attitudes. The arithmologist V. G. Alekseev noted that the attack on the Darwinian notion of the continuity of organic evolution was a frontal attack on atheism.[70] The philosopher L. M. Lopatin congratulated the arithmologists for making freedom a "philosophical imperative."[71] He claimed that the existence of discontinuities in the "inner" or "mental" domains of life provided the best proof for the presence of a force in nature that is not limited by the laws of the physical world.[72] Biology, unlike physics, must be mindful of the unlimited powers of a supreme intelligence. A. I. Vvedenskii, another philosopher, greeted the arithmological crusade as a "turning point" in the growth of modern mathematics.[73] The theologian S. S.

Glagolev commended arithmologists for showing the full unity of theology and mathematics, the most perfect of the sciences.[74]

While a strong wing of metaphysicists accepted arithmology with utmost enthusiasm, the leading mathematicians chose to ignore it. The mathematicians wanted no part of an orientation that identified the defense of free will with the defense of autocratic values. Aside from the heavy and obvious ideological entanglement, for which Alekseev and Nekrasov were awarded high government positions, arithmology represented a sensible but insufficiently elaborated effort to meet the modern need for systematic work on the mathematics of discrete or discontinuous functions.[75]

The primary battle against Darwin was carried out on ideological rather than on scientific grounds. The ideologues used scientific arguments to achieve ideological results. In the 1880s—as typified by N. Ia. Danilevskii—they criticized Darwin's theory because it "violated" the cardinal principles of Newtonian continuity and Laplacian causality. Darwin's random variation was no match for the majestic sway and calculable regularity of Newtonian gravitation. At the beginning of the twentieth century the critics aimed their fire in the opposite direction: now Darwin's fault was not in ignoring the guidelines of Newtonian explanation but in fitting himself too snugly into the Newtonian fold. At the same time he was found guilty of dogmatic adherence to the supreme authority of causal explanations and of disregarding the element of discontinuity in natural processes.

Neovitalist criticism of Darwin's theory flared up during the 1890s. After 1900 it survived in the scientific community primarily as a component of the organicist orientation, which claimed historical roots in, but not full identity with, the philosophical views of Aristotle, Leibniz, and Schelling, and which professed close ties with the ideas of Bergson and Driesch. The chief representatives of the organicist orientation were the Moscow University histologists I. F. Ognev and V. P. Karpov, teacher and disciple. Ognev started the process of developing the new orientation in a critical review of neovitalism published in 1900. While rejecting the monopoly of mechanical models and interpretations in biology, he at the same time criticized neovitalism for unpardonable lapses into mechanistic patterns of scientific explanation. Mechanism, he contended, was clearly on its way out as a unitary orientation in biology, but neovitalism was in no position to replace it. Neovitalism did not become a fully developed and independent orientation; it did not achieve a full emancipation from mechanism.[76]

Ognev did not expect neovitalism to replace mechanism fully, but he was ready to announce that mechanism had already surrendered its prerogatives as a unitary orientation in biology and had become "only one of the possible methods of studying nature."[77] The period of the absolute reign of the mechanistic world outlook was giving place to a "period of synthesis" of different views. Ognev felt that he lived in a period of transition: "The altars of the old gods have been destroyed; fire is still burning in them, but their gods have only a few believers. New gods have not yet arrived to take the place of old gods." He made no effort to anticipate or to elaborate the details of the philosophy of the new gods. This did not prevent him from making it abundantly clear that he regarded Darwinism as a system of theoretical and methodological principles encased in a mechanistic world view and much in need of a broad philosophical and scientific recasting.

In a later essay, Ognev endorsed Emil du Bois-Reymond's classic discussion of the cognitive limitation of science in general and of Newtonian mechanism in particular. He stated that the laws of organic nature are incompatible with the mechanistic world outlook, and that Darwin's effort to read "purposiveness" into natural selection is far from convincing.[78] He reaffirmed his belief that the future of biology is neither in the supremacy of mechanistic views nor in the reign of vitalism, but in a new orientation that would be equipped conceptually and methodologically to handle the universal aspects of both the structure and the evolution of life. Unsure about the course of the future development of neovitalism, he was only certain that the days of mechanism were numbered.

Vladimir Karpov gave the organicist effort wider scope and more depth in a series of studies from 1908 to 1913.[79] After a detailed and critical analysis of the ideas of Driesch, Reinke, Hartmann, and Bergson, he decided that the future of biology lay in a "higher synthesis" of a thesis (mechanism) and an antithesis (vitalism).[80] He was also influenced by Ostwald's energeticism, a unique criticism of mechanistic views, and by the neo-Kantianism of the Baden school. The future "general biology," as he saw it, should consist of two branches: one identified as "natural science," the other as "history." As a natural science, biology should concentrate on cosmic characteristics that unite organisms with other natural objects, whether they be celestial bodies or drops of water. This branch calls for an ahistorical or "geometrical" approach. It deals with organisms as holistic entities and as states of equilibrium in the constant flow of organic energy.[81] As history, biology concentrates on organic evolution. Karpov may be counted among the Russian pioneers

of the structuralist orientation in biology—a unique reaction to the supreme reign of the evolutionary approach.

Karpov may also be counted among the pioneers of organicism in
Russian biological and philosophical thought. Insofar as it applied to
his theoretical views, organicism had two meanings: it stressed the mor-
phogenetic and holistic unity of organisms, and it suggested the use
of the organism as a model for the explanation of the internal unity of
all cosmic objects and of the cosmos itself. Karpov could not accept
Bergson's emphasis on fundamental differences between living and inorganic nature.[82]

Karpov envisaged the "natural science" branch of biology as a discipline concerned with the organism as a network of reversible processes.
He viewed the historical branch of biology as a special discipline concentrating on irreversible processes. Under the spell of the Baden school
of neo-Kantianism, he was ready to treat the first branch of biology as a
nomothetic discipline, and the second branch as an idiographic science.
The future of biology, he said, was in synchronizing a study of the structural stability of living forms with the study of evolutionary dynamics.
He tried to cast biology within a framework that embraced a general
theory of organization and a concern with evolution. Combined, the
two orientations covered not only genetic ties but also formal similarities between organisms.[83] His arguments carried one unmistakable
message: the Darwinian or any other evolutionary study of the forms of
life was not sufficient for a full understanding of the structure and dynamics of life.

The organicists seldom mentioned Darwin and avoided a direct confrontation with his theory. Their general orientation, however, worked
against Darwinism on three strategic fronts. In the first place, it emphasized the need for abandoning the Darwinian tradition of keeping teleology out of biology. Indeed, it relied heavily on teleological metaphors
borrowed from neovitalist literature. In the second place, it favored a
reversal of the historicist trend in biology by shifting the central emphasis from the instability of evolution and transformism to the stability of
the structural features of life. In the third place, it encouraged close cooperation with relevant currents in philosophical thought, even when
these lapsed into metaphysical speculation of the most tenuous kind. At
one point, Karpov hinted at the possibility of advancing a philosophy
that would treat the universe as an organism.[84] While organicism did
not present a major threat to Darwinism, it did contribute to the growing pressure to bring Darwin's legacy closer to the conceptual and meth-

odological challenge of new developments in experimental biology—
and to the new ties between biology and physics.

Karpov's argumentation lacked precision and any consistent use of
philosophical terminology. Buried in the dim passages of his unwieldy
discussion was a two-pronged criticism of Darwinism. In the first place,
Karpov admitted the evolutionary approach to life only as a subsidiary
of the organismic or holistic approach. In the second place, he thought
that Darwinism could survive as a viable theory only by separating itself
from the models of Newtonian mechanics. Karpov found neovitalism
open to criticism as well. He thought that the neovitalists erred in push-
ing the boundaries of biology beyond their natural limits. In his opinion, the neovitalists concentrated too much on proving that an organism
is not a machine, even though the answer to the question was obvious
and required no lengthy discussion. The neovitalists, he charged, exaggerated the role of the psychological factor in biology.[85]

Organicism was a mere episode in the history of Russian attitudes
toward Darwinism. Ognev and Karpov did not occupy positions high
enough on the scale of academic prestige and honors to attract influential followers. Nor did they give their analysis sufficient depth and requisite precision. Both Ognev and Karpov eventually abandoned their
efforts on behalf of organicism—and holism—and returned to more
traditional academic work. Karpov's new activity included the writing
of a basic textbook in histology and translations of ancient Greek medical texts. He also produced a Russian translation of Aristotle's *Physics*.

In the scientific community, anti-Darwinism took on many forms
and appeared in various intensities, levels of organization, and types of
scholarship. It seldom took the form of a pure and detached scientific
inquiry. All sweeping refutations of Darwinism depended more on philosophical convictions than on carefully adduced and examined empirical
facts and scientific arguments. The anti-Darwinists in the scientific community were not ideologically united. A. A. Tikhomirov and the arithmological school made the attack on Darwinism part of a determined
effort to contribute to the preservation of autocracy and the value system clustered around it. Bachinskii looked forward to the triumph of an
ideology that rejected Laplacian determinism because of its philosophical affinity with political absolutism. He found Darwin guilty of a
Laplacian bias. Borodin criticized the mechanistic bent of nineteenth-
century science (which included Darwinism) and the positivism and materialism of Russian nihilism. His attack on nihilism did not make him a

defender of autocratic ideology; he belonged to the academic intelligentsia involved in a search for political moderation. In 1905 he was among 342 scholars who signed the famous "Note" to the government demanding extensive reforms in the educational system.[86] V. P. Karpov made neither laudatory nor unfavorable statements about Darwin's theory; what he inferred, however, was that the time had come for biology to sail to a new ocean, rich in challenges that eluded Darwin. He did not want to do away with Darwin's contribution: he merely wanted to reduce its authority to a more modest size.

CHAPTER EIGHT

Strategies for Retrenchment

The revolutionary stirrings in biology at the beginning of the twentieth century created new challenges for Darwinism—for its method, theory, and general outlook. Two groups of scholars played major roles in helping Darwin's scientific legacy survive the crisis and find a functional place in the complex world of the new biology.

The first group grew within the community of Darwinian scholars. Its members were willing and determined to press for a retreat from the more orthodox—and less defensible—positions of Darwinian thought, which, they felt, limited the scope of evolutionary exploration, followed misdirected paths of inquiry, and perpetuated critical lacunae in the systems of theoretical principles. Determined to soften the paralyzing rigidity of orthodox constraints, they did not abandon their basic loyalty to Darwin. They were "unorthodox Darwinists."

The second group consisted of biologists who responded favorably to drastic changes in theoretical and experimental approaches to heredity and to the rise and phenomenal growth of genetics, a science built on Hugo de Vries's mutation theory and Gregor Mendel's mathematical explanation of the mechanism of heredity. Immersed in biological developments outside the classical framework of the Darwinian tradition, these scholars pressed for a redefinition of the nature and evolutionary role of variation and heredity.

These two groups, each in its own way, represented the most fertile developments in Russian biology during the first two decades of the twentieth century. They played primary roles in the ongoing search for a new and more comprehensive theory of evolution. Both merit more detailed scrutiny. This chapter concentrates on the major representatives of each group, particularly on their suggestions for synchronizing the twentieth-century developments in biological thought and the Darwinian tradition, and for formulating a broader approach to organic evolution.

Unorthodox Darwinism

Unorthodox Darwinists, a clearly identified and rapidly growing group, agreed that, despite their general scientific validity, Darwin's principles required extensive amplification, refinement, and modification in the light of new advances in biology. In "useful" contributions of modern experimental biology they saw much more than the products of research Darwin had anticipated or recommended but did not pursue. In no way, however, were they ready to abandon the idea of natural selection as a decisive factor in the evolutionary process. They shared two distinctive characteristics: they did not work actively in experimental research related to variation and heredity, and they labored primarily on protecting and consolidating the preeminence of Darwin's theory. They helped stem the tide of anti-Darwinism propelled by the more extreme, and rashly generalized, claims of experimental genetics and various branches of neo-Lamarckism.

The representatives of unorthodox Darwinism, like the representatives of strict orthodoxy, regarded Darwin's theory as the quintessential explanation of organic evolution. Unlike the defenders of strict orthodoxy, however, they placed strong emphasis on the urgent need for bringing Darwin's views into a closer working relationship with major developments in modern biological theory. They recognized that modern experimental biology had made significant discoveries in the domains of life that Darwin had left unattended, and that it had produced theoretical insights contradicting some of the leading principles built into Darwin's theory. Darwinism, in the view of these scientists, faced two vital tasks: to absorb the new knowledge that filled in the lacunae in evolutionary thought; and to modify some of its own views in order to facilitate their accommodation to advances in experimental biology contradicting Darwin's theory.

Among the supporters of unorthodox views in the community of Darwinian scholars, four were particularly active: N. A. Kholodkovskii, I. I. Mechnikov, V. M. Shimkevich, and V. I. Taliev. All were united by a firm loyalty to the basic principles of Darwin's theory, by a conviction that the survival of Darwinism depended on a cautious search for cooperation with new stirrings in evolutionary biology, and by a strong distaste for neovitalism and psychological orientations steeped in metaphysical speculation. Although united in their general views on the current strategy of Darwinian research, these biologists acted strictly as individuals, guided by their specific theoretical ideas and general philosophical attitudes. Some individuals gave the cooperation of Darwinism and selected developments in the experimental studies of heredity a particularly high priority; others favored closer working contact between Darwinism and various branches of neo-Lamarckism.

Nikolai Aleksandrovich Kholodkovskii, professor of biology, first at the Forest Institute in St. Petersburg, and then at the Academy of Military Medicine in the same city, was known for his literary gifts. A translation of Goethe's *Faust* made him a national celebrity and earned him the coveted Pushkin Prize. He also translated Erasmus Darwin's *Temple of Nature* into Russian and wrote his own poetry. A noted specialist in entomology and parasitology, he wrote numerous technical and popular articles, some of them reissued in a volume entitled *Biological Essays*. His popular articles covered such topics as "parthenogenesis," "animal instincts," "human reason," "cell and nucleus," "protective coloration among animals," and "the social life of ants." He dealt extensively with the development of post-Darwinian evolutionary thought.[1]

Kholodkovskii did not present a systematic discussion of the contributions of de Vries and Mendel to the experimental study of heredity. The meteoric growth of genetics at the very beginning of the twentieth century did not attract his attention. His heavy concern with the rapidly mounting involvement in the theoretical and experimental aspects of variation and heredity did not go beyond the age of Weismann. In fact, he was the first Russian zoologist to undertake a methodical and critical scrutiny of the scientific work of Kölliker, Nägeli, Weismann, Eimer, and other leading evolutionary biologists of the early post-Darwinian era. Written in a popular style, without excessive technical involvement, his essays appealed particularly to general readers interested in the more exciting developments in modern science.

Kholodkovskii thought that Darwin's theory was more comprehensive and more deeply grounded in empirical substance than that of any other naturalist. But he was also convinced that since Darwin left some basic questions unanswered, there was an imperative need for blending his theory with selected contributions by more recent biologists. Darwin, for example, did not look into the origins of variation: he worked exclusively with variation as ready-made raw material which natural selection either rejects or integrates into the course of evolution. More recent studies, however, showed the close relationship of the mechanisms of heredity to the dynamics of variation. While Darwin asked the general questions related to the origin of species, experimental biologists showed a particular interest in the specific questions related to the origin of variation. Kholodkovskii even welcomed the rebirth of Lamarckism because of its possible complementary relation to Darwinism.[2]

Kholodkovskii combined a firm adherence to Darwin's views with a careful search for an objective and minute appraisal of the theoretical vistas opened by post-Darwinian scholars. He noted in 1895 that the time had come to admit that the idea of evolution had entered a "new phase of development."[3] The current—and rapidly growing—interest in the physiological aspect of evolution, the experimental study of the evolutionary role of the environment, the limitless potential of the physicochemical analysis of ontogeny, and the promise of experimental morphology attracted much of his attention.[4] He wrote in 1898, on the eve of great discoveries—and rediscoveries—that laid the foundations for genetics as a distinct discipline:

> Experiment, previously considered a tool used mainly by physiologists, is beginning to find a place in anatomy and embryology. In his study of animal cells, a modern researcher not only is interested in morphological characteristics . . . but tries to reduce all vital activity and every morphological structure to a physicochemical base (Otto Bütschli); he studies the influence of various chemical agents on the nucleus and on the protoplasm; he endeavors to isolate physiological activities (Demoor); and, finally, he devotes particular attention to the process of cell division, depending on the special activity of the centriole, and to the chromatin base of the nucleus.[5]

Caught in a whirlpool of scientific crosscurrents, Kholodkovskii sensed and recorded the multiple lines of development in the new biology. In 1891 he thought that phylogenetic morphology, tracing its roots to the first generation of Darwinian scholars, would continue to reign supreme for a long time. Seven years later he was ready to admit that the phylogenetic approaches, based on the intricate techniques of observation, had surrendered their preeminent role in evolutionary biology to experimental studies.[6] In all this, he operated on the assumption that in order to preserve its reigning authority in biology, Darwinism needed methodological retooling and broader theory. In comparison with new developments in biology, Darwinism had a definite edge because it covered a much broader universe and was simpler in conceptual design and more directly related to the concrete facts of nature.

Kholodkovskii, who knew more about the developments in biology during the 1890s than any other Russian professor of the life sciences, had a keen understanding and appreciation of the threads of thought that led to the emergence of new biological sciences—such as cytology and genetics—and to new research strategies. Greeting the growing specialization in biology, he welcomed at the same time the growing cooperation between various branches of biology, both old and new.[7] Such an integration, he thought, would achieve two positive results: it would recognize the growing advantage of experimental techniques in modern biology, and it would give new strength and broader meaning to Darwin's theory.

Skeptical about its current achievements, Kholodkovskii was cautiously optimistic about the future of the experimental branches of biology. He showed a particular interest in three orientations that were in the ascendant during the 1890s: Weismann's neo-Darwinism, Wilhelm Roux's "developmental mechanics" (*Entwicklungsmechanik*), and various experimental schools rooted in the Lamarckian tradition. In all cases, he was interested primarily in relating individual orientations to the strengths and weaknesses of Darwinism.

Kholodkovskii dealt in some detail with Weismann's theory of heredity as the backbone of the general theory of evolution. This theory appealed to him because it included natural selection as a bridge connecting the notion of germ plasm, the immutable and "potentially immortal" hereditary matter totally independent of external influences, with the idea of organic transformation. This did not stop him, however, from criticizing Weismann's exaggerated emphasis on natural selection. It is not advisable, he wrote in 1897, to view natural selection as "the only or almost the only factor responsible for such a complex phenomenon as organic development."[8] The rejection of panselectionism was a common trait of Russian representatives of unorthodox Darwinism.

In Kholodkovskii's view, Weismann deserved much credit for two contributions to modern biology. First, he laid the foundations for the modern scientific concern with heredity as a basic mechanism of organic evolution. His "germ plasm," a synthesis of Galton's "stirps" and Nägeli's "idioplasm," served as a basis for the modern recognition of the "material foundations of heredity."[9] Second, he showed conclusively that the theory of the inheritance of acquired characteristics had no empirical support. Kholodkovskii had two objections to Weismann's theory: it is laden with impenetrable conceptual structures—and improvisations upon structures—difficult to translate into manageable designs for empirical research, and it cannot account for parthenogenesis.

Kholodkovskii also showed a keen interest in the ambitious effort of the German professor Wilhelm Roux of Halle University to reduce experimental biological data to the universal laws of mechanics, an undertaking that Nägeli initiated in 1884 on a purely theoretical level. He thought that, at least in one respect, Roux's effort was unrealistic and rather pretentious: at the end of the nineteenth century, he contended, science was not yet ready to establish firm links between experimental data and the laws of mechanics.[10] Roux received credit, however, for advancing a corpuscular theory of heredity based on principles very similar to Weismann's. He "erred" in proposing a shift of emphasis from morphology (Darwin's base of operation) to physiology as the main science of organic evolution, and in exaggerating the advantages of "experiment" over "observation" as the main source of biological knowledge. Far from disparaging the role of experiment in biology, Kholodkovskii merely wanted to note that experiment had not yet become an efficient research tool in the study of organic evolution. Nor did physiology reach a level of perfection that would allow it to occupy a commanding position in unraveling the secrets of the evolutionary process. Darwinism continued to be a dominant approach in the study of evolution for the simple reason that no other orientation was strong enough to take its place.

Lamarckian research in heredity was the third major branch of experimental biology that attracted Kholodkovskii's critical attention. Here, too, his analysis explored the possibility of closer ties between Lamarckian and Darwinian theoretical views and research methods. In his critical review of observations by Max Standfuss in Germany and Shmankevich in Russia, he concluded that, at the current level of the development of biology, it was impossible either to prove or to disprove the direct environmental influence on the emergence of heritable characters.[11] The fruitful cooperation of Darwinian and Lamarckian approaches was feasible, but it clearly belonged to the undetermined future. Both orientations deserved credit for recognizing the stimulating environment as an important factor of evolution. Kholodkovskii favored the development of Lamarckism and Darwinism as complementary theories. While Darwinism, he thought, held the keys to an understanding of the origin of species, Lamarckism tackled the more difficult question of the origin of variation. He cited two reasons for considering

Darwinism a more productive biological theory than Lamarckism: first, it operated within a narrower frame of reference, which helped it avoid "unanswerable questions"; and second, it had no need for assuming the existence of an inscrutable innate drive that renders evolution an ordered progress.

While hailing the possibility and necessity for at least a partial fusion of the Darwinian tradition and the new branches of experimental biology, Kholodkovskii refused to accept neovitalism as a stream of thought contributing to a grand synthesis of biological thought. In this respect, he acted as a true representative of unorthodox Darwinism. In the 1890s, he thought, neovitalism offered the most belligerent and sustained opposition to Darwinism as a general and comprehensive interpretation of the natural mechanism of organic evolution. He contended that the strongest neovitalist opposition to Darwinism came from Hans Driesch. Unlike Eduard Hartmann, who defended neovitalism from metaphysical positions, Driesch was a respectable scientist who, at one point in his career, decided to widen the cognitive horizons of biology by employing neovitalist tools. He borrowed the Aristotelian notion of entelechy to describe the essential powers that separate life from the inorganic world and to help in understanding the purposiveness of the vital processes. Kholodkovskii gave Driesch's ideas a fair airing in Russia, but he also joined a large group of biologists who opposed every "metaphysical" interference with biological thought. Teleology, according to Kholodkovskii, is a "philosophical notion," and natural scientists must treat it as "unnecessary" and "useless." [12] His firm opposition to neovitalism helped create a firmer base for his dedicated search for closer functional ties between Darwinism and burgeoning genetics.

The second—and most distinguished—representative of unorthodox Darwinism was I. I. Mechnikov. He, too, recognized not only the fundamental "correctness" of Darwin's theory but also the need for widening its empirical and theoretical base. Evolutionary theory, as he interpreted it, needed more than the mere addition of details to Darwin's views: it needed new and more precise perspectives in the study of heredity, variation, and psychological adaptation. Mechnikov worked within a broad framework of evolutionary thought: his research and publications covered comparative embryology, comparative pathology, the theory of evolution, the history of biology, and the relations between biological evolution and social change.

During the entire course of his scholarly career, Mechnikov wrestled with the problem of evolution in general and with the Darwinian theoretical legacy in particular. His position was generally clear: he accepted evolution as the key to a true understanding of the structure and the processes of life, but he contended that no evolutionary theory—not even Darwin's—could explain by itself all the key aspects of evolutionary dynamics. Darwin's theory, he maintained in 1871, cannot claim the attribute of universality, for there are many aspects of the evolutionary process that it cannot account for. For example, it cannot explain why, in rare cases, "very similar" animals, living under the same environmental conditions, undergo different transformations.[13] Nor can it explain how the numerous species of the phylum Nematoda can preserve the same body organization even though they live in a wide variety of environments.[14]

At first Mechnikov concentrated on studies in comparative embryology: many of his objects of inquiry belong to the rich fauna of the Mediterranean Sea. It was also at this time that he developed a habit of writing critical surveys of current developments in zoology. His doctoral dissertation, which earned him academic employment at the University of Odessa, dealt with the embryonic growth of the crustaceans. During the 1870s Mechnikov continued to work in evolutionary embryology, but he also tried to reach a wider audience with popular discourses on the questions of evolution that he found most puzzling and, at the same time, most important for the formulation of the basic principles of a historical view of the living universe. A perceptive historian of science, he now tried to elucidate the intellectual roots of the notion of biological evolution. He confronted the question of evolution outside the traditionally defined domain of biology. The riddle of the sociological meaning of the struggle for existence as an evolutionary mechanism of human society did not escape his serious consideration, nor did he overlook the complex relation between the biological causation of conflict and the moral imperative of cooperation. While he did not completely rule out conflict as a mechanism of social evolution, he had little use for Malthus's demographic formula. He also wrote about the complexities and dilemmas of extending Darwinian evolutionism to the domain of anthropology, at the time one of the most rapidly developing disciplines. All these writings carried the same message: "Science commands so few facts on the struggle for existence that . . . it is forced at every step to rely on indirect proofs and logical deductions."[15] Still, he missed no opportunity to remind his readers that, despite its glaring imperfections and need for extensive recasting, Darwin's theory was more promising than any other transformist theory.

During the 1880s Mechnikov's identification with Darwin's theory became stronger and more direct. The gradual solidification of his allegiance to Darwinism appeared at the time of his growing engagement in work on the phagocytic theory of inflammation and, in general, on the embryological foundations of evolutionary pathology. He presented the struggle between pernicious bacteria and protective phagocytes as a special case of the universal struggle for existence.[16] Mechnikov devoted much attention to explaining the adaptive mechanisms of phagocytes and bacilli during various phases of the ongoing war.

Phagocytic reactions, as intracellular defense mechanisms, are much older than secretions of extracellular antitoxins, at least according to Mechnikov. By injecting toxins into various animals, he claimed to have established that invertebrates were incapable of producing antitoxins, an activity that begins with reptiles.[17] Mechnikov drew an evolutionary conclusion about the phagocytic defensive reactions: they are both the oldest and the most basic mechanisms for protecting multicellular organisms from pernicious bacilli. He claimed that the comparative-evolutionary method unveiled not only the general characteristics of inflammation that are shared by animals and human beings, but also special characteristics, representing evolutionary stages and growing complexities of animal organization. His effort to place immunology into an evolutionary framework was a bold effort to carry Darwin's theory to a new field of biological research. He wrote at the end of the century: "The phagocyte theory, based on the principles of Darwin's and Wallace's transformism, can serve as a particular basis for the study of the organic world. [By studying inflammation] we participate, as it were, in a struggle and observe natural selection at work."[18]

In 1916 Mechnikov received the Albert Medal of the Royal Society of Arts, an English learned body, for his scientific achievement. The citation placed a particular emphasis on his work on phagocytes—and on the nature of immunity to infectious diseases. It noted that his observations of the activity of the mesoderm cells in the embryonic organs of echinoderms led to the knowledge that "the white blood-cells of phagocytes devour the invading microbes in vertebrates also," thus showing "the universal applicability of his generalization."[19] The citation stated that Mechnikov had contributed to the control of infectious diseases "more than any other living being." It was this achievement that earned Mechnikov the Nobel Prize in 1909—which he shared with the German pathologist Paul Ehrlich—and the Copley Medal of the Royal Society of London.

As Mechnikov saw it, Darwin's theory of organic evolution made a strong impact on medicine. "Illnesses," he wrote, "which are not an exclusive privilege of mankind or of a limited number of higher animals, are regulated by the great laws Darwin had discovered. The recognition of this truth has laid the foundations for comparative pathology, a branch of biology concerned with pathological processes that take place in all the phases of the entire organized world."[20]

Triumphant in his scholarly achievement and deeply involved in elaborating a philosophy of "rationalism," Mechnikov was now inclined to be much more direct and categorical in recognizing the revolutionary sweep of Darwin's theory of evolution. In 1909 he was among the noted scientists who gathered in Cambridge to observe the one-hundredth anniversary of Darwin's birth and the fiftieth anniversary of the publication of the *Origin of Species*. In a short speech at the celebration he noted that Darwin's theory made it possible for pathology to reap bountiful harvests from an evolutionary approach.[21] In a summary of the main points of his speech, Mechnikov noted:

> The truth that man is a blood relative of the animal world has become the basis of comparative pathology. With the help of the study of lower organisms it has become possible to prove that inflammation is not a manifestation of illness, but a reaction of an organism to pathogenic activities. The theory of the origin of species offers suggestions for the study of the most difficult problems of medicine, among which the questions of malignant tumors (cancer and sarcoma) occupy the first place. Medicine has no room for the assumption that . . . tumors develop from faulty starts of germ layers. On the contrary, Darwinism has shown that among lower animals that develop from germ layers tumors are always caused by parasites, which suggests a similar origin of cancerous tumors among men.[22]

E. Ray Lankester had no difficulty in detecting a strong strain of Darwinism in the phagocyte theory. He pointed out that "inflammation, as a creative process, depended on special mechanisms established by natural selection." He viewed the "Struggle of the Organism against Microbes," Mechnikov's initial report on phagocytes, as an evolutionary study guided by Darwin's ideas. The phagocyte theory, he wrote, was a major step in the advancement of modern medicine.[23]

While enriching many disciplines, Darwin's theory acquired a broader and firmer empirical support.[24] It also acquired a more precise and complete conceptual structure. In Mechnikov's view: "While it is correct to state that many points of the theory of evolution have not yet been fully explained, it is perfectly obvious that the foundations Darwin had built can be regarded as fully established."[25] This assertion, however, did not make Mechnikov a champion of Darwinian orthodoxy. Unlike Timiriazev, he welcomed the new developments in the experimental study of heredity and variation and readily accepted the possibility of a cross-fertilization of various theoretical orientations in evolutionary biology. He openly criticized "the purveyors of orthodoxy who would have us believe that science has made no progress since the time Darwin published his works." "Orthodox Darwinists," he wrote, "do not recognize the incontrovertible fact that changes in living forms are sometimes sudden leaps that skip intermediate phases."[26]

The Darwinian commemorative sessions in 1909 at Cambridge University helped strengthen Mechnikov's admiration for Darwin's scientific contributions. They also convinced him that Darwin's legacy could only be strengthened by a refashioning of its basic principles in the light of the evolutionary ideas advanced by experimental biology.[27] He was much impressed with de Vries's mutation theory, but never before was he closer to Darwin's ideas on the evolutionary role of the struggle for existence and natural selection. He viewed Darwin's theory as a turning point in the history of modern science, but he publicly chastised E. Ray Lankester for his attack on de Vries's mutation theory and on "Bateson's research in heredity based on Mendel's theory."

During the closing decades of his active engagement in scholarship, Mechnikov relied on Darwin's ideas as the most comprehensive and most viable evolutionary theory of his age. He treated Darwin's theory as a rich source of suggestions for reconciling and unifying the apparently discordant theoretical products of many branches of experimental biology. By taking Darwin's principles to a new realm of inquiry, he helped establish comparative pathology as a distinct discipline. Darwinian thought occupied the central position in his world outlook, based on a firm conviction that the growth of rationalism was the most significant index of cultural progress. A true Darwinist, he found no use for neo-vitalist inroads into biological thought.[28]

Vladimir Mikhailovich Shimkevich was a professor of zoology at St. Petersburg University.[29] His enormous scientific output dealt primarily with the morphology, embryology, and systematics of a wide representation of invertebrates, many from the phylum Arthropoda. Particularly noted were his studies of the embryonic development of individual organs or complexes of functionally related organs. Together with his assistants at St. Petersburg University, he made a sustained effort to combine advanced techniques of observation with experimental research. At the end of the nineteenth century the zoological laboratory at his university was known as the most advanced Russian center for experimental study of organic evolution. Shimkevich earned wide reputation for his essays in biological theory and for his critical surveys of current developments in the leading branches of experimental biology.

From the beginning to the end of his active engagement in scientific work, Shimkevich was a loyal supporter of Darwin's theory. In the 1890s—before the emergence of de Vries's mutation theory and the revival of Mendel's theory of heredity—he was happy to see Darwinism not only as a culminating point in the history of evolutionary ideas but also as an inexhaustible source of research perspectives. He wrote: "All research activity in embryology, comparative anatomy, and paleontology of the post-Darwin era consists of efforts to meet the goals stated in Darwin's legacy. Darwin lived to see the evolutionary theory, built on biological facts, become part of sociology, linguistics, and, finally, philosophy." [30] The future of biology, as Shimkevich saw it in the 1890s, lay in building upon the basic principles of Darwin's theory. This did not stop him from welcoming research in the mysterious world of heredity. In an essay dealing with the Spencer-Weismann dispute he looked optimistically at the ongoing engagement of experimental biology in the microscopic study of the molecular processes of heredity.[31] He reasoned that this branch of biology would start a new phase in the history of evolutionary thought.

During the early years of the twentieth century, Shimkevich conceded that the future of evolutionary biology lay in broad and progressive adjustments of Darwinism to the ideas of various new trends in experimental biology. At no time, however, did he abandon his firm conviction that Darwin's theory was the foundation of modern biology. Relying on a dispassionate style of writing and on cautious analysis, Shimkevich gave support to an orientation that saw the future of biology in clustering modern theoretical advances around a core of Darwinian thought. Although he favored the Darwinian side of the evolutionary equation, Shimkevich was ready to treat the Mendelian theory of heredity and de Vries's mutation theory as fundamental components of a general theory of organic evolution. While interpreting Weismann's views on heredity as a revival of the theory of preformation, he anticipated the search for a meeting point between epigenesis and preformation to occupy a central position in the future development of genetics. He warned, however, that the problems of epigenesis were much more accessible to experimental study than were the problems of preformation.[32] The *Biological Foundations of Zoology*, his major university textbook, introduced students to the intricacies of Mendelian laws.[33] It also presented the mutation theory, analyzing both its strengths and its alleged weaknesses.[34] The textbook gave a limpid and accurate account of Korzhinskii's and de Vries's theoretical views. Shimkevich played a major role in resisting Timiriazev's campaign to slow down the flow of ideas generated by modern genetics through the channels of scientific communication.

In an effort to unify Darwin's and de Vries's evolutionary views, Shimkevich preferred to divide all heritable characters into three categories: those that emerged suddenly, those that emerged gradually, and those that emerged either suddenly or gradually. In a special study he tried to show that the most pronounced mutations belonged to the realm of teratology. He thought that generally there is no fundamental difference between Darwin's "slight modifications" and de Vries's "mutations" except that individual species prefer one or the other method. At one time he thought that mutations are not a reality sui generis, but a specific kind of Darwin's variations.[35] Not questioning the reality of mutations, he was inclined to view "slight modifications" as the most common mode of transformation in the organic world.[36]

Despite his generally favorable outlook on current developments in experimental biology, Shimkevich protected Darwinism from basic criticism. He fought against "repeated efforts" in contemporary Russia to discredit the struggle for existence as a factor of evolution without taking note of the full spectrum of meanings Darwin attributed to it. "Every biologist knows," he wrote, "that the word struggle denotes not only rivalry, in the strict meaning of the term, but also a predatory mode of life, passive defense, and resistance to external agents—briefly, an entire network of complex relations that determine the survival potential of plants and animals."[37] He made no effort to minimize the role of the struggle for existence as one of the prime movers of human society.

Darwinism, according to Shimkevich, had three particularly strong scientific assets: it was firmly anchored in empirical knowledge; it offered promising perspectives for new research; and it received strong support from the history of biological thought. A typical Darwinian scholar at the beginning of the twentieth century dealt exclusively with the first two assets. Shimkevich belonged to the small group of biologists who studied the history of their discipline with the explicit purpose of presenting Darwinism as a high point in the gradual growth of evolutionary thought.[38]

In his university textbook on zoology and comparative anatomy, Shimkevich expressed an unwavering allegiance to the cause of Darwinism. At a time of trial and crisis in evolutionary thought, this book helped reinforce and consolidate Darwinian tradition in Russia. It helped soften the pressure generated by various new orientations that either challenged Darwin's basic principles, such as natural selection, or gave "unexpected" answers to questions Darwin raised but did not answer, particularly in the domains of variation and heredity.

Shimkevich acted as a consistent defender of Darwinism, regardless of the source of attack. In 1901 the Ministry of Public Education ordered secondary public schools to emphasize anti-Darwinian views in teaching the life sciences. Shimkevich joined a group of educators who appealed to the authorities to rescind the ordinance and who received a relatively favorable response to their plea.[39]

Shimkevich occupied a borderline position between the defenders of Darwinian orthodoxy and the defenders of unorthodox views. He deserved to be identified as an unorthodox Darwinist because he recognized that Darwinism must find a way to coexist with the great discoveries that made genetics one of the most progressive biological disciplines. Moreover, he played a major role in drawing the attention of both the scientific community and the general public to the new biological disciplines. On occasion, however, Shimkevich acted more as an orthodox Darwinist. In some papers, he interpreted the contributions of Mendel and de Vries in such a way as to make them fit into a larger pattern of Darwinian thought. He made no systematic or sustained effort to accommodate Darwin's theory to the demands of the more recent developments in evolutionary thought. One is often left with the impression that he was concerned much more with protecting Darwinism from the heretical ideas unleashed by genetics than with proposing a working synthesis of the two strands of evolutionary thought.

The list of noted biologists who supported Darwinism but opposed Darwinian orthodoxy and who contributed to the search for a more comprehensive theory of organic evolution should also include Valerii Ivanovich Taliev. A docent at Kharkov University from 1900 to 1919 and then a professor at the Moscow Agricultural Academy, he worked in several fields but was particularly noted for his studies in botanical geography and the theory of evolution. His most important contributions to botanical geography included volumes on the plant life in Kharkov province and in Crimea, enriched by a systematic analysis of the influence of human activity on the succession of vegetation covers and on the impoverishment of soil.[40] He added fresh insights to the strong interest of Russian botanists in the evolution of "plant communities," or "plant associations," in S. I. Korzhinskii's terminology. The biotic process responsible for transforming dense forests into marshland, on the one hand, and into rocky steppe, on the other, attracted much of his attention. He helped inaugurate a systematic study of the ecological effects of combined natural and artificial selection.

Taliev likened Darwin to a mountain stream. Tumbling down the precipitous gorges, a mountain stream destroys everything that stands in its way and absorbs numerous little creeks along its course. Once it reaches the flatland the stream becomes a river, rich in water but without mountain stimuli to give it the power to crush all the obstacles it encounters. As a mountain stream, Darwinism, bursting with power, destroyed every opposition, partly by converting enemies into supporters. At the end of the century it had become a river, large in size but too sluggish to sweep away the mounting challenges from unfriendly critics.[41]

In assessing the strengths of Darwinism Taliev was guided by two premises. First, he gave Darwin full credit for making one unchallengeable contribution to modern biology: he "completely destroyed the notion of the immutability of species and replaced it with the fertile idea of evolution, making it a solid basis of modern science."[42] Second, all Darwinian evolutionary principles must be reexamined and reassessed in the light of new empirical evidence and theoretical insights. Such questions as variation and heredity, whose formative processes and mechanisms Darwinism did not broach with sufficient vigor and determination, must be given high priority in evolutionary research. The time had come for Darwinism to reconcile or blend its theoretical principles with those of modern evolutionary theories that grew outside—and often in direct opposition to—the Darwinian tradition.

In "Biological Ideas during the Second Half of the Nineteenth Century," published in 1900—before the resurrection of Mendel's theory and the emergence of de Vries's mutationism—Taliev concentrated on the most common challenges to Darwinism, particularly during the 1890s. Whether he discussed the teleological and panpsychic orientation of neovitalism, the neo-Lamarckian defense of the idea of the inheritance of acquired characteristics, or the neo-Darwinist negation of any role of the environment in generating heritable variation, his primary interest was to survey the basic questions that a modern theory of evolution could no longer ignore. His analysis made it clear that neovitalism, neo-Lamarckism, and neo-Darwinism were successful in pointing out a number of critical problems that required a reorientation in biological thought, but not in providing satisfactory answers. The time had come, he said, for the formation of new approaches to the compounded dilemmas of evolution, particularly in the domain of heredity. He was optimistic that the time for new Darwins would come in the near future.[43]

At this time, Taliev sensed that the new evolutionary theory would depend on Spencer's ideas to build a bridge between Darwinism and various branches of neo-Lamarckism, and on Weismann's ideas as a link between Darwin's thought and the swelling streams of scientific knowledge that led to the founding of genetics as a distinct discipline. In his opinion, the future of the theory of evolution belonged to the study of heredity, which in turn concentrated on both internal and external sources of heritable characteristics.

Taliev returned to the idea of organic evolution in several studies.[44] In 1902 he published an essay entitled "On Purposiveness in Nature" in which he elaborated his criticism of the teleological orientation of neovitalism and defended Darwin's interpretation of purposiveness in the organic world as a factor operating strictly within the scope of natural causation.[45] In 1907 he published *Lectures on General Botany*, the first Russian university textbook to be cast completely within an evolutionary framework.

In December 1909 Taliev read a paper, "Darwinism, Lamarckism, and the Theory of Mutations," in which he acknowledged "the broad scope of the revolution in science wrought by Darwinism."[46] He rejected de Vries's alleged effort to transform the mutation theory into a general biological theory, but he was ready to admit that the idea of "saltatory variation" was a strong addition to the theory of evolution and that it introduced an "important corrective" into Darwin's theory of slight modifications. He added that mutations are not the only source of variation. Future research, he said, should illumine the role of the changing environment—a Lamarckian factor—in making mutations part of a progressive process in the evolution of the forms of life. He tried to harness both mutationism and Lamarckism in support of Darwinism as a general theory of evolution.

Taliev's doctoral dissertation, *A Study of Speciation in Living Nature,* defended at St. Petersburg University in 1915, was actually a collection of essays grounded in empirical research and concerned exclusively with the theory of organic evolution. Now he was clearly less optimistic about the promise of genetics. At the very beginning of the study, he announced that he found experimental biology a rather lim-

ited source of information on the development of species.[47] He admitted
that he found the research based on the mutation theory particularly
wanting. Nor was he favorably disposed toward the rapidly growing
Mendelian research. In his opinion, organic evolution is such a slow and
intricate process that for a long time it will continue to be beyond the
reach of the experimental method. Instead of the experimental method,
he opted for the "comparative-morphological" inquiry, the Darwinian
method, as the main source of data for his essays.

Taliev did not ignore the mutation theory altogether. He thought,
however, that it could be used most profitably as an integral component
of a larger complex of Darwinian thought. He also thought that the difference between Darwin's minute modifications and de Vries's mutations were more in semantic interpretation than in naked reality. Mutations, like natural numbers, could be interpreted either as discontinuous
(discrete) phenomena or as units of continuous processes.[48] Darwin did
not exclude "leaps" (or "sports") from his view of organic evolution as
a continuous augmentation of minute modifications. In general, Taliev
stuck to the view that evolution is a result of "infinite changes in existing variations."[49] Not mutation as a discontinuous change outside
"existing variations," but the continuous and ever-deepening process of
fission within "existing variations," is, in Taliev's view, the primary
wheel of evolution. He was ready to admit, however, that the acceptance or rejection of the mutation theory must wait for more systematic
and thorough testing.

Along with Tsinger's work on the false flax (*Camelina linicola*), Taliev's study of speciation was the only empirical study in Russian botany
devoted in its entirety to the problems of evolution. Although he
thought that "the evolutionary theory has become the meeting ground
of all branches of biology,"[50] he was not sure that experimental genetics
was an evolutionary discipline in the first place. In fighting mutationism,
in particular, he fought Korzhinskii much more than de Vries, for it was
Korzhinskii who relied on the theory of heterogenesis to conduct an
open war on Darwinism. In his references to Mendel and Mendelism
Taliev was more fragmentary but equally unreceptive. Endorsing Timiriazev's attack on Western scientists who tended to "exaggerate" the
contributions of genetics, he argued that Mendel's approach to heredity
substituted "algebraic schemes" for "an analysis of concrete factors."

His pronounced skepticism about the usefulness of Mendel's and de
Vries's contributions did not mean that Taliev had decided to fall back
into the fold of Darwinian orthodoxy. It meant only that he looked in

other directions for theoretical and methodological ideas that could inject new blood into the Darwinian tradition. For example, he used every opportunity to suggest the need for exploring the avenues of cooperation between Darwinism and Lamarckism. Several times he returned to the idea of evolution as a process of orthogenetic adaptation, which came out of the Lamarckian tradition. Darwinism, he thought, needed a toning down of its exaggerated emphasis on the struggle for existence and natural selection as mechanisms of evolution. It should also join the modern efforts to study organisms as physicochemical systems.

Taliev's book received wide attention in Russia. The *Nature* reviewer objected to its negative attitude toward the mutation theory and to its heavy dependence on Darwin's theory of the divergence of characters.[51] In the newly established *Journal of the Russian Botanical Society,* V. L. Komarov criticized the author for disregarding Mendel's theory of heredity, itself not without imperfections.[52] The main criticism came from R. E. Regel' in a twenty-four-page review in the *Proceedings of the Bureau of Applied Botany.*[53] Regel' obviously used the opportunity not only to provide a detailed rebuttal of Taliev's methodology and theoretical criticism but also to explain the basic principles and recent achievements of genetics. The review marked the first systematic and comprehensive effort in Russia to mount a counteroffensive against the foes of Mendel, de Vries, and their followers. Regel' pointed out that to study the process of speciation by disregarding the vast resources of genetics was to ignore some of the most dramatic achievements of modern biology. He made it clear, however, that he was far from claiming perfection for all information that genetics had produced. His critique of Taliev's work gave Regel' a splendid opportunity to make up for the sin of early geneticists: instead of harping on the alleged flaws in Darwin's thinking, he admitted that the foundations which Darwin placed under biology had continued to be indestructible.[54]

The supporters of unorthodox Darwinism, represented by Kholodkovskii, Mechnikov, Shimkevich, and Taliev, were remarkably consistent in their general views on the major theoretical orientations in modern biology. All recognized the vast potential of Darwinism as the fundamental framework for evolutionary theory. All expressed a firm conviction that Darwinism must come to terms with the new ideas in biology, not by abandoning its guiding principles but by making them more flexible and less authoritative. Although they were not always judicious in presenting the relative strengths and weaknesses of competing theories, they helped open the gates for new concerns and ideas in evolutionary biology. They agreed with Lamarckism on the powerful role of environment in the evolutionary process, but they contended that no empirical evidence supported direct influence of the external environment on the transformation of species. Shimkevich was not alone in acting as if he would not be surprised if future research produced evidence in favor of such an influence, at least on a limited scale. In one of his major papers on heredity, he talked much more extensively about the Lamarckian experiments conducted by Standfuss, Przibram, and Kammerer than about the achievements of genetics.[55] No unorthodox Darwinist saw the possibility for fruitful cooperation with psycho-Lamarckism, represented by Raoul France and August Pauly in Germany. Nor did they envisage any points of contact with neovitalism.

Darwinism and Genetics

In addition to unorthodox Darwinists, there was another group of biologists who exercised a strong influence on the development of Darwinism at the beginning of the twentieth century. This group consisted of scientists who labored in the new field of genetics. While unorthodox Darwinists were strongly inclined to look at the leading ideas of genetics from Darwinian positions, the geneticists preferred to look at Darwinism from the theoretical heights of their discipline. Both groups saw the future of evolutionary biology in a combination of Darwinian heritage and the ideas of genetics and related branches of experimental biology.

At the beginning of the twentieth century, two capital developments in experimental biology—the mutation theory and the rediscovery of Mendel's laws of heredity—laid the foundations for the emergence of genetics, a new science involved in the study of two cardinal aspects of organic evolution: variation and heredity. The representatives of the new discipline presented a vigorous challenge to Darwin's theory of organic evolution. But they also inspired a sustained search for a working reconciliation of Darwin's evolutionary views with the new theoretical insights. Only to isolated interpreters in the changing panorama of biological thought did the new theories appear as harbingers of an irreversible collapse of Darwinian thought.[56] One such interpreter was Evgenii Shul'ts, who favored the experimental method of the new biology over Darwin's emphasis on observation and was impressed with Hans Driesch's neovitalism. He anticipated that "the new biology" would move away from historicist orientations in biology, which, in his view, made it exceedingly difficult to design research methods capable of producing reliable empirical data.[57] Representing an isolated position in Russian biological thought, Shul'ts made no serious effort to elaborate his antihistoricist arguments.

In 1899 S. I. Korzhinskii published a paper entitled "Heterogenesis and Evolution" in which he presented an extensive survey of examples illustrating what he thought to be the key role of saltatory changes—or "sudden divergences"—in the origin of species.[58] He anticipated the mutation theory, Hugo de Vries's bold new step in the development of biological evolution. Indeed, Korzhinskii provided the first systematic and elaborate effort to advance and systematize the mutation theory. He died in 1900, before he had completed the second—theoretical—part of the study. Korzhinskii did not base his study on his own empirical—or experimental—work; he merely went through a mass of literature in search of illustrative material for his thesis. The collected material convinced him that there were two categories of change in organic nature: gradual change producing variation within the type, and heterogenetic change, made of drastic deviations, leading to the formation of new types.[59] Heterogenetic changes, in Korzhinskii's view, are fully independent of, and are not induced by, the external environment, and they are hereditary. He admitted, however, that he was in no position to identify the internal causes of saltatory changes. Heterogenesis produced both "progressive" and "regressive" evolution. The mechanism of evolution is mainly what Darwin said it is not. Korzhinskii gave his theory of heterogenesis a clear and direct explanation:

> Regardless of what their real causes may be, heredity and variation are two forces hidden in the organism; they are two tendencies in a state of mutual antagonism. Under normal conditions—that is, in races that are well established and unimpaired—heredity has absolute power in producing identical generations. Variation, by contrast, is not continual. During many generations it must, so to speak, gather energy for the purpose of overcoming the power of heredity and of giving birth to new races.[60]

One year after Korzhinskii's death, Hugo de Vries published the first volume of *The Mutation Theory*, based on extensive and prolonged experiments with *Oenothera lamarckiana*, which provided a full confirmation of Korzhinskii's theory of heterogenesis and was quickly recognized as a turning point in the history of evolutionary thought. De Vries acknowledged the universality of Korzhinskii's heterogenetic interpretation of speciation and view of natural selection as a "conservative process," which preserves the existing forms and prevents further changes.[61] Almost paraphrasing Korzhinskii, he asserted that there are two types of variation—"individual" and "mutational"—which are fully independent of each other, and that only mutational variation can overstep the species limits.[62] De Vries did not try to minimize Darwin's contribution to modern biology; he merely chose to place strong emphasis on one of the factors of evolution which did not attract Darwin's primary attention. This did not prevent him from giving Darwin credit for making organic transformation the major subject of empirical study in biology.[63] He supplied anti-Darwinists with new ammunition, but he refused to join the anti-Darwinian movement.

Mutationism, in brief, challenged Darwin's theory on three grounds. First, Darwin claimed that the continuous acquisition of minute variation is the natural process of speciation; Korzhinskii and de Vries recognized only discontinuous or discrete leaps or mutations as the sources of new species. Mutationism rejected not only Darwin's canon of *natura non facit saltum* but also the mechanical law of the continuity of motion, the paramount principle of Newtonianism. De Vries's strong emphasis on discontinuity in evolutionary processes took place in the same year that Max Planck discovered the discontinuity of black-body radiation. Second, while Darwin placed the primary emphasis on the struggle within the species, the mutationists emphasized the struggle between the species. Third, Darwin viewed natural selection as a positive process leading to the origin, development, and perfection of species; de Vries and his followers viewed it as a negative process—a process that destroys species burdened with inadaptive mutations. Outside the scientific community, mutationism was widely regarded as a total denial of Darwinism, and because of its formal semblance to creationism it was widely heralded by theological writers.

Most Russian Darwinists welcomed the new theory, but they also voiced strong reservations. A typical critic made it known that de Vries incorporated two Darwinian ideas into the mutation theory: the struggle for survival and natural selection.[64] Most biologists accepted mutation as a source of new species, but they tended to treat it merely as a minor factor of the evolutionary process. The mutation theory, they thought, complemented, rather than contradicted, Darwin's theory. When de Vries conceded that the struggle for existence determined which mutations would survive, he received commendations from many Darwinists, even though he made little effort to elaborate his views.

In 1903 O. V. Baranetskii, a well-known plant physiologist from Kiev University, published a lengthy review of several recently published studies critical of Darwin's theory, including Korzhinskii's "Hetero-genesis" and de Vries's *Mutationstheorie*. He noted that organic evolution, as Darwin portrayed it, "occupies the same place in history that axioms occupy in mathematics." Current "criticism," he said, was directed not at evolution but at various misinterpretations of its inner mechanisms.[65] By implication, Korzhinskii and de Vries did not deny the "axiom" of evolution but gave it a more precise and accurate interpretation. A scrupulous analysis of the new ideas did not prevent Baranetskii from concluding that the idea of evolution continued to be unclear and imprecise. From Darwin to de Vries, he said, the biologists had been busy trying to reduce the evolution of life to a "material substratum"—to explain the laws of organic nature in terms of the laws of inorganic nature. Baranetskii sided with "the growing number of serious scholars" inclined to recognize a special vital principle that explains the most fundamental characteristics of life.[66] He was not sure, however, that "the natural course of scientific progress" would lead to the triumph of this principle.

Commentators on the relationship of new developments in experimental biology to Darwin's theory followed many lines of reasoning. Most of them occupied a position somewhere between full acceptance and full rejection of the complementary roles of mutationism and Darwinism. Baranetskii, for example, hinted that Darwin's and de Vries's theories of evolution were not comparable empirically and operated at different levels of abstraction. While de Vries, in his opinion, operated with "critically" examined and precisely established empirical facts, Darwin operated with data that were not adequately authenticated. De Vries's theory is an empirical generalization; Darwin's natural selection is a "theoretical construction," devoid of a solid empirical base. De Vries, however, had much more difficulty in staying within the realm of natural causation in interpreting mutations as factors of organic evolution. Darwin's random variations are "slight" and occur in endless profusion; de Vries's "random mutations" represent drastic changes and occur very rarely. For these reasons, Baranetskii argued, it is much easier to accept natural selection as a mechanism of evolution. Darwin's theory retained its popularity because it did not transgress the boundaries of natural causation. It preserved its authority because it offered "simple" and "logical" explanations.[67] This did not mean, however, that it did not require major improvements. In general, Baranetskii considered both theories imperfect. What biology needed, in his opinion, was a general principle that could explain all the manifestations of life.

He was obviously under the spell of the proliferating neovitalist literature. This did not stop him, however, from considering de Vries's theory
the beginning of a new era in biology.

A. S. Serebrovskii admitted that de Vries's mutation theory contradicted Darwin's views on the mechanism of evolution; he noted, however, that the new theory harbored too many ambiguities and uncertainties to present a serious threat to Darwin's ideas.[68] Unlike Timiriazev, he
voiced a generally optimistic view of the future role of the mutation theory in the development of evolutionary biology. The future, he said, belongs to the harmonizing of "progressive mutations" with "progressive
evolution."

At first Korzhinskii's and de Vries's theories received only perfunctory
attention in Russia: they attracted attention more as challenges to
Darwin's theory than as new designs for empirical research. Their departure from Darwin's theory was so radical that most biologists required time to digest their basic research implications. Korzhinskii attracted wide attention in Russia primarily as a critic of the mechanistic
foundations of the evolutionary theory.[69] Timiriazev, Menzbir, Taliev,
and other Darwinists expressed particularly strong criticism of his views
that brought him close to neovitalism. The critics of Darwinism, typified
by I. P. Borodin and A. S. Famintsyn, were pleased with Korzhinskii's
antimechanistic stance. The first generation of Russian geneticists, represented by Regel' and Filipchenko, wasted no time in recognizing
Korzhinskii as a true pioneer of the "new biology."

Neo-Darwinism and mutationism made the rediscovery of Mendel's
theory, presented and promptly forgotten in 1865 and 1867, an unavoidable event. The rediscovery of Mendel's scientific legacy in 1900
marked a high point in the growing effort to give the evolutionary theory an experimental base. William Bateson, who coined the term *genetics* in 1906, deserved the major credit for placing the new science on
two pillars: Mendelism and mutationism. Mendelism showed that all
normal variations resulted from mathematically predictable combinations of the particulate units of heredity. Mutationism showed that
all "unexpected" variations came from internally caused and sudden
"leaps" in the makeup of the cytological base of heredity. The new science challenged the two basic principles on which Darwin's theory was
built: Mendelism challenged the interpretation of evolution as a random
process; the mutation theory challenged the idea of evolution as a gradual process. The new biology agreed with neo-Darwinism in rejecting
the notion of the inheritance of individually acquired characteristics as a

factor of organic evolution. It differed from neo-Darwinism in rejecting the notion of natural selection as a primary mechanism of evolution.

The generally cautious reaction of Russian biologists to Mendel's theories was far from uniform. In 1903 the popular journal *God's World* (*Mir bozhii*) gave a clear presentation of Mendelian laws and expressed a guarded view of their future role in the development of the theory of organic transformation.[70] Despite his enthusiastic endorsement of Mendel's ideas, I. P. Borodin, the author of the article, thought that they did not present a universal law of nature.[71] Nor did he hide his skepticism about de Vries's emphasis on mutation as the prime mover of organic evolution.[72] He devoted much more space to spelling out the substance and the logic of the new theory than to giving his criticism more depth and precision. The article carefully avoided references to the specifics of Darwin's theory. In the third edition of his textbook *The Biological Foundations of Zoology*, V. M. Shimkevich devoted a special section to a careful summary of Mendel's theory.[73] M. A. Menzbir's textbook on zoology and comparative anatomy made no mention either of Mendel or of his theory.

In 1909 N. Iu. Zograf, a Moscow University professor, produced a Russian translation of *Experimental Zoology* (1907) by Thomas Hunt Morgan. The translated work was a welcome addition to the growing literature on the state of the new biology. This was the first serious scholarly work to give a critical review of salient developments in experimental biology during the first decade of the twentieth century. It summarized the strong empirical arguments against the Lamarckian theory of the inheritance of acquired characteristics. It presented the points of contact between Mendelism and mutationism as a most promising prospect for the future development of evolutionary biology. Morgan admitted that the complexity of the process of evolution made it imperative to exercise extreme caution in relating the new stirrings in biology to Darwinism. To survive as a viable scientific theory, Darwinism, in Morgan's view, must take into account the new developments in experimental biology. He thought that Darwin made his theory so general that it could easily be interpreted as covering de Vries's view on the role of mutations in the evolutionary process.[74] All this did not mean that Morgan accepted Mendel's and de Vries's views without criticism. In current developments he saw the beginning of a new science, which would give Darwin's theory more precision and broader latitude. Implied in Morgan's arguments was the idea that, in order to realize its potential, the Darwinian notion of evolution must be modified in the light of new advances in science. The main value of Morgan's volume was in recording the questions that must be answered before the theory of organic evolution could be raised to a higher level of scientific abstraction.

Thanks to the energetic and realistically ambitious efforts of R. Regel', the Bureau of Applied Botany became the first Russian center of genetic research. In "Selection from a Scientific Point of View" (1912), Regel' offered a long and sympathetic discussion of the new frontiers of knowledge opened by the Mendelian laws.[75] He looked forward to a future synthesis of different strands of evolutionary thought. In his view: (1) mutations are the only source of primary—nonderivative—hereditary characters; (2) Mendelian laws show that the external environment cannot alter hereditary characters and that, under normal conditions, only the crossing of different races can produce secondary—derivative—variability; and (3) the struggle for existence alone determines which of the new combinations have a survival potential. "Scientific breeding," according to Regel', depends on a functional interdependence of organic processes described by de Vries, Mendel, and Darwin. In all this, the struggle for existence is strictly a conservative factor; it contributes to the preservation of living forms adapted to the environment.[76]

It is fair to say that to Regel', as to Darwin, evolution depended on three basic factors: variation, heredity, and natural selection. It is equally fair to say that he gave these factors a more modern explanation, placing a particularly strong emphasis on such achievements of modern experimental biology as Mendel's laws of heredity and de Vries's mutation theory. Although he used every opportunity to pay homage to Darwin, he dealt primarily with the contributions of genetics to the study of transmission and combination of hereditary factors. He worked hard to neutralize the current inclination of the new biologists to consider genetics an anti-Darwinian branch of biological knowledge. Regel' went so far as to claim that the notion of "species" is too ambiguous to serve as the basic taxonomic unit and that it should be replaced by a new unit, which he identified as "race" or, following de Vries, as "elementary species," based on the genetic criteria of Mendel's laws.[77]

In his appraisal of the general factors of evolution, Regel' noted the basic differences between Darwin's views and those of modern geneticists. To Darwin, he said, variation in living forms is caused by external factors, crossing, or "undetermined" changes of internal origin. Since in his time there were no experimental data to study the internal causes of variation, Darwin, according to Regel', was compelled to leave this problem for future generations. At long last, the advances in genetics during the first decade of the twentieth century made this possible. These advances, he thought, marked a full collapse of the Lamarckian environmental bias built into Darwin's theory, and a triumph for Mendel's mathematical approach to the study of hybridization. They produced extensive improvements in the physicochemical analysis of organic processes, and a confirmation of natural selection as a mechanism of organic evolution. As he saw them, mutations appear only in individual organisms. His contention that the "individual" origin of mutations and the cumulative effects of natural selection are the keys to an understanding of the transformation of living forms placed him among the early ancestors of the idea of synthetic evolution.[78]

At the end of the first decade of the twentieth century, the new theories showed clear signs of taking root in Russian science. In 1910 Mendel's classic was translated into Russian, and its ideas began to spill into the larger community. A year later N. I. Vavilov presented a paper at the Golitsyn's Women's Courses in Agriculture, reviewing the salient points of Mendel's laws of heredity, Korzhinskii's and de Vries's "mutationism," and Johannsen's theory of "pure lines." In his opinion, the rise of genetics in Russia was closely related to the rapid growth of selection stations in southern and eastern regions and to the emergence of breeders as a professional group.[79]

In 1912 a group of scientists began to publish *Nature* (*Priroda*), a monthly journal devoted to making the knowledge of current developments in science accessible both to specialists and to the educated public. The journal placed particularly strong emphasis on current developments in genetics. It published popular and relatively favorable reviews of Russian translations of Western books dealing with the mechanism of heredity and its relation to organic evolution.[80] A group of enterprising young biologists published a series of books, entitled *Bios*, which featured studies in experimental biology, especially genetics. In 1913 a leading theological journal published a detailed article on Mendelian genetics, greeting it as a step toward a full reconciliation of religion and modern science—and as a major defeat for Darwinism.[81] In the same year Iu. A. Filipchenko, a budding geneticist, was appointed instructor at St. Petersburg University with the assignment of offering an introductory course in modern genetic theory.[82] While universities continued generally to be the strongholds of Darwinism, the new institutions of higher technical education and new specialized research centers, less bound by tradition and vested academic interests, were much more hospitable to the new theories.

In 1914 the British *Journal of Genetics* published a paper on the genetic basis for the immunity of certain cereals to fungous diseases, written by Nikolai Vavilov, a young Russian geneticist. The paper considered the physiological study of fungal reactions as a source of information on the genetic relations of plant forms.[83] It clearly indicated that the Russian scientific community was ready to make original contributions to genetics. The *Proceedings* of the Bureau of Applied Botany added a new dimension to its deep commitment to the advancement of the new science of variation and heredity: it began to wage an open war on the more belligerent enemies of genetics.[84]

E. A. Bogdanov, professor of animal husbandry at the Moscow Agricultural Institute, digressed from the main line of his scholarly interest in 1914 to produce a thick volume on Mendel's theory—on its scientific content and practical applicability. Loosely organized and unwieldy, the tome did not attract the attention it deserved. A broad and astute defense of Mendel's contributions to modern science was its most distinctive feature. "Mendel's notion of evolution," Bogdanov wrote, "relies primarily on fact, observation, and experiment, and on the elimination of errors and careful verification of conclusions." Bogdanov barely touched on the interrelation of Mendelism and Darwinism. He criticized "orthodox Darwinists" for their unwillingness to recognize important points of contact between the two theories. While recognizing the universal role of selection in the evolutionary process, he stuck to Mendel's views that denied random variation as a source of new species. Bogdanov did not use Mendel's theory as an anti-Darwinian weapon; what he wanted was a recognition of genetics as a key link in the realm of Darwinian thought. Obviously referring to Timiriazev, he lamented the strong role of authority in Russian biology, which stood in the way of critical thought, the cognitive fountain of science. To dramatize his point, he linked orthodox Darwinists to D. I. Pisarev, "the idol of the youth" and the culture hero of the nihilist intelligentsia in the 1860s, who relied on the authority of Ludwig Büchner in dismissing Pasteur's bacteriological theory as a legitimate part of science.[85]

Without mentioning his name, Bogdanov criticized Timiriazev, the living link with the Pisarev era, for his unfair and abusive treatment of Mendel. He referred scornfully to Timiriazev's assertion that Darwin clearly anticipated the basic points of Mendel's theory, and that Mendelian genetics was a unique expression of "pan-Germanic aspirations" and a weapon of clericalism.[86] If genetics was a German conspiracy, why, he asked, did it develop most intensively in the United States and Great Britain? He hinted that, in his blind devotion to Darwinian orthodoxy, Timiriazev displayed a kind of hero worship that was most injurious to science. His defense of Mendel, however, did not prevent Bogdanov from expressing genuine admiration for the greatness of Darwin's achievement.

Small in number and weak in influence, the geneticists worked on two fronts: they took on the challenging task of advancing and popularizing the new theories, and they used every opportunity to stress their affinity with the Darwinian tradition. Even when individual geneticists reduced natural selection to a secondary position in the evolutionary theory, they did not challenge the gigantic proportions of Darwin's contributions to the great triumphs of modern science. Darwin, after all, made biology an empirical science, and, as Hermann Helmholtz had pointed out, he replaced metaphysical teleology by naturally evolved purposiveness—he chased the ghosts out of the realm of biology.

In their views on the factors of evolution, Russian geneticists were now clearly divided into two groups: the group attaching primary significance to natural selection, and the group placing the central emphasis on mutations and Mendelian laws of heredity.

N. K. Kol'tsov was the most dynamic and most articulate spokesman for the first group. His lavish praise of Darwin's contributions to evolutionary biology was part of a strategy to appease K. A. Timiriazev, a powerful member of the scientific community and a bitter foe of serious tampering with Darwinian thought. True to his appeasement strategy, he wrote:

> Among Russian naturalists, the name of K. A. Timiriazev has been profoundly respected for a long time. More than one generation has learned biology from his books, and his numerous students and admirers have seen in him one of the pioneers of Russian Darwinism, who has contributed more than any other person to the advancement of the evolutionary idea. . . . Naturally, the merciless criticism of Mendelism coming from such a high authority has made various segments of Russian society suspicious of the new theory: many persons now view Mendel's and de Vries's theories as anti-Darwinian orientations. In his published papers and public speeches, this writer has fought for a long time against such a position.[87]

Despite the most favorable attitude toward Darwin and Darwinian tradition, Kol'tsov never abandoned the idea that genetics stood an excellent chance of adding new substance and new theoretical insights to the very core of evolutionary thought. While paying homage to Darwin and his contributions to modern biology, he chose the powerful challenges of genetics for the central concern of his professional involvement in science. Not only did he not turn against the "physicochemical analysis" as a sure scientific path to the secrets of life, but, on the contrary, he became an internationally recognized pioneer in laboratory work that led many years later to the discovery of the macromolecule of heredity and to the triumph of molecular biology.

Kol'tsov made no secret of his profound dedication to the cause of modern genetics and experimental biology in general. This dedication did not stop him from placing natural selection, which he identified as "Darwin's law," on the same plane with Newton's law of gravitation.[88] Just as Newton's law made it unnecessary to rely on miracles in interpreting the order of the physical universe, so "Darwin's law" took the whim of miracles out of living nature. "Darwin's law," however, did not reduce the importance of Mendel's and de Vries's theories for an understanding of organic evolution: these theories opened the previously inaccessible mysteries of living nature to scientific analysis and harnessed the tools of experimental study in support of evolutionary theory. Kol'tsov viewed evolutionary theory as both a world outlook and a scientific axiom. Darwin's major contribution, he thought, was in bringing about the triumph of evolutionary theory as a world outlook; the major contribution of Mendel and de Vries was in giving firmer footing to the scientific theory of evolution.[89]

Iu. A. Filipchenko was the most eloquent and the most advanced representative of the other extreme in the views of Russian geneticists on the place of Darwin in modern evolutionary biology: he thought that natural selection was not the primary wheel of the evolutionary process. He was firmly convinced that the study of the mechanism of heredity held the primary key to the understanding of organic evolution (which was not true in Darwin's case) and that the basic answers must be sought outside the framework of Darwinian theory. He contended that the causes of variation—the building blocks of evolution—are inside organisms rather than in the outside environment; he was an unswerving advocate of autogenesis.[90] Nor was "simple [nonmutational] variation," which Darwin had in mind, the real cause of evolution, for it was not hereditary. Relying on the experiments of Mendel, de Vries, and Johannsen, Filipchenko concluded that neither natural selection nor "slight modification" was a prime mechanism of the transformation of species.

He admitted, however, that all this was not sufficient ground for turning against Darwin; it merely meant a shift of emphasis to the more general aspects of Darwin's contributions. The more the idea of natural selection and "slight modification" retreated to a secondary plane of evolution, the more biologists were inclined to give Darwin credit primarily for transforming biology into an inductive science, for recognizing natural causality as the exclusive mechanism of scientific explanation in the life sciences, and for establishing evolution as the supreme integrative principle of biological knowledge.

Filipchenko represented the biologists who contended that "natural selection," "the struggle for existence," the stochastic nature of variation, and the continuity of change were not Darwin's major claims to fame in the first place. Darwin's major contribution, as he saw it, was in making evolution an unchallengeable fact of science and in elevating biology to the level of a true natural science. Even some of the most ardent Darwinists now admitted that the theory of natural selection did not preclude the possibility of alternative or parallel theories of biological evolution. Darwin's theory produced an indisputable result: it brought forth the "permanent and irrevocable" downfall of the idea of the static nature of living forms.[91]

In *Heredity*, published in 1917, Filipchenko consolidated his firm allegiance to modern genetics based on autogenesis and Mendelian laws of heredity. This did not deter him from paying homage to Darwin as the main contributor to the rise of evolutionism in biology. Changing his strategy somewhat, now he selected two of Darwin's principles for particular emphasis: variation as a source of organic evolution, and heredity as the mechanism of generational transmission of variations. Filipchenko pointed to the *Origin of Species* as a pioneering study of the sources of variation and to *The Variation of Animals and Plants Under Domestication* as a detailed explication of pangenesis, the mechanism of heredity. This did not mean that he accepted Darwin's interpretation of variation and heredity: it meant, however, that he was ready to concede that the interpretation of these principles by modern experimental biology was not a negation but an elaboration and a refinement of Darwin's scientific ideas. While admitting that de Vries's "intracellular pangenesis" is far from a simple recasting of Darwin's theory, he was eager to point out a genetic similarity of the two views. Filipchenko reminded his readers that the particulate carriers of heredity, variously named by contemporary biology, were descendants of Darwin's gemmules.[92] He accepted all the digressions of modern genetics from Darwin's postulates, but he also worked hard not to be labeled an anti-Darwinist.

In 1917 Filipchenko noted that the study of variation and heredity, the subject matter of genetics, had made more progress during the past ten years than during the previous one hundred years. Impatient with biologists who ignored the pressing need for integrating the knowledge of genetics into the general theory of evolution, he criticized *Les théories de l'évolution* by Y. Delage and M. Goldsmith, which appeared in a Russian translation in 1916, for its failure to relate the developments in genetics to the theory of evolution. He considered the book an "anachronism."[93] Nor did he see how a modern discussion of natural selection could overlook Johannsen's experimental study of "pure lines," or how a modern biologist could consider Weismann's "anachronistic" studies the most substantial contribution to the theory of heredity.[94] The function of genetics, according to Filipchenko, was not to "liquidate" the theory of evolution but to build it on firmer foundations. In 1916 and 1917 *Priroda* carried his annual bibliographical analysis of major developments in various branches of genetics.[95]

The sparse ranks of Russian geneticists received wholesome and beneficial encouragement from isolated scientists working in other fields of biology. The most noted among these scientists was D. I. Ivanovskii, the discoverer of the virus of tobacco mosaic. In 1908 he delivered the annual "university lecture" in Warsaw, where he taught microbiology and plant physiology. The annual "university lectures" were festive occasions at all Russian universities and honored the outstanding members of the academic community. In his address, Ivanovskii argued that the experimental method is indispensable to the study of organic evolution, but that it alone cannot answer all the questions. The future of evolutionary biology, he said, is in a combination of Darwin's legacy and recent products of experimental research. He thought that experimental biology rested on three principles: a reaffirmation of the paramount evolutionary role of the external environment (he obviously had in mind a branch of Lamarckian experimental biology); a recognition of the primacy of mutations in the origin of species; and a full acceptance of natural selection as the chief mechanism of organic transformation. In drawing his conclusions, Ivanovskii was strongly inclined to view the new biology as a combination of Lamarckian environmentalism, de Vries's mutationism, and Darwin's theory of natural selection. For unspecified reasons, he made no mention of revived and reinvigorated Mendelism. He noted that de Vries's mutation theory was a product of refined observation techniques rather than of the experimental method.[96]

Ivanovskii admitted that the future development of evolutionary research would be inseparable from advances in experimental biology. He noted, however, that the new evolutionary theory was still too fluid to be conducive to successful empirical testing. This interpretation did not prevent him from concluding that the time for a general reevaluation of the factors of organic evolution was approaching fast. He thought that this reevaluation would result in downgrading natural selection as a "negative factor" and in upgrading variation and heredity as "positive factors."[97]

The New State of Darwinian Biology: A General Picture

In the West, more than in Russia, the architects of the new orientations in experimental biology gave prominence to strongly stated anti-Darwinian views, creating the impression of a rapidly approaching collapse of Darwin's theory. The threat to Darwinism, however, appeared much stronger than it actually was. The anti-Darwinian camp lacked unity and consolidation. Bateson, for example, was very reluctant to tie Mendel's laws of heredity to the burgeoning chromosome theory, a challenging and most promising development at the time. He represented those geneticists who "tended to see the particulate theory of inheritance as bordering on preformation."[98] In general, the conflict between Pearsonian biometricians and Mendelians was bitter and intense. As a result of his experiments with the fruit fly *Drosophila melanogaster* in 1910, Thomas H. Morgan accepted Mendelism but also became convinced that many mutations were neither "jumps" nor parts of heritable characteristics.[99] In a modified form, he went back a distance toward Darwin's "slight modification" and the law of natural selection. The Dutch biologist I. P. Lotsy attracted considerable attention with an original theory that recognized the gene theory while rejecting mutationism. While some geneticists fought against the "materialistic" bias of the chromosome theory, others thought that the main strength of the new genetics was in its unveiling of the secrets of the corpuscular basis of heredity, the main stumbling block to understanding organic evolution.

No new development was strong and comprehensive enough to replace Darwinism as a general evolutionary view of life. Mendelism was generally thought of as a brilliant development that strengthened the scientific study of heredity while leaving the other aspects of transmutation generally unattended. The mutation theory faced similar criticism. It was widely thought that, at best, mutations could be regarded only as one source of evolutionary change. Neo-Lamarckism was diffuse and far removed from the most promising and rigorous domains of experimental research. Various neovitalist schools were too preoccupied with building metaphysical enclaves to give serious attention to questions of a methodological nature.

In 1907 V. L. Kellogg offered a pertinent summary of the situation on the biological front:

> The basic truth is that the Darwinian selection theories, considered with regard to their claimed capacity to be independently sufficient mechanical explanations of descent, stand today seriously discredited in the biological world. On the other hand, it is also fair to say that no replacing hypothesis or theory of species-forming has been offered by the opponents of selection which has met with any general or even considerable acceptance by naturalists. Mutations seem to be too few and far between; for orthogenesis we can discover no saltatory mechanism; and the same is true for Lamarckian theories of modification by cumulation, through inheritance of acquired or ontogenetic characters.[100]

There was also another development that worked in favor of the Darwinian positions: the pronounced inclination of the scientific community to look at most new advances in biology as realizations of fertile ideas Darwin had presented in his numerous published works. Darwin's theory of pangenesis, for example, was now interpreted as an ancestor of the involvement of modern genetics in the nuclear basis of heredity. Individual commentators noted that Darwin did not exclude saltatory evolution as one of several possible modes of change in living forms.

The anti-Darwinism of the new genetics peaked quickly and began to show signs of receding. In 1909 de Vries and Bateson were among those who gathered at Cambridge University to commemorate the one-hundredth anniversary of Darwin's birth, to celebrate the fiftieth anniversary of the publication of the *Origin of Species,* and to reaffirm their respect for the scientific legacy of the English naturalist. These and most other leaders of the new science were eager and ready to note that modern transformism owed a primary debt to Darwin and that the future of biology belonged to a synthesis of diverse theoretical ideas among which Darwin's contributions occupied a preeminent position. De Vries, widely heralded as the most powerful enemy of Darwin's theory, said that whether variations arise from "simple fluctuations," as Darwin thought, or from mutations, as he believed, natural selection "will multiply them if they are beneficial, and in the course of time accumulate them," producing a great diversity of organic forms. Darwin may have been only partially correct when he identified minute digression as the real stuff of evolution; but he was absolutely correct in making natural selection the real factor determining the course of evolution.[101]

Darwin helped prepare the way for modern biological advances not only by suggesting topics for future research, but also by deliberately making the basic principles of his general theory open-ended and amenable to exceptions. Bateson found it appropriate to remind his Cambridge listeners that Darwin never abandoned his early conviction that "natural selection has been the main but not the exclusive means of modification."[102]

Soon the representatives of Darwinian and Mendelian biology were united in recognizing natural selection as a mechanism of evolution. To be sure, there were differences: whereas to the Darwinists natural selection was a primary mechanism, to the Mendelian scholars it was a secondary mechanism. Nevertheless, individual biologists hinted at the possibility of unifying Darwinian and post-Darwinian biology. The theory of evolutionary synthesis, however, did not triumph until the late 1920s and the early 1930s. The emergence of population genetics in the mid-1920s gave biology inner order and theoretical consistency, without which no evolutionary synthesis could be profitably undertaken.

The upsurge in biological thought gave new direction to evolutionary research. It achieved this by improving experimental techniques, by advancing more effective procedures of field investigation, by adopting theoretical models from the rapidly advancing post-Newtonian physics, and by soliciting help from such previously detached sciences as mathematics and psychology. The excellence of the new evolutionary research was not in producing new ideas but in giving a sense of urgency and concreteness to a number of general notions that coexisted with—and were in opposition to—Darwin's theory. In one form or another, the idea of "sudden jumps" in the evolutionary process had been part of standard criticism of Darwin's theory since the early 1860s, when Kölliker introduced the notion of heterogenesis. The idea of the independence of heredity factors from the environment received strong support from Bateson's research almost a decade before the rediscovery of Mendel's laws. Somewhat earlier it became a key part of Weismann's elaborate logical constructions related to the cytological basis of heredity. The much-heralded orthogenesis of Eimer and Cope stood in direct opposition to Darwin's idea of the random nature of variation. The pressure for a systematic study of the psychological dimension of evolution came from many sides, particularly after Wundt's bold efforts to tie experimental psychology to the evolutionary idea. Romanes's work on physiological evolution had an appreciative and growing audience, as did the new vitalistic metaphysics (Henri Bergson) and vitalistic biology (Hans Driesch).

The appearance of the mutation theory and the rediscovery of Mendel's heredity laws did not precipitate immediate and extensive retooling of biological research. The process of translating the new theories into research strategies was rather slow. It was not until the end of the decade that the designs for new research acquired tolerable precision and that carefully conceived and organized inquiries were in full swing. The work of Thomas G. Morgan in the United States, William Bateson in Great Britain, and W. Johannsen in Denmark, and the founding of the first two journals in genetics—the *Zeitschrift für Vererbungslehre* (1908) in Germany and the *Journal of Genetics* (1910) in Great Britain— played a major part in helping the new research to acquire a solid foothold in the scientific community.

As it developed in Russia, the new world of biology had four distinguishing national characteristics. First, the new biology did not receive much help from the most distinguished national scientific centers— from the St. Petersburg Academy of Sciences and from St. Petersburg and Moscow universities. Its main support came from the new institutions, still involved in rapid growth and unencumbered by rigidly defined curricula or by vested academic interests. Indeed, the leading universities continued to be the bastions of Darwinism, pure or tempered. Second, it was the botanists, rather than the zoologists, who produced the chief architects, defenders, and articulators of developments in biology that went against the cardinal principles of Darwin's theory. The fact that an overwhelming majority of Western geneticists came from a botanical background had much to do with this phenomenon. Third, a majority of Russian scientists actively engaged in genetic research belonged to the younger generation of scholars. Employment insecurity and a low standing in the hierarchy of academic ranks influenced them to exercise restraint in their criticism of Darwinism, fervently defended by the luminaries of the older generation. Fourth, the Russian biologists avoided extremist philosophical positions. They favored a middle-of-the-road synthesis of opposing orientations. A typical biologist was willing to listen to neovitalist arguments only as a critique of the more extreme claims of mechanical philosophy. More than ever before, philosophical discourse occupied a strategic position in the critical examination of new developments in the leading branches of biology. This discourse, however, displayed neither particular depth nor a propensity for dogmatic system-building.

The triumph of genetics, the unmistakable sign of the new age, compelled the Darwinian scholars to give serious thought to recasting the ideas of random variation, heredity, and natural selection as an adaptation to new advances in biology. It made geneticists determined not to discard Darwin's legacy but to make it part of the general onrush of modern scientific currents. Geneticists gained strength on both philosophical and practical fronts. On the philosophical front, they benefited from mounting challenges to the reigning ideas of the mechanistic orientation in science—challenges that came from both ontological and epistemological criticism. On the practical front, they had succeeded in alerting the rapidly growing ranks of scientific breeders to the enormous potential of the new techniques of crossing domestic varieties and species.

Although the rise of experimental biology challenged some of the cardinal principles of Darwin's theory, it did not seriously affect the preeminent position of Darwinian thought within the general context of evolutionary biology. M. V. Arnol'di, professor of botany at Kharkov University, gave what may be considered the most typical assessment of the current state of Darwin's contributions. He wrote in 1904 that all new theories actually concentrated on expanding and clarifying Darwin's ideas. He was not surprised when de Vries identified himself as Darwin's follower. In his view, every effort to explain organic evolution "must begin with Darwin."[103]

A new encyclopedia referred to Darwin as "the Copernicus or Newton of the organic world." It noted that by proclaiming man a member of the animal world, Darwin brought "the sciences of man" into close contact with the natural sciences and made the "genetic method" the main research tool of many fields of scientific inquiry.[104] By categorizing man as an animal, he vastly expanded the scientific study of human society and culture. Expressing an unduly optimistic view, the anonymous writer noted that the enemies of Darwin's ideas had been silenced long ago. Published in 1902, the article appeared before the challenge of genetics had begun to take on the proportions of an avalanche. Another article in the same encyclopedia noted that the rejection of individual propositions of Darwin's theory should in no way be construed as a rejection of the theory as a whole.[105] The author had no reservations about viewing Darwinism as an "indestructible" scientific theory.

CHAPTER NINE

Darwinian Anniversaries

Ceremonial Recounting of Achievement

In 1909 a group of leading European and American evolutionary biologists gathered at Cambridge University to commemorate the one-hundredth anniversary of Darwin's birth and to observe the fiftieth anniversary of the publication of the *Origin of Species*. The distinguished participants in the academic proceedings represented the full spectrum of evolutionary thought ranging from Haeckel's version of Darwinian orthodoxy and Weismann's neo-Darwinism to de Vries's mutation theory and Bateson's genetics built upon Mendel's mathematical theory of heredity. The speakers differed in their choice of Darwin's strengths for special emphasis, but they were united in considering the English naturalist the architect of one of the most fateful turning points in the history of scientific thought. The dual anniversary was marked by ceremonial gatherings sponsored by scientific institutions, learned societies, and universities throughout the civilized world.

Taking the anniversary dates with the utmost seriousness, the Russian scientific community sponsored a wide variety of commemorative activities paying homage to the man who, by changing the course of science, changed the world view and self-image of modern man. The high point in the solemn expressions of gratitude to the great man was the publication of *In Memory of Darwin* in 1910.[1] Consisting of new and previously published essays, all by Russian scholars, the volume was dominated by a single theme: the vast and continuously growing scope of Darwin's influence on modern scientific thought and world outlook.

With minor exceptions, the writers paid no attention to the formidable challenges coming from the achievements and stirrings in experimental biology.

Ivan Pavlov and Il'ia Mechnikov were the most noted contributors to *In Memory of Darwin*. Pavlov's essay, originally presented at the Twelfth Congress of Russian Naturalists and Physicians in December 1909, did not refer specifically to Darwin or his theory. By inference, however, Pavlov drew attention to the close relationship between the evolutionary process, as Darwin saw it, and the adaptive functions of neurophysiological processes, as he saw them. The presence of Pavlov's name made both the symposium and the Darwinian celebration all the more appealing, impressive, and authoritative. In his contribution Mechnikov was particularly eager to point out the essential role of comparative material in the study of inflammation and immunity. He predicted the growing role of microbiology in the search for specific causes of variation in plants and animals, and of the origin of new species.[2] Referring to the experimental work of Louis Pasteur and his assistants, he also envisaged an increasing participation of microbiology in the general study of heredity, the key to the understanding of organic evolution. The high status of Pavlov and Mechnikov in the scientific community made it all the more difficult for scientists who were critical of Darwin's legacy to give full sway to their heretic ideas. Most of these scientists found it advantageous to accommodate their heterodox ideas to the substance and the spirit of Darwinian evolutionism.

In their contributions to *In Memory of Darwin*, Pavlov and Mechnikov looked at Darwin from the vantage points of their scientific specialties. K. A. Timiriazev, by contrast, offered a general analysis of the Darwinian revolution. During the fifty-year period after the publication of the *Origin of Species*, he wrote, continuous efforts to "denigrate that book" had been unsuccessful. The *Origin* had continued to serve as the only sound "philosophy of biology." "It alone gives the key for the full understanding of the general structure of organic nature, and it alone continues to serve as the North Star for modern biologists whenever they look beyond the narrow vision of daily work in hope of catching a new glimpse of the biological universe."[3] Timiriazev took this opportunity to assure his readers that the developments in experimental biology triggered by the work of Mendel and de Vries were minor developments in comparison with the universal significance and permanence of Darwinian contributions. He resented Bateson's view of genetics as a "new province of knowledge" rather than as a "branch of physiology."[4]

By attacking the legacy of "the golden age" of biology—the age of Darwin—Bateson and Bergson, each in his own way, hoped to return to "the dark ages of scholasticism and irresponsible thought."[5]

Nor did Timiriazev's wrath bypass the two main neo-Lamarckian groups: one looking for a "physical explanation" of the emergence of new forms of life, the other linking the idea of organic evolution to an "immanent teleology of the purposefully acting environment," "a purposefully directed process of the development of organisms," or "a conscious protoplasm envisaged by the supporters of German panpsychism."[6]

Learned societies of the most diverse allegiances and interests were particularly active and articulate in paying homage to Darwin and his contributions to modern science. At the end of 1908 the Moscow Society of Naturalists held a special meeting at which N. A. Umov, a physicist with a strong philosophical predisposition, was the featured speaker.[7] Evolution, he thought, reinforced the idea of the unity of nature. Every component of nature—from the lowly atoms to the most elaborate structures of human thought—occupied a place in the cosmic sequence of natural objects. He treated human psychology as a unique mirror of cosmic order and harmony. In the mental growth of man he saw both a recapitulation and an extension of the evolution of the universe. To study evolution, as he saw it, was to study the ascending levels of the organization of nature. The cosmic scope of the psychological foundations of evolution formed the only sound basis for the study of the most sublime questions of ethics. Discerning listeners recognized in Umov's address an effort to blend a strong interest in psychology with Newtonian atomism and Darwinian evolutionism. Umov wanted to show that the study of evolution stood at the threshold of a new era, which promised to give the Darwinian theoretical legacy a cosmic meaning and a psychological essence.

D. N. Anuchin, a popular anthropologist and geographer, read a paper before 1,200 members and guests of the Moscow Society of the Friends of Natural Science, Anthropology, and Ethnography in which he recounted the achievements of Darwinian biology.[8] Not all biologists, he said, were convinced that Darwin's explanation of organic evolution was necessarily correct. But no scientist would deny that Darwin gave science "a powerful impulse for broader and more penetrating research undertakings, methods of inquiry, and modes of explanation." Thanks to Darwinism, systematics embraced new concerns, and many areas of biology, biogeography, and biometrics were transformed from secondary to primary scientific concerns. Paleontology became a major discipline concerned with phylogeny, and medicine recognized new research perspectives and challenges. Anthropology acquired the status of a major science. The studies of culture and behavior also gained a stronger scientific footing. Darwin's theory invited an empirical study of the moral foundations of human society.

Nor did Anuchin overlook the distinctive features of Russian Darwiniana. Darwin's ideas, he noted, were received most favorably in Russia. In addition to bolstering the empirical base of Darwin's theory and to popularizing Darwin's ideas, the leading Russian biologists and paleontologists were also engaged in translating Darwin's works into Russian. This group included V. O. Kovalevskii, I. M. Sechenov, K. A. Timiriazev, M. A. Menzbir, A. N. Beketov, and A. P. Pavlov. He noted that the struggle between Darwinists and anti-Darwinists in Russia was much milder than in Germany, where the conflict between Haeckelians and anti-Haeckelians was particularly intense and bitter. The attacks on Darwin in Russia were motivated primarily by "moral" and "social" considerations.

On March 21, 1909, the St. Petersburg Society of Naturalists needed the largest hall at St. Petersburg University to accommodate the crowd that came to take part in the Darwin celebration. All widely recognized experts in evolutionary theory, the speakers covered the main topics of Darwinian thought. V. M. Shimkevich spoke on "the historical significance of Darwin's contributions to zoology." He took the opportunity to lament the conditions that prevented the rise of Russian Darwins.[9] V. A. Vagner spoke about "Darwin and the social sciences," and V. L. Komarov about Darwin's sizable contributions to botany.

Much smaller in attendance but equally impressive because of the topics discussed was the celebration sponsored by the student Circle of the Friends of Nature at Kharkov University. In addition to the "fundamental questions of Darwinism," the speakers concentrated on the relationship of Darwin's ideas to current developments in evolutionary biology: N. F. Belousov spoke on the theme of "Darwin and Weismann," and V. I. Taliev on "Darwin and de Vries."

Vladimir Vagner, the leading Russian expert in the psychological aspects of organic evolution, was the featured speaker at the Darwinian convocation sponsored by the Pedagogical Museum of Military Schools. A comparison of the *Cours élémentaire d'histoire naturelle,* a high school textbook written by Milne-Edwards in 1857, with the *Origin of Species* led Vagner to point out Darwin's original contributions to the

study of the origin and variability of instincts, of genetic links between
the mental characteristics of man and animals, and of the elementary
reasoning capacity not only among higher mammals (which Milne-
Edwards was ready to recognize) but also among the invertebrates of
"almost all levels of classification." [10] Vagner criticized comparative psychologists for their one-sided and generally inadequate exploration of
Darwin's legacy. While staying too close to the ideas presented in the *Origin of Species*—and to the Darwinian emphasis on the struggle for existence and natural selection—the comparative psychologists were unpar-
donably lax in exploring *The Descent of Man,* particularly the chapters
dealing with moral sentiment as a foundation of social solidarity and
cultural values. [11] In *The Descent* Vagner saw an invitation to explore
the links between the theory of organic evolution and social psychology
as an academic discipline.

The succession of commemorative events ended in late December
1909, at one of the sessions of the Twelfth Congress of Russian Naturalists and Physicians. On that occasion, V. I. Taliev presented a paper
entitled "Darwinism, Lamarckism, and the Mutation Theory," which
was followed by a lively discussion. Emphasizing the interdependence of
current orientations in evolutionary biology, he placed particular stress
on the complementary relations of the "experimental method" of modern biology and "careful scientific speculation of the old school," which
depended on comparative-morphological facts and other data obtained
with the help of systematic observation. He also emphasized the complementary relations of "leaps" and "slight modifications" in the evolutionary process and looked optimistically toward a theoretical reconciliation of autogenesis and Lamarckian sources of evolution. The
future of biology, as he saw it on this occasion, lay in a recognition of
the multiple aspects of the evolutionary process and of the interdependence of various theories. Darwin deserved major credit for accelerating
the growth of modern biology and for opening up limitless avenues to
new research. [12]

Addressing the same congress, A. N. Severtsov chose to defend Darwinism from the attacks mounted by the supporters of neovitalism led
by Driesch and strongly espoused by Rádl, a widely read historian of
biology. [13] The neovitalists, he said, considered purposiveness not only
an essential aspect of life but also a "constant" of nature, which cannot
be analyzed into component parts and therefore cannot be subjected to
scientific treatment. Darwinism, by contrast, regarded purposiveness as
a phylogenetic development made up of simpler and scientifically scru-

table components. It closed the door on metaphysical interference with the work of science.

The St. Petersburg Academy of Sciences did not hold a special convocation to honor Darwin. It did delegate two eminent biologists—the plant physiologist I. P. Borodin and the embryologist V. V. Zalenskii—to participate in the Cambridge celebration.[14] In its official greeting to the Cambridge gathering, the Academy paid homage to the grandeur of Darwin's contributions to science. A catalyst of epochal developments in biology after 1859, the *Origin of Species,* according to the Academy's salutatory message, stimulated the growth of a strong tradition in Russian embryology, "a science closely allied with evolution."[15]

An impressive number of journals joined the anniversary celebrations by publishing essays on Darwinian themes. The journals *Education* and *Natural Science and Geography* marked the occasion by publishing a Russian translation of the commemorative speech Richard Hertwig had delivered before the members of the Munich Society of Naturalists. Hertwig presented a warm and human picture of personal strengths that contributed to the greatness of Darwin's scientific stature. He mentioned Darwin's intellectual perseverance, disciplined adherence to the rules and perspectives of the inductive method, mild temperament, high moral standards, sense of justice, and exemplary modesty.[16]

The journal *Natural Science and Geography* published also a Russian translation of Haeckel's paper on the world views of Darwin and Lamarck, presented in Jena in observance of the one-hundredth anniversary of Darwin's birth. In addition to reaffirming his unlimited devotion to the extraordinary power of Darwinian thought, Haeckel called Lamarck and Goethe the chief contributors to the triumph of evolutionary thought, which opened "wide cosmological perspectives," carried the searching human mind "far beyond the limits of space and time," and brought full victory to the idea of the unity of nature. Darwin's destruction of anthropocentrism belonged to the same level of scientific accomplishment as Copernicus's destruction of geocentrism.[17] To appear more realistic and judicious, *Natural Science and Geography* also carried a sample of articles that were somewhat less ecstatic about Darwin's scholarly acumen. In an excerpt from the writings of Raoul Francé, a German representative of psychological Lamarckism, Darwin is portrayed as a scholar who could be labeled a "researcher" but not a "thinker."[18] Darwin's frame of mind and temperament made analysis, rather than synthesis, his main concern. He was too involved in a study of the parts of the organic universe to take a closer look at the harmony, unity, and symmetry of the infinite manifestations of life. He described the substance of the living world, not its architecture. On his part, Francé belonged to the group of biologists who sought a modern synthesis of biological theory and idealistic metaphysics.

The editors of the *Herald of Knowledge,* deeply involved in the popularization of current developments in science, added to the festivities by dedicating the February 1909 issue of the journal to Darwin's memory.[19] Seven of the eight commemorative articles were translations from the German language. They were selected with the view of touching on the main areas of Darwin's scholarly activities and of providing a general assessment of Darwinian positions in evolutionary thought.[20] Some invited criticism of Darwin's more tenuous ideas. All provided insightful details on Darwin's style of thought, exemplary dedication to scholarship, and high standards of moral behavior. In 1910 the journal published and distributed an "inexpensive" edition of the *Origin of Species* as well as the part of *The Descent of Man* dealing with sexual selection. In the same year it asked its subscribers to name their favorite writers and scientists. The answers showed that the subscribers favored Tolstoy, Turgenev, and Dostoevsky, among writers, and Darwin among natural scientists.[21]

The *Herald of Europe* marked the celebration in a most auspicious manner: it published Timiriazev's two papers offering a defense of Darwinism and a description of the Cambridge festivities respectively. "Charles Darwin," the first paper, was quickly reprinted in *In Memory of Darwin.* It provided both the most glowing and the most dogmatic accolades for Darwin's achievement. The second paper, "Cambridge and Darwin," gave a detailed account of the commemorative ceremonies at the old English university. Timiriazev was particularly pleased with— and recounted in some detail—the paper read by Ray Lankester, a strong Darwinist from Oxford University. Lankester argued that fifty years after the publication of the *Origin of Species* Darwin's evolutionary principles continued to be indestructible despite all efforts to undermine them.[22] Natural selection, the main principle of Darwin's theory, continued to be incontestable as the main agent of organic transformation. Mendel added a statistical approach to the study of heredity, but in no way did he challenge the accuracy of Darwin's thought. Darwin rejected the idea of mutations (he called them sports) as the basic—or the only—source of evolution, but only after he subjected them to serious scrutiny. Timiriazev was obviously displeased with the favorable attention Korzhinskii's mutationist ideas received in the Cambridge proceedings. Nor was he satisfied with the state of evolutionary biology at Cambridge University, where William Bateson's "anti-Darwinism" was rapidly becoming a dominant factor.[23]

In yet another commemorative article, published in the newspaper *Russkie vedomostii* on January 30 and 31, 1909, Timiriazev described his visit to Darwin at his home in Down in early February 1877. Timiriazev thus became the first Russian who not only visited Darwin but also made his impressions accessible to the reading public. He noted Darwin's familiarity with the development of the theory of evolution in Russia. Darwin, he recorded, recognized the enormous scope of the contributions of the Kovalevskii brothers to the evolutionary theory. Timiriazev also noted that Darwin considered the contributions of Vladimir to paleontology more valuable than the contributions of Aleksandr to comparative embryology.[24] Darwin informed Timiriazev that his and his family's "sympathies" were with Russia and that he was particularly pleased with the favorable reception of Buckle's, Lyell's, and his own works in Russia.

The *Herald of Europe* did not stop with Timiriazev's articles recounting the highlights of Darwin's contributions to modern science. In the May 1910 issue it portrayed Darwinism as a high point in the development of modern science and as a cultural force of immense magnitude. In addition to causing a revolution in science, it placed the mainstream of modern thought on a new course. By showing that the differences between human beings and animals are differences of degree rather than of kind, it showed that man's intellectual capacities, esthetic feelings, and moral rules are products of a long and inexorable evolutionary process, propelled by natural causes. Darwin's theory has shown that animals are our "little brothers"—that they too "have nerves and feelings of joy."[25] The journal stated that Darwinism, "interpreted correctly," posed no threat either to religious beliefs or to religious morality. In fact, it added new strength to both. It noted that the theologians could benefit enormously by relying on Darwinism in their study of the natural sources of "altruism, sense of duty, and conscience."[26] The author of the article expressed gratitude to Darwin for laying the foundations for a new world outlook based on a firm unity of modern science and traditional culture expressed in intellectual, moral, and religious values.

Theologians and theological journals did not overlook the rising tide of pro-Darwinian sentiment. S. S. Glagolev, one of the more erudite theologians, marked the occasion in 1909 by publishing an article in the *Theological Herald* dealing specifically with the evolutionary theory of the origin of man. Bitterly critical of Haeckel's interpretation of *Pithe-canthropus erectus* as a link between man and ape, and unbending in his rejection of the cardinal principles of the theory of natural selection, he did take time, however, to make a salutatory comment on Darwin's dual anniversary. In Darwin's *Origin*—as well as in Lamarck's *Philoso-phie*—he saw the kind of books that "do not appear every decade" and that contain "valuable observations and ideas that have become a perma-nent property of science." [27] After paying homage to Darwin's science in general, he proceeded to argue against the role of human paleontology in adducing anticreationist arguments. Glagolev spared Darwin by di-recting his heaviest arguments against Haeckel. Saddened by the threat of Darwinian menace, he stated with a strong dose of rhetorical exag-geration: "If a person would appear in Moscow or St. Petersburg with a public lecture on 'the evolution of man from donkey' he would face a full auditorium. If, in the same cities, a speaker would offer to talk about 'the scriptural description of the creation of man' he would face an empty auditorium." [28]

The publication of two monographs made the Darwinian festivities richer and more appealing. In *Charles Darwin: The One-Hundredth Anniversary of His Birth*, V. I. Taliev emphasized the commanding posi-tion of Darwinism in the swelling stream of contemporary evolutionary thought. The struggle for existence, he wrote, continued to be the only satisfactory way to apply the principle of causality to the study of or-ganic evolution as a unique expression of the universal harmony of na-ture. [29] The book offered one of the earliest bibliographic surveys of Rus-sian studies in Darwinism. A. A. Ostroumov, professor of zoology at Kazan University, published *The Origin of Species and Natural Selec-tion: Fifty Years of Darwinism*, a collection of empirical and logical ar-guments in favor of the "granite stability" of the main principles built into Darwin's theory. [30] With a minimum of technical involvement, both books were clearly expected to appeal to a diverse and rapidly expand-ing reading public.

The anniversary celebrations received strong support from yet an-other momentous development: a new publication of Darwin's collected works in a Russian translation. The new edition, richly illustrated, brought together all the books Darwin had written, some in new trans-lations. Published in eight volumes, the new edition was the result of a deliberate and carefully planned effort to comb out errors that marred earlier translations. Timiriazev, Menzbir, and the geologist A. P. Pavlov deserved the main credit for making the new collection a noted success both academically and commercially.[31] During the tsarist era thirty-five thousand copies of the *Origin of Species* were published in various Russian translations.

The festive atmosphere of the celebration encouraged the participating scholars and popular writers to take a broad view of the inner makeup and scientific potential of the evolutionary idea in biology. The generalizing mood helped make the difference between orthodox and unorthodox Darwinism precise and fixed. Speaking for the rapidly shrinking group of orthodox Darwinists, K. A. Timiriazev argued that Darwin became a giant of modern biology because he not only identified evolution as a universal attribute of organic nature but also was successful in explaining the general mechanisms of the evolutionary process. Without his explanations of the struggle for existence and natural selection he would not have been recognized as the founder of modern biology. Lamarck, he said, had failed because he did not match his pronouncement of the universality of organic evolution with acceptable explanations of the mechanisms of the evolutionary process.[32]

In 1909 the journal *Russian Wealth* published a seventy-page essay on Darwin's key contributions to evolutionary biology. The author of the paper was Sinai Chulok, a Ukrainian native associated with the Zurich Polytechnic Institute, first as a student and then as a privatdocent. In 1910 he published *Das System der Biologie in Forschung und Lehre*, a noted effort to present a comprehensive and integrated picture of the methodological and theoretical foundations of the biological sciences. The *Russian Wealth* essay was intended to be part of Darwin's anniversary festivities. In response to the attacks mounted by geneticists, Chulok noted that the victory of the evolutionary view in biology rather than the recognition of natural selection as the basic mechanism of evolution was Darwin's primary—and unchallengeable—contribution to modern science.

In Chulok's view Darwin took three steps that placed him far ahead of his predecessors.[33] First, he concentrated strictly on the origin of species, for he knew that this question must be answered independently of that on the origin of life. This enabled him to reduce the intensity of philosophical interference with his scientific concerns. It also gave him a decided advantage over his predecessors, who were often handicapped by asking questions that scientific wisdom was not yet in a position to tackle successfully. Second, Darwin was the first scientist who asserted that the successful construction of a genealogical tree required not only an accumulation of relevant empirical data but also the support of a general law of the transformation of species. The work on a universal genealogy must synchronize inductive and deductive approaches. Third, Darwin was the first to show that the facts of systematics, biogeography, and paleontology would be meaningless as building blocks of science without a thorough consideration of evolution as a process of organic transformation. He showed that change in form and function is the essence of life.

No Russian writer had written more extensively about the general theoretical fermentation in biology than V. V. Lunkevich. He held no academic position and published all his biological articles in popular journals. The *Basis of Life*, his major work, appeared in several editions. It treated every trend in modern biology, ranging from Le Dantec's effort to reconcile Lamarckism and Darwinism on a physicochemical basis to Erich Wasmann's bold effort to reconcile Darwinism and Christian theology. He expressed two general judgments about the state of the field. First, Darwinism explained some of the basic factors that accounted for the transformation of species. It showed the evolutionary role of the struggle for existence, adaptation, and divergence of characters. Second, Darwinism did not explain some of the other pivotal aspects of organic evolution. Above everything else, it did not explain the growing "complexity" and the "progress" of living forms.[34] In order to tackle these problems successfully, the evolutionary theory of the future must be built upon a synthesis of Darwinism and the streams of new theories advanced by various branches of experimental biology. Lunkevich gave a clear impression that a future synthesis of evolutionary thought would assign Darwinism a role of primary significance.[35]

In general, the commemorative literature was more implicit than explicit in recognizing the serious need for blending Darwinism with Mendelian genetics. A careful reading of the voluminous material published in connection with the celebration suggests that a typical speaker or essayist felt that much preparatory work was needed to set the stage for a successful reconciliation of apparently discordant views held by the Darwinists and geneticists. In the festive atmosphere, the writers dealt much more with the heights of Darwinian achievement than with the ways of resolving the conflict between Darwinian principles and the current products of experimental studies in biology.

Lunkevich's comparison of Darwinism and de Vries's mutationism favored the former. The mutation theory, as Lunkevich interpreted it, was sound in its full acceptance of two Darwinian principles: the struggle for existence and natural selection. In its digression from the Darwinian

course, however, it showed strong weaknesses: its basic propositions
were doubtful, its logic was full of strained interpretations, and its argumentation followed a path of distorted Darwinism.[36] De Vries's entire
system of theoretical considerations rested on the observation of a single
plant—*Oenothera Lamarckiana*. Almost as an afterthought, he noted
the primary significance of the idea of mutation in the modern scheme of
evolutionary principles. No doubt, Lunkevich's writing at this time was
influenced more by the celebration of Darwin's anniversary dates than
by the desire to undertake a careful scrutiny of the achievements in a
vital branch of modern biology.

Darwin and Lamarck

The participants in the anniversary celebrations frequently compared
Darwin and Lamarck. Published in a Russian translation in 1910,
Haeckel's essay "The World View of Darwin and Lamarck" stimulated
a lively and sustained discussion of the place of Lamarck in the pre-
Darwinian evolutionary tradition. Haeckel considered Lamarck Darwin's truest and most formidable predecessor: both were thorough, consistent, and categorical in claiming that only natural causes can solve the
riddle of evolution. Lamarck's theory of evolution generated little enthusiasm: its labyrinthine structure could not find support in the empirical data available at the time. Lamarck asked more questions than
science was ready to answer. Darwin, by contrast, was eminently successful because the rapid accumulation of empirical information invited
and made possible a serious concern with the theory of evolution.
Darwin succeeded because he answered all the basic questions he undertook to answer.

Haeckel helped reinforce the view of Lamarck as a central link in the
chain of developments that culminated in the appearance of the *Origin
of Species*. His intent was to recognize the great historical value of *Zoological Philosophy* without taking anything away from the *Origin of
Species*. Darwin, he said, did not negate Lamarck's work but carried it
to a successful completion. The intent of Haeckel's address was not only
to pay respect to the French scientist on an important anniversary but
also to fight the rising tide of neo-Lamarckism, a heresy that considered
Darwin guilty of digressing from the evolutionary path Lamarck had
taken and that pleaded for a return to Lamarck's original thought. By
praising Lamarck, Haeckel was protecting the interests of Darwinism.

Ieronim Iasinskii, a popular writer with a long list of publications, honored not only the Darwinian commemoration but also the one-hundredth anniversary of the publication of Lamarck's classic *Zoological Philosophy*. In "Darwin and Lamarck," published in the journal *New Word*, he relied on Félix Le Dantec, the well-known French scholar who combined a profound involvement in evolutionary physiology with a strong flair for philosophy. Le Dantec recommended a combination of the idea of evolution as a random process, Darwin's main concern, with the idea of evolution as a universal law of nature, the center of Lamarck's attention. Iasinskii recommended the search for regularities hidden behind the random occurrence of variation in plants and animals. In other words, he, too, recommended a synthesis of Lamarckian and Darwinian approaches to organic evolution. Although Darwin's pangenesis, a unique search for universal regularities in the living universe, did not lead to a formulation of empirically verifiable laws of nature, it was a solid move in the right direction. While the full development of evolutionary theory belonged to the future, "the basic principles built into the great Darwinian theory of the origin of species" will continue to be as "unassailable" as the laws of Kepler and Newton.[37] Implicit in Iasinskii's argument were two ideas: first, the future of evolutionary theory lay in building upon Darwinian foundations; and second, the future of Darwin's theory lay in closer cooperation with the Lamarckian tradition. In making his plea, Iasinskii, like Haeckel, referred to Lamarck, not to various branches of neo-Lamarckists.

Speaking at a session of the Twelfth Congress of Russian Naturalists and Physicians at the very end of 1909, V. I. Taliev found a new task for Darwinism: to serve as a conciliatory force between the extreme positions of renewed Lamarckism and the mutation theory, the former stressing the primacy of environment as a factor of evolution and the inheritance of acquired characteristics, the latter minimizing the evolutionary role of environment and rejecting the idea of the inheritance of acquired characteristics. He believed in the possibility of bringing the contradictory evolutionary theories under one roof, but he argued that the new unity could be forged under the auspices, and through the well-established intellectual equipment, of Darwin's theory.

The zoopsychologist Vladimir Vagner presented a unique and rather amusing comparison of Darwin and Lamarck.[38] In terms of their scientific work, the two men, he wrote, were more similar than dissimilar. They relied on essentially the same general views of organic nature and arrived at basically the same evolutionary conclusions. Despite these similarities, they represented two distinct kinds of scholars. Darwin was a strict empiricist, operating within a relatively narrow but carefully and precisely delimited framework. Lamarck, by contrast, had a mind prone to speculation unrestrained by empirical considerations and constantly searching for answers to problems of universal significance. Darwin allowed science to set the limits to his generalizations; Lamarck depended on philosophical imagination when the facts of science happened to be in short supply. Vagner expressed a firm conviction that evolutionary biology was so much richer because of striking differences in the mental makeup, temperament, and style of work of its two most eminent architects. The evolutionary theory achieved great victories because Lamarck was a philosopher in science and Darwin was a scientist in philosophy.

In a lengthy essay published in *Russian Wealth* in 1910, V. V. Lunkevich compared Darwinism with Lamarckism and mutationism. He noted that various branches of neo-Lamarckism, particularly of a "psychological" variety, did not have much in common with Lamarck's scientific legacy: unlike true Lamarckists, they violated the modes and canons of scientific reasoning and verification and allowed themselves to sink in the morass of metaphysical elaborations. Relying heavily on the address Delage had delivered in the Jardin des Plantes in Paris on the occasion of the unveiling of a statue of Lamarck in celebration of the centennial of *Philosophie zoologique,* he noted that, despite their basic differences, the theories of Lamarck and Darwin mutually reinforced each other. He also added that, had Lamarck lived in the twentieth century, he would most probably have accepted Darwin's idea of transformation.[39]

Most biologists who commented on the theoretical views of the two illustrious celebrants treated Lamarck's ideas as integral parts of the Darwinian evolutionary framework. P. F. Lesgaft represented a small group of biologists who took an opposite position: he thought that Lamarck's theory was more advanced and more general than any other theory, including Darwin's. In a polemical article published in 1909 in connection with the one-hundredth anniversary of the publication of *Philosophie zoologique,* Lesgaft gave a forthright expression to his evaluation of the two theories. In comparison with Darwin's views, he thought that Lamarck placed a stronger and more explicit emphasis on the unity of inorganic and organic nature.[40] He created a more solid basis for making biology an integral part of Newtonian science. By recognizing spontaneous generation, Lamarck linked the theory of the origin of life with the theory of organic evolution.

General Views and Assessments

The commemorative literature was not only an important part of the celebration of Darwin's scientific achievement but also a parade of Darwinian forces in Russia. Although it did not produce a critical and objective appraisal of the current state of Darwinian studies, it did create a strong and basically accurate impression of the vast respect for Darwin in both the scientific community and the general public. The remarkable feature of this literature was that only in very rare cases did praise of Darwin's achievements involve an attack on the mushrooming biological orientations outside the Darwinian tradition. It was most successful in linking the philosophical foundations of Darwin's biology to the accelerated growth of the scientific world outlook in Russia.

Nor was the commemorative literature without major flaws. In the first place, it missed a golden opportunity to make a methodical survey of Russia's direct contributions to the evolution of Darwinian thought. Expansive in their references to Western sources, the Russian biologists were generally reluctant to cite their compatriots. In the second place, the commemorative literature made only isolated and fortuitous efforts to comment on recent advances in genetics and their relevance for Darwinism. Russia did not follow the model of the Cambridge commemorative celebration, where the likes of de Vries and Bateson were invited to speak on Darwinism from the vantage point of the burgeoning field of genetics. Since genetics was rather slow to take root in Russia, the participants in Darwin's ceremonies did not feel sufficient pressure to present it as a critical problem requiring special scrutiny. In his introductory remarks to the Russian translation of T. H. Morgan's *Experimental Zoology*, published in 1909, the translator, N. Iu. Zograf, noted that it was a "bitter experience" to realize that the country that "only recently had produced such giants in science as Mendeleev, Butlerov, Aleksandr Kovalevskii, and Mechnikov" had chosen to overlook the rise of genetics and related branches of experimental biology.[41]

The commemorative literature was basically realistic in portraying Darwinism as a reigning orientation in evolutionary thought. University textbooks in zoology and related fields, which customarily provided the most detailed discussion of theoretical orientations in evolutionary biology, expressed this attitude vividly.[42] Typical textbooks—such as the *Biological Foundations of Zoology*, by V. L. Shimkevich, *Textbook in Zoology and Comparative Anatomy*, by N. A. Kholodkovskii, *Short Course in Zoology*, by N. M. Knipovich, and *Short Course in Zoology*, by Ia. P. Shchelkanovtsev—supported Darwin's theory with three sets of arguments.

First, they treated the *Origin of Species* as a turning point in the history of modern biology. Darwin received the primary credit for giving biology a historical orientation and for moving the idea of organic evolution from the realm of "hazy hypotheses" to the solid ground of established scientific disciplines. Darwinism was considered the only general theory of organic evolution; all other theories covered only specific problems of the vast reality of evolution.

Second, they stressed the need for the utmost caution and fairness in interpreting Darwin's legacy. They pointed out specific cases of deliberate misrepresentation of Darwin's ideas, particularly by anti-Darwinists writing "popular books." Although he was falsely criticized for it, Darwin never claimed that natural selection caused modifications in plants and animals, which in due time produced new species. Darwin stated explicitly that natural selection has nothing to do with the origin of variation; it works solely to preserve the existing variations that have proven useful to organisms in their struggle for survival. In Darwin's words, natural selection means "the preservation of favorable individual differences and variations, and the destruction of those which are injurious."[43]

Third, they presented the current advances in the scientific study of the nature and mechanisms of organic evolution as essential components of the larger system of Darwinian thought. Most textbook writers, for example, interpreted the birth of genetics as a development called for by the Darwinian paradigm of evolutionary biology. They relied on Darwinism to legitimate the streams of new biological knowledge—to make various branches of experimental biology a certified component of established knowledge. In one breath, they paid homage to Darwin and recounted the achievements and promises of modern experimental biology. They saw the future of biology in a symbiosis of Darwinism and the most promising branches of experimentally oriented studies of life. In this symbiosis, however, they were sure to assign Darwinism a position of strategic importance.

The *Progress of Science,* a luxurious multivolume survey of current developments in the major sciences, published in 1912, provided a more guarded but less theoretical assessment of the current state of Darwin's principles. The authors of the papers on biological themes, all recognized scholars, shared three general ideas. First, they rejected all "metaphysical" orientations, represented most typically by neovitalism and psycho-Lamarckism. Second, they recognized Mendel's theory and the mutation theory as firmly established components of modern biology. Consistently, however, they treated these theories as separate and unrelated bodies of knowledge, rather than as interdependent pillars of genetics. In no instance did they place these theories in opposition to Darwinism. Third, they viewed Darwin's theory as the mainstay of evolutionary biology. The triumph of the idea of evolution, in turn, made embryology, comparative anatomy, and paleontology "sciences of genetic relations within and between groups of organisms."[44] All contributors recognized the acute need for expanding the research domain of organic evolution beyond Darwinian limits.

Among the contributors to the *Progress of Science,* V. L. Komarov, professor of botany at St. Petersburg University, and N. M. Knipovich, senior zoologist at the Zoological Museum of the St. Petersburg Academy of Sciences and professor of zoology at the Psychoneurological Institute, expressed the most favorable views on Darwin's contributions. Komarov referred to Darwin's overall contribution when he stated: "Highly convincing generalizations and the skillful use of illustrative material from the life of plants and animals have produced brilliant victories for Darwin's ideas."[45] He represented most biologists when he said that the advances in evolutionary biology after 1859 consisted mainly of improvements in methodology, substituting more exact research techniques for Darwin's "descriptive method."[46] Knipovich referred to the struggle for existence, the key notion of Darwin's evolutionary theory. He stated: "The struggle for existence in all its different forms . . . is the most general mode of biological relations, embracing the entire organic world. There is not a single organism that stands outside that struggle."[47]

The longest and most profusely illustrated essay dealt with the origin of man. Written by the geographer-anthropologist D. N. Anuchin, it summed up the current evidence supporting the idea of the anthropoid origin of man: the salient clues distilled from the history of the idea of organic evolution from Lamarck to Darwin, Huxley, and Haeckel; the relevant embryological and anatomical data; and the information presented by paleontology. Darwin received ample recognition for advancing the first "complete" theory of organic evolution from which the idea of the anthropoid origin of man followed as a logical conclusion.[48] Anuchin presented a mass of information supporting the thesis Darwin elaborated in *The Descent of Man.* Numerous lacunae in paleontological knowledge and complex controversies in interpreting the available data led him, however, to conclude that science was not yet in a position to draw incontrovertible conclusions on the origin of man. In this essay—as well as in his article on the same subject contributed to the popular Brockhaus-Efron *Encyclopedic Dictionary*[49]—he helped confirm the idea that the empirical base of Darwinism was becoming more substantial and more encompassing. It should be noted, however, that Anuchin showed much stronger interest in evolution as a body of empirical facts than as a system of theoretical principles.

Anuchin's surveys of human paleontology kept Russian readers well informed about the growing list of hominid fossils and about the general anthropological interpretations of the evolution of man. For example, Anuchin surveyed the significant details of Hermann Klaatsch's involved effort to advance a polytypic interpretation of human ancestry, which claimed the dominance of different anthropoid strains in various human groups.[50] To play it safe, Anuchin assumed that the evolution of human society and culture should in no way be interpreted as a correlate of biological evolution. Of all Darwinian studies in Russia, the work on the biological evolution of man was easily the least advanced. This was the branch of science that experienced the most vigilant policing by the guardians of traditional values. Despite all the restrictive pressure, the educated segment of the population took the idea of the anthropoid origin of man as a foregone conclusion.

The commemorative literature, the zoology textbooks, and the *Progress of Science* volumes were different ways of recording a deep allegiance to Darwin's theory. The differences between the individual categories of surveys were minor: they showed up primarily in the intensity of sentiment expressed and in the choice of rhetoric. This entire complex of evolutionary literature elaborated and defended unorthodox Darwinism, the strongest theoretical orientation in contemporary Russian biology. It saw that the future of biology lay in combining the new advances in experimental biology with Darwinian tradition. The emphasis was not on the search for a certified evolutionary theory but on the recognition of complementary contributions of particular orientations.

Nikolai Konstantinovich Kol'tsov: The Prophet of Evolutionary Synthesis

As the memories of the celebration of 1909 receded, Darwinism faced a formidable challenge from the rising power of genetics, a new science that opposed some of the basic principles built into Darwin's theory. Timiriazev met the challenge by changing his strategy from a defense of Darwinism to a furious attack on the main representatives of genetics—particularly on Mendel, Bateson, Johannsen, de Vries, and Korzhinskii. A. N. Severtsov worked on giving Darwin's theory more precision and conceptual unity as a prerequisite for forging more productive relations with the avalanches of developments in the key branches of experimental biology. A rapprochement of the two wings of evolutionary thought could take place only after Darwinism had put its own house in order. N. K. Kol'tsov, a junior member of the academic community, distinguished himself as the main advocate of the immediate search for the unity of Darwinism and genetics.

A graduate of Moscow University, Kol'tsov was deeply imbued with the spirit of Darwinism. Selected as a "professorial candidate" by his university, he received sufficient funds at the turn of the century to continue postgraduate studies in Western universities of his choice. On this tour he spent some time in the university laboratories at Kiel and Heidelberg and at the Russian marine research station in Ville Franche, near Nice, France.[51] It was at this time that he initiated and completed a study of metameric features of the vertebrate head. His Western travel took place at a time of rapid reorientation in biology signaled by the rise of genetics. Deeply moved by the promises of new biology, Kol'tsov became convinced that to study the inner structure and dynamics of the cell is the same as to study the first spark of life and the prime motor of organic evolution. To study the cell, in turn, is to study the ties of life with the physical universe. He agreed with Umov's prophetic announcement that the study of life should be covered by the "third law of thermodynamics," expressing the work of natural forces opposite to entropy.[52] To study the cell is to study it as a "molecule of life." In the study of the physics and chemistry of the cell Kol'tsov saw the safest path to a scientific study of the foundations of biology.

In 1903 Kol'tsov returned to Russia and was immediately appointed a zoology instructor at Moscow University. His extensive and too conspicuous participation in the 1905 revolution on the side of the forces opposed to the government and his unorthodox scientific orientation prevented him from climbing professionally above the lowly rank of privatdocent. He even took time to write a book on police terrorism in the university community in 1905. In 1909 M. A. Menzbir, annoyed with Kol'tsov's criticism of the national government's university policies, denied him access to Moscow University's zoological laboratory. In 1911 Kol'tsov was among the professors and instructors who resigned from Moscow University in protest against the efforts of the Ministry of Public Education to curb academic autonomy. Menzbir, who previously occupied a high position in the university administrative hierarchy, was also among the professors who chose to resign. In 1914 Kol'tsov established a small laboratory for experimental zoology at Shaniavskii University, a recently founded private institution. The laboratory emphasized research in genetics.[53]

Before the October Revolution, Kol'tsov published a series of papers reporting the results of his inquiry into the principal forms of cells. Influenced and inspired by the German biochemist O. Bütschli, he was guided by the idea that twentieth-century biology should combine the comparative method and observation, standard research tools of nineteenth-century biology, with the analytical method of modern physics and chemistry. The introduction of the modern experimental method in biology, he wrote, did not imply a turn in an anti-Darwinian direction: on the contrary, it meant making Darwinism a twentieth-century scientific orientation. His writing and, particularly, his leading role in editing the journal *Nature*—which placed strong emphasis on developments in genetics and related disciplines—made him known as one of the most ardent supporters of new developments in experimental biology, including genetics. At the same time, his consistent and eloquent defense of Darwin's theory made him a model Darwinist.

Describing the situation in Russia during the 1910s, Kol'tsov noted the bewildering diversity of evolutionary views. He also noted that all Russian biologists were evolutionists, which, in his opinion, made them Darwinists as well. "It appears to me," he wrote, "that Darwin's great contribution—the high-level generality of his theory and the wide scope of the influence of his ideas on modern biologists—allows us to claim that there is no difference between Darwinists and evolutionists."[54]

Kol'tsov went on to say that although Darwin had established evolution as a universal feature of the living world, he did not explain the mechanisms of the transformation process. He noted two major efforts to answer the question of transformation. One group of evolutionists built on Darwin's view of evolution as a process of accumulating small random variations in heritable characteristics. The second group made Darwin's view of "genes" the key to the understanding of evolution. The interpreters of "genes" as carriers of the evolutionary process were, in turn, divided into two subgroups: while de Vries and his followers considered "genes" mutable in the same way that atoms of radioactive elements are mutable, the pure Mendelians undertook the difficult task of combining the notion of evolution with the notion of the immutability of "genes."

Kol'tsov made his view unambiguous and unwavering: "Genes, Mendelian characteristics, mutations, pure lines, and isogens, all these are important discoveries of our time that do not in any way contradict the foundations of Darwin's theory, but, on the contrary, give it more scope and more depth. These notions play the main role in artificial experiments in which the scientist controls crossbreeding—the main factor of evolution."[55]

Behind Kol'tsov's reasoning was a quiet recognition of two complementary facts of the riddle of Darwinism. First, Darwin did not give a clear and incontrovertible explanation of heredity and variation as mechanisms of evolution. His theory was much in need of further clarification and amplification. In this respect, Darwinian scholarship found it necessary to welcome the help of research perspectives presented by experimental genetics and related disciplines. Second, Darwin, more than any other scholar, contributed to making evolution the pivotal idea of modern biology. In this sense, every modern biologist, whether he admitted it or not, was a Darwinian scholar. In this respect, it was the recognized duty of all biologists to build on the tradition Darwin had originated. Kol'tsov noted, however, that the major task of these scientists was not merely to refine Darwinian thought but to add new research areas, new methods of inquiry, and new challenges to traditional thought.

Kol'tsov did not intend to say that "evolutionism" was not different from "Darwinism." He meant to say that under all conditions Darwin's ideas would continue to form a strategic component of the general notion of organic evolution. More than most of his colleagues, he fully realized that the future of Darwinism qua evolutionism was primarily in advancing the frontiers of genetics, a science well on its way to answering questions that, despite their great importance, did not attract Darwin's attention. He dedicated his scholarship to strengthening the cause of Darwinism by helping genetics develop deeper roots in Russian culture. His papers on "the molecules of life," published in the 1920s, may rightfully be considered an important step in the series of developments that laid the foundations for molecular biology.

There was the strong possibility that Kol'tsov used Darwinism as a shield to protect the burgeoning research in genetics. An established orientation, Darwinism drew strength not only from its own scientific and philosophical triumphs but also from the vested interests in the leading universities and from popular support, so clearly manifested during the anniversary celebrations in 1909. By making genetics a component of the Darwinian scientific framework, Kol'tsov hoped, first, to placate the leaders of Darwinian orthodoxy, who looked with the utmost suspicion at most new developments in experimental biology, and second, to accelerate the acceptance of genetics as a legitimate academic field. He knew that the inordinate strength of Darwinism in the major universities was one of the major reasons for pronounced sluggishness in the early development of Russian genetics. Kol'tsov's strategy helped to mitigate the initial conflict between genetics and Darwinism and to create a tradition working in favor of an evolutionary synthesis.

Praiseful recountings of Darwin's contributions to science, rekindled by the anniversary celebrations in 1909, soon gave way to more realistic assessments. The rising criticism, however, did not prevent the scientific community from continuing to credit Darwin with making biology a true science, a study of natural causes operating in the world of life. Darwin received credit also for making the historical method the main tool of biological analysis. In all other respects, Darwin's legacy invited varied, and often contradictory, assessments. In the maze of interpretations of the place of Darwinism in the mainstream of biology, four names were clearly in the limelight. (1) Timiriazev represented a small but influential group that considered genetics a passing aberration that was in no position to challenge the dominant status of Darwin's theory. (2) Kol'tsov saw the future of evolutionary biology in a synthesis of Darwinism and genetics—a synthesis in which Darwinism would provide the basic frame of reference. (3) Filipchenko also was inclined to look toward a synthesis of Darwinism and genetics, but a synthesis in which genetics would provide the basic framework and Darwinism would play an auxiliary role. (4) Komarov thought that evolutionary biology had entered a new state of development, marked by the coexistence of independently functioning theoretical and methodological orientations. He emphasized the autonomy, rather than the synthesis, of various biological orientations. In his view, Darwinism was destined to provide only one of the many possible approaches to the scientific study of the organic world. He wrote that the reign of "universal theories" belonged to the past and that modern biology preferred specific theories explaining separate categories of problems.[56] He appreciated the gigantic proportions of Darwin's contributions to the triumph of the evolutionary view in biology; at the same time, however, he was firmly convinced that answers to many key questions on the origin of species were outside the competence of the theory of natural selection.

CHAPTER TEN

Darwinism and the Radical Intelligentsia

To talk about the radical intelligentsia in Russia is to talk about partially or fully formed ideological movements dominated by elaborate beliefs in the inevitability of social progress and in the coming doom of the autocratic regime. Three of these movements—populism, anarchism, and Marxism—stood out because of minute elaborations of distinct sociopolitical philosophies, basic criticism of the existing structure of Russian society, and strong roles in shaping and directing political opposition. They did not reach the state of full growth until the second half of the 1880s. The leaders of these movements were strongly influenced by the nihilist intelligentsia, which peaked during the early 1860s.

The three movements shared a strong belief in science as the main source of social progress and humanitarian values. All acknowledged the gigantic proportions of Darwin's contributions to modern science as a body of positive knowledge and a most notable achievement in emancipating the human mind from the tyranny of prejudice and superstition. They also shared serious doubts about certain aspects of Darwin's theory, particularly about the applicability of the biological principles of evolution to the social and cultural fabric of human existence.

Populism

Like nihilism, populism was a product of social and cultural stirrings that made the reforms of the 1860s a historical watershed. It was more stable and less bellicose than nihilism. In the philosophical and sociological elaboration of its ideological program it differed from nihilism in four notable respects.

First, its chief architects firmly rejected all efforts to ally positivism with materialism. While accepting the positivist view of science as the major source of social progress, they refused to give allegiance to materialistic ontology.

Second, the populists did not agree with the nihilist claim that a new Russian society could be built only on the ruins of the old social system. Any future Russian polity, they argued, must be built on the democratic residues of traditional national institutions rather than on Western models. By relying on selected elements of national tradition, Russia could even elude capitalism.

Third, unlike the nihilists, the populists rejected every effort to treat nonscientific modes of inquiry—philosophy, art, and ethics—as mere subsidiaries of science. Nor did they accept the nihilist elaboration of Buckle's view of science as the only true fountain of the democratic spirit. All this did not prevent the populists from applauding the great strides science had made in their time and to think seriously of the enormous potential of a scientific study of social dynamics and of a scientific approach to social reforms.

Fourth, unlike the nihilists, the populists rejected the idea of social sciences as theoretical and methodological extensions of the natural sciences. They anticipated the emergence of the Baden school of German neo-Kantianism, which concentrated on logical and methodological differences between natural and social sciences.

N. K. Mikhailovskii, a major contributor first to *Fatherland Notes* and then to *Russian Wealth* and a generally prolific writer on social themes, is usually considered the spiritual leader of populism. His writing is not only a rich display of populist criticism but also a most reliable indicator of turning points in the evolution of populist ideology. By his own admission, Mikhailovskii started his long and distinguished writing career under the influence of the journalistic activity and world outlook of N. D. Nozhin. Beginning as a promising embryologist, Nozhin lost no time in making social criticism his main concern. His articles in *Book Herald*—where Mikhailovskii started his writing career—evoked favorable response from contemporaries critical of orthodox thought but weary of nihilist excesses.

By his own admission, Nozhin conducted biological research for the purpose of gaining insights into the inner dynamics of human society. He wrote: "I direct the attention of the reader to the fact that the main goal of my anatomic and embryological studies is to discover the laws of the physiology of society." [1] Nozhin was eager to point out not only the strengths but also the limitations of biological models in social theory. He did not object to making the idea of transformism the guiding principle of sociology; he demanded, however, that it be adapted to the specific features of society as a reality sui generis.

Nozhin taught Mikhailovskii to recognize both the strengths and the weaknesses of Darwin's theory. The strength of Darwin's evolutionary theory lay in its making "the transformation of organisms" the central problem of biology. According to Mikhailovskii: "At the present time we can fathom only a small part of the monumental significance of [Darwin's] powerful thought. In due time, a fuller study of this theory will undoubtedly prove that never before did man have at his disposal such an exact, broad, and fertile idea." This idea is covered by the label "transformism," which Mikhailovskii preferred to "evolution"; "evolution," in his view, does not cover all the manifestations of "transformism." He thought, for example, that processes that lead to the emergence of new species include both gradual (evolutionary) and saltatory (revolutionary) changes.

The weakness of Darwin's theory, as Nozhin saw it, lay in its viewing the struggle for existence as the chief mechanism of evolution. Mikhailovskii had no reservations about endorsing Nozhin's claim that "Darwin did not see that the struggle for existence was not an instrument of development but only a source of pathological phenomena," and that "for this reason, Darwin's entire theory can be termed the theory of a bourgeois-naturalist." [2] In opposition to Darwin, Nozhin formulated his own law, according to which closely related animals are united by common interests and cooperation: they are not split by a division of labor and competition. This "law" served as the pivotal point of Mikhailovskii's sociological criticism of Darwin's evolutionary theory. [3]

S. N. Iuzhakov described the chief dilemma of populist thought when he noted that Mikhailovskii gave prominence to two opposite assessments of Darwin's evolutionary view. On the one hand, he treated Darwin's contribution as a scientific feat of the same magnitude as Copernicus's and Newton's achievements. On the other hand, he viewed the principles of the struggle for existence and natural selection as totally unacceptable because they denied the role of "cooperation" and "solidarity" as a factor of organic and social progress. [4] Mikhailovskii felt no need to compare and examine critically and methodically the two disparate appraisals of Darwinian thought.

Mikhailovskii used many occasions to state his views on Darwin's theory.[5] While rich in incisive comment and broad in substantive coverage, his discussion tended to be fragmented and diffuse. In most cases, his analysis of Darwin's scientific notions belonged to a sustained effort to articulate the finer points of populist ideology. On the issue of the anthropoid ancestry of man, pregnant with ideological implications, Mikhailovskii stood firmly on Darwin's side:

> The folk tradition of all peoples ascribes a more or less high origin to man. Darwin is perfectly correct in asserting that the folklore imputation of a divine or semidivine descent of man is only an illusion that does not flatter the human species; what flatters man immensely more is the idea that he has risen from lower spheres—from the depths of nature. In fact, this is the only viewpoint that allows for the advancement of man; all other views assume that man has fallen and disgraced his ancestors.

Mikhailovskii recognized two independent and rather discordant steps in Darwin's elaborate handling of the struggle for existence. The first step consisted of making a specific feature of capitalist society an attribute of the living universe—of transforming a unique type of social behavior into a general law of organic evolution. Mikhailovskii thought that the origin of the struggle for existence was in "the moral and political conditions of contemporary Europe"—in capitalist values. The second step consisted of extending the struggle for existence, as a universal law of organic evolution, to the structure and historical dynamics of human society in general—of transforming a biological idea into a sociological law. The struggle for existence was first a specific feature of capitalist society, then a biological principle, and finally a universal law of both biology and sociology.

Mikhailovskii expressed high regard for Darwin as a man and a scientist. He credited Darwin with placing biology on a solid and reliable path to full citizenship in the realm of positive knowledge. He praised Darwin for making biology an empirical and inductive science and for emancipating it from mysticism and metaphysics. In his view, evolutionary biology made sociologists aware of the high standards of scientific inquiry: rigorous methodology, tight verification procedures, and precise theoretical organization and integration of empirical data. It alone provided the necessary instruments for the study of human social life at the animal level.

Mikhailovskii found much in Darwin's theory to disagree with on both biological and sociological grounds. On biological grounds he echoed the criticism advanced by some of the leading critics of Darwin's views. He favored both Kölliker's notion of heterogenesis, which challenged Darwin's principle of *natura non facit saltum,* and Nägeli's view of evolution as a process moving in a definite direction. In his view, Darwin and Darwinists erred in making random or "perturbational" change the main source of variation in plants and animals and in identifying all variation as an unfailing path to the evolving complexity of organisms.[6] Mikhailovskii joined the critics of Darwin's theory who claimed that natural selection is essentially a conservative process: even if it conserves and improves the variations that survive in the struggle for existence, it cannot originate new variations. He was equally skeptical about the Darwinian identification of organic evolution as a progressive process: "Changes a species undergoes as a result of struggle, selection, and useful adaptation can go in all possible directions. In accordance with the theory of probability, chances for unilinear-progressive change . . . are not stronger than chances for a unilinear-regressive development, that is, in the direction of complete extinction."[7]

Mikhailovskii was willing to recognize, on a limited scale, the struggle for existence as a factor of the evolutionary process. He was unwilling, however, to accept Darwin's explanation of the struggle for existence as the sole or the main factor responsible for evolution as a progressive process.[8] In the growing "physiological division of labor," rather than in the struggle for existence, Mikhailovskii saw the key to an understanding of "biological progress." He identified the evolutionary regularity of this division as "the law of development" (*zakon razvitiia*). The evolution of the physiological division of labor, as he saw it, led either to the emergence of new organs or to the increasing functional differentiation of existing organs—and, as a result, to enhanced adaptability of organisms to the environment. Since the struggle for existence exercises only a "perturbational" influence on the development of organs, it interferes with the normal line of "organic progress."[9]

Mikhailovskii's biological criticism of Darwin's theory was fragmentary and rather perfunctory. By contrast, his sociological criticism was comprehensive and much more methodical. Like the leading sociologists of his age, particularly before the 1890s, Mikhailovskii dealt primarily with the idea of social progress. Opposed to every kind of biological reductionism in sociology, he viewed human society as a reality *sui generis,* often involved in processes unknown, or contradictory, to the biological universe. In "What Is Progress?" he rejected Spencer's effort to develop a grand sociological theory by depending solely on organic analogies. He rejected not only the organic theory in sociology, which viewed human society as an organism, but also Darwin's evolutionary theory as a source of sociological models. Both Spencer and Darwin, he said, emphasized the biological notions of the division of labor and of competition as prime movers of social organization and dynamics.[10]

Spencer and Darwin, in Mikhailovskii's view, relied heavily on Karl von Baer's law, which defined evolution as a growing differentiation and specialization of life processes—as a growing division of labor in organisms. What Mikhailovskii opposed specifically was the sociological use of von Baer's "law": the treatment of the growing division of labor in society as the most reliable indicator of social progress. In Mikhailovskii's view, the differences between organic and social progress are fundamental: the progress of organisms is measured by the increasing division of labor among organs; the progress of society is measured by the decreasing division of labor among individuals.[11] In the gradual realization of the wholeness, integrality, inner unity, or harmonious development of the individual Mikhailovskii saw the infallible sign of social progress.

The wholeness of the individual, as Mikhailovskii portrayed it, could not be achieved by competition between various social groups or social strata—it could come only from solidarity and cooperation, the twin pillars of sociology. Not the struggle for existence but the "struggle for individuality," propelled by solidarity and cooperation as primary conditions of social unity, is, according to Mikhailovskii, the true source of social progress. On the level of organic evolution, progress is indicated by a growing dissimilarity between organs; on the level of human society, progress is indicated by a growing similarity between individuals. In defending the similarity of individuals as an index of social progress, Mikhailovskii defended an ideology that considered the *obshchina*, based on simple cooperation and minimal division of labor, the most favorable model for a future Russian society.

In essence, Mikhailovskii's criticism of Darwin's theory was part of an effort to marshal sociological arguments favoring a utopian blueprint for a society prepared to bypass the "capitalist phase" in the grand scheme of evolutionary change. In the process of advancing his arguments, he also concentrated on presenting Darwin's theory as a special reflection of industrial-capitalist values. "Darwinism," he wrote in 1876, "could appear only in our time, not because such an array of biological knowledge did not exist earlier . . . , but primarily because our time is the time of the triumph of such Darwinian principles as the

struggle for existence, divergence of characters, and adaptation."[12] He
stated: "Darwinism not only does not provide weapons against inequality and privilege in general, but, on the contrary, it gives them a new and
firmer basis. . . . Darwinism is not only undemocratic in its essence but
it moves sharply and resolutely in making inequality and the struggle for
higher positions in society the cornerstone of ethics and politics."[13] In
natural selection and the struggle for existence, treated as sociological
categories, Mikhailovskii found the main barriers on the path to *obshchina* socialism. In all this, he made it clear that he criticized, not
Darwin, but the flagrant and quite common abuses of his theory by
those who claimed to be his followers. Some of these "abuses" came
from Darwin himself, particularly in the "sociological" sections of *The
Descent of Man.*

In dissecting Darwin's legacy, Mikhailovskii operated on the assumption that all that is scientific is not necessarily moral. He classified
Darwin's theory among those components of scientific knowledge that
belied Buckle's claim that scientific knowledge, by its very nature, is a
most reliable indicator of social progress.[14] Even if Darwin's theory met
scientific standards, it did not always meet moral expectations. Mikhailovskii had no reservations about viewing Darwin's theory as the primary intellectual source of Social Darwinism, the backbone of a new
"sociology of inequality." To bolster his critical stance, however, he
found it compelling to challenge Darwin both on scientific and on moral
grounds. Despite fortuitous concessions, he considered natural selection
indefensible both as a scientific concept and as a source of moral precepts. Critical of Darwin's theory, Mikhailovskii was blindly uncritical
of the arguments put forth by anti-Darwinian biologists. Among other
arguments against Darwin's biological reasoning, he claimed that the
struggle for existence is not an unfailing path to organic progress—that
the stronger groups are not always victorious in natural selection. In
other words, survival in the struggle for existence is not always a step
upward on the evolutionary ladder. This argument found support among
some of the leading biologists, including Karl von Baer.

Mikhailovskii both accepted and rejected Darwin on ideological
grounds. On the one hand, he saluted Darwin for his appreciable role in
the struggle for the full emancipation of modern thought from mysticism and metaphysical doctrines. He viewed Darwin's science as a
source of powerful arguments against the sacred values of the ancien
régime. Most of all, he had high respect for Darwin's contribution to the
triumph of the historical view of nature and human society. On the

other hand, Mikhailovskii criticized Darwin as a man whose scientific theory supported an ideological position that contradicted the populist dream of a society based on *obshchina* principles of equalitarian cooperation. It was this view that prevented him from undertaking a dispassionate, systematic, and logically consistent analysis and interpretation of Darwin's scientific legacy. In marshaling anti-Darwinian arguments, Mikhailovskii did not think about the welfare of science; he merely protected his version of populist ideology. He was an ardent spokesman for the scientific world view so long as it did not stand in the way of the populist vision of a better society. To a nihilist, science was socially beneficial by its very definition; to Mikhailovskii, as a representative of populism, science was socially beneficial only on a selective basis—only when its products did not violate moral principles.

Darwin's notion of organic evolution assumed an upward, progressive course in the transformation of living forms—a movement toward increasing morphological and physiological efficiency of organs. Mikhailovskii's notion of social evolution implied a retrogression to rural existence—to technological simplicity and reduced division of social labor. In *obshchina* he saw a higher *type* of social organization at a "lower stage of development." The populist goal was to elevate this particular *type* of social organization to a higher *stage* of development. This could be achieved only by improving the social mechanisms of "simple cooperation," that is, cooperation unobstructed by a complex division of labor.[15]

Mikhailovskii's populism consisted of two disparate ideas: the utopian dream of an ideal society, dominated by the equalitarianism of a rural community built on simple cooperation,[16] and the positivist belief in the intellectual supremacy of the scientific attitude as a most desirable path to social progress. As Mikhailovskii and most populists saw him, Darwin lost points as an "ideologue" of capitalist values; but he more than made up for this loss by a gigantic contribution to the triumph of scientific thought in modern society. His work presented a crystalline expression of the intellectual superiority of inductive thought, secular outlook, and critical judgment. On the practical level, for example, Darwinism helped open the institutional vestiges of the feudal era to devastating criticism.

Populism was a generic term for a wide range of theoretical designs for social change in Russia. Populist writers believed that human societies do not necessarily follow the same line of development. Every society is presented with various developmental possibilities. Performing his duty as a scholar, the sociologist is expected to point out the most promising and most desirable possibilities for the progressive development of individual societies. He is also expected to avoid heavy reliance on natural science models and on nihilist treatment of science as a panacea for all social ills. The similarities in the basic social philosophies of populist leaders were matched by strong differences in personal styles of argumentation, temperaments, intellectual preferences, and modes of operation.

In their plans and dreams for a perfect society, populist leaders operated strictly within a Russian framework. They wanted a society built on Russian values and ideals. They did not hesitate to present Darwin as a product of English society and an expression of English values and ideals. Mikhailovskii wrote in 1871 that Darwin adhered to a "narrow but powerful English emphasis on the practicality of knowledge." He treated the struggle for existence solely as an instrument of "useful adaptations" of plants and animals to their environment. Like Bacon, Hobbes, and Bentham, he also showed a strong inclination to immerse himself in details and to apply a limited number of principles to as many ramifications of life as possible.[17] Mikhailovskii described Darwin's work as "brilliant, sharp, consistent, and masterful," but also "too narrow and slow-moving." In all this Mikhailovskii wanted to show that Darwin's thought was a product and an expression of values not necessarily related to Russian culture.

Despite its broad sweep and unusual richness in insightful details, Mikhailovskii's discussion of Darwin's ideas did not exercise a strong influence on the evolutionary controversy in Russia. The reasons for this must be sought in Mikhailovskii's generally ambivalent and contradiction-ridden attitude toward Darwin. Mikhailovskii went on record as a bitter foe of the use of organic analogies in social theory, but this did not prevent him from making sociological use of Haeckel's hierarchy of six types of morphological "individualities" conceived purely as organic units. Two mutually exclusive notions formed the conceptual core of his sociological theory: "the struggle for individuality," which treats the individual as the centrifugal unit of human society, and the evolution of "solidarity," a centripetal social force. He made no effort to reconcile the notion of "individuality" as a self-contained and self-sufficient personality with the notion of "solidarity" as a chief factor of social integration.

Mikhailovskii's attitude toward the cultural value of Darwin's work in science was equally ambivalent. In some of his writings he praised Darwin's contributions to the harnessing of science for a sustained war against prejudice and ignorance; in other writings he attacked the unethical implications and the undemocratic orientation of Darwin's evolutionary principle. To him Darwin was both an equal of Newton as an intellectual giant and a direct contributor to the moral pathology of Social Darwinism. A similar ambivalence is built into Mikhailovskii's general views on science. On the one hand, he made laudatory statements on the culture of science as a configuration of values that emphasize a constant war against dogmatic thought. He praised objective standards for the certification of knowledge and the practical utility of accumulated wisdom. On the other hand, he could not but relate science to technological progress and specialization directly opposed to his dream of "simple cooperation" as a measure of an ideal society.

Next to Mikhailovskii, Petr L. Lavrov and N. I. Kareev contributed most to the translation of populist ideology into an elaborate sociological theory. At the time Mikhailovskii had begun his sociological writing Lavrov was already a well-known social theorist, working on a general synthesis of his ideas. Despite Lavrov's great erudition, Mikhailovskii is generally considered the chief architect of populist ideology.[18] Lavrov was not only an ideologue but also an intellectual historian of the first order. He spent the last three decades of his life in western Europe, where a personal acquaintance with the leaders of various revolutionary movements played a key part in shaping his ideological orientation. Kareev, by contrast, performed an important role in carrying the populist ideas to Russian universities; he made bold efforts to integrate populist ideas into modern sociology and the philosophy of history. He was widely acclaimed as the first historian in Russia and abroad to write a systematic and comprehensive study on the state of the French peasantry during the revolutionary period.

Lavrov considered Darwin "one of the greatest minds of our time" and readily acknowledged the singular role of the evolutionary theory in shaping both modern biology and a modern world outlook. Darwin's basic contribution, he thought, was in reinforcing the idea of the unity of nature:

> From 1848 to 1864 the bankruptcy of the old order had become increasingly more evident. The theory of evolution and particularly of Darwinian transformism has given a new unifying principle to all spheres of thought and life. This has happened at the time when the conquests of synthetic

chemistry have brought the organic world closer to minerals, and when spectrum analysis . . . has laid the foundations for the idea of the scientific unity
of physical and chemical study and of the entire universe.[19]

Lavrov readily acknowledged the extensive role of evolutionary theory in shaping modern biology. In addition to those of Darwin, he acknowledged the contributions of Wallace, Weismann, E. Ray Lankester,
and Romanes to the triumph of the evolutionary idea in modern biology.[20] He thought that Haeckel's elaboration of Darwin's theory enriched science as a special mode of inquiry, an instrument of enlightenment, and a world outlook. The *Origin of Species*, in Lavrov's view,
built the foundation for a scientific approach to animal systematics; it
achieved this by placing primary emphasis on the genetic unity of animal forms.[21] Lavrov made his acceptance of organic evolution clear and
unconditional:

> All leading modern scholars think that the evolution of organisms and the
> animal origin of man are as firmly established as the conservation and trans-
> formation of energy in physics, and as gravitation in astronomy. Despite
> their general scarcity, paleontological finds show consistent uniformity in the
> evolution of living forms. Fritz Müller has clearly established that the knowl-
> edge of the embryonic growth of individual organisms, which appears as an
> abbreviated and simplified repetition of the development of species, makes it
> possible to reconstruct the successive stages in the development of the or-
> ganic world, inaccessible to us in any other way. A more careful study of
> atrophied and rudimentary organs, of the struggle for existence between
> various parts of an organism, and of direct links between separate forms of
> domestic plant or animal life will provide new confirmations for the grand
> picture of mankind evolving from the primordial residues of life.[22]

One of the most eloquent and consistent Russian exponents of the
grand theory of social evolutionism, Lavrov built his evolutionary view
on an "anthropological principle": he was interested in the evolution of
plants and animals only insofar as it contributed to a fuller understanding of the evolution of man. This principle consisted of several intellectual layers, beginning with solid ingredients of Comte's positivism and
ending with Sir Edward Tylor's theory of the universality and unilinear-
ity of cultural evolution. Lavrov's "anthropological principle" designated first a general scientific approach to the study of man, then a philosophical orientation, and finally a sociological theory.[23] The primary
goal of all these strategies was to establish the "normal order" of successive stages in the evolution of human society. Lavrov wrote: "The
most important task of the study of history is to discover the law of the

normal succession of stages in the evolution of social life, individual groups, and mankind as a whole."[24]

Lavrov argued fervently that no natural science model—even if it came from evolutionary biology—can answer all the key questions of social evolution. The scientific approach to social evolution must concentrate not only on the universal aspects of the succession of historical phases, but also on the idiosyncratic historical features of individual societies. It must also study the evolutionary significance of the purposive nature of human action. The teleological element separates social evolution from organic evolution; it makes social evolution both an objective and a subjective reality. In Lavrov's anthropological orientation man is simultaneously an object and a subject of inquiry. Man is both a product and a creator of society. In human thought and action, "the subjective truth of ethics is no less important than the objective truth of physics." Sociology is a synthesis of the subjective method of ethics and the objective method of physics.

Mikhailovskii equated social evolution with the development of harmonious, or symmetrical, personality.[25] The division of labor in society could produce only disharmonious, fragmented personality. This was why he placed such a strong emphasis on the "wholeness" of personality as the most dependable index of social evolution. Lavrov, by contrast, equated social evolution with the development of "critically thinking" individuals.[26] Under the spell of Auguste Comte's positivist philosophy, he believed that the growth of "positive knowledge"—or science—was the real force behind the progressive meaning of social evolution.

In Lavrov's thinking there is yet another index of social progress: the growth of cooperation, harmony, and mutual aid as the propelling force of social integration. Mikhailovskii, too, believed in the evolution of cooperation as the essence of social evolution, which led him to dispute Darwin's emphasis on the evolutionary role of competition—the unending and fateful struggle for existence. Unlike Mikhailovskii, Lavrov showed many signs of willingness to consider the evolution of cooperation as a unique manifestation of the struggle for existence.[27] He stressed mutual aid and cooperation without abandoning struggle, conflict, and competition. To satisfy the Darwinian tradition, he did not argue that solidarity, a condition of social cooperation, replaced the struggle for existence, a condition of social competition; he merely noted that solidarity gradually became the primary mechanism of the struggle for existence.[28] In the course of evolution, the struggle for existence takes on many forms: cooperation is one of them.

Lavrov wrote that many writers considered the struggle for existence not only a law of the organic world but also a law of all natural objects from celestial bodies to the sounds of human language. The fact that the struggle for existence is universal does not mean that, in the long run, it may not change so much as to "differ from its primordial appearance in the same way as an insect differs from the larva from which it developed. An unending process, the struggle for existence may give birth to new features that differ radically from its previous characteristics."[29] Small wonder that in the long run the process of evolution may transform competition into an effective way of safeguarding social solidarity and cooperation.

In his discussion of cooperation as a mechanism of social integration, Lavrov was much more realistic than Mikhailovskii.[30] He rejected Mikhailovskii's apotheosis of "simple cooperation—based on a negation of the division of labor in society"—which amounted to presenting social evolution as a "regressive phenomenon." To assure a triumph for simple cooperation, he thought, would be nothing short of a return to the prehistorical phase in the history of man. Mikhailovskii's "formula for progress," which emphasized a reversal of the trend of the growing division of labor in society, meant a retreat into hopeless obscurity.[31] Mikhailovskii, he said, preferred a "pleasant" but subjective view of social evolution to an "unpleasant" but "true" interpretation. Like Mikhailovskii, Lavrov accepted the *obshchina* as a model of higher achievement in social organization, but, unlike Mikhailovskii, he saw in it a combination of "simple" and "complex" cooperation.

Lavrov's *Anthropological Life* presents a massive reconstruction of the earlier phases of human society viewed in the light of biological and social evolution of solidarity.[32] He recognized three basic "sociological" phases in the evolution of the animal world. During the first phase, prior to the formation of firm social ties, the struggle for existence is dominated by the Hobbesian principle of *bellum omnium contra omnes*. During this phase the biological or instinctive base for solidarity has no opportunity to express itself. During the second phase "unconscious solidarity" becomes the most important mode of the struggle for survival. This solidarity has an instinctive basis. During the third phase solidarity becomes a conscious—a deliberately chosen—weapon in the struggle for existence and is strictly a human characteristic. This does not mean, however, that the evolution of conscious solidarity is an automatic process in human society. History has known social systems that, in Lavrov's opinion, negated conscious solidarity and gave unlimited sway to the Hobbesian war as a universal principle of social relations. These social systems represent pathological digressions from the normal course of social evolution.

In comparison with Mikhailovskii, Lavrov was much more favorably disposed toward Darwin. He built a philosophy of history on the idea of social evolutionism, an offshoot of Darwin's biological theory.[33] While working on the general theory of evolution, he kept up with and learned from the advances in biology. He reasoned that, in order to survive as a major strategy for a scientific study of the organic world, Darwin's theory must experience incessant elaborations, refinement, and expansion. The Lamarckian tradition, he thought, should be much more firmly integrated into Darwinian thought. Moritz Wagner's suggestive ideas on the evolutionary role of geographical isolation appealed to him, as did George Romanes's views on "physiological selection."[34]

As Lavrov saw it, social evolution is made up of two parts. The grand history of cooperation or solidarity, the basis of human society, is the first part. Gradual but inexorable growth of useful knowledge is the second part. Clearly under the spell of Comtian positivism, Lavrov considered the evolution of scientific knowledge the most significant indicator of social progress. In the Darwinian revolution he saw a triumph of modern science and scientific spirit. Darwin gave mankind "a theory of evolution and transformism that has given new unity to the entire world of thought."[35] The rise of Darwinism coincided with two "remarkable" developments in the modern world: the advent of "scientific socialism" in the West and the rise of realism in Russian art and literature—the two major participants in the struggle for social and political reforms.

While Mikhailovskii concentrated on laying the foundations for a general sociology, and Lavrov worked on a positivist and evolutionary version of the philosophy of history, N. I. Kareev made an effort to unify the two concerns. A professor of history, first at Moscow University and then at St. Petersburg University, Kareev presented his arguments carefully and systematically, and he tried to keep his text clear of obvious ideological pronouncements. Not directly involved in defending the cause of populism, he was much impressed with Mikhailovskii's and Lavrov's discourses on the subjective nature of social theory. In *Introduction to the Science of Sociology* (1897) he viewed the unity of "objective" causality and "subjective" purposiveness as the true method of sociology.[36]

During the second half of the nineteenth century, the theory of progress, as Kareev saw it, came from three distinct intellectual sources:

positivism, transformism, and evolutionism. Positivism, a creation of
Comte's, explained human society and culture. Transformism, particularly in Darwin's version, explained the living world in general. Evolutionism, a product of Spencerian thought, explained the entire universe.[37]
Taken together, the three orientations elucidated the sociological, biological, and philosophical sides of progress, a reigning idea of nineteenth-century thought. Kareev contended that Spencer's cosmic law of
progress was too general to be useful outside the domain of philosophy.
He was primarily concerned with showing that Darwin's "law" of biological evolution (and progress) not only did not coincide with Comte's
"law" of social evolution (and progress) but, in some of its basic features, was contrary to it. Darwin's biological theory, in Kareev's view,
invites three kinds of major criticism.

First, it represents a kind of transformism that has little sociological
value. Biological transformism, applied to society, can shed light on
only a limited number of questions: the origin of man, the development
of the most elementary mental capacities and social instincts, and the
emergence of community life.[38] The notion of organic transformation
applied to man as an animal, but not to human society as a creator and
a product of culture.

Second, it views the evolution of organic forms exclusively as a mechanical approach that allows no room for purposive action. Even if the
mechanical approach to organic evolution is recognized as perfectly
valid for the organic world, this does not mean that it is useful also for a
study of social development. The psychological factor, the very essence
of human society, is irreducible to mechanical explanations. According
to Kareev, Darwin erred in attributing little value to the mental factor as
a vehicle of organic evolution.[39]

Third, it reduces plant and animal evolution to two principles: the
struggle for existence and natural selection. Kareev, by contrast, contended that this view invited a mistaken identification of the biological
category of "species" with the sociological category of "society." He not
only collected a mass of arguments refuting the use of biological analogies in sociology, but also opposed the use of sociological analogies in
biology. Like Mikhailovskii, he noted that before Darwin made the
struggle for existence a biological principle Malthus and Ricardo had
made it a key sociological concept.[40] Human society is built on solidarity, and solidarity has an organic and a cultural foundation. The cultural foundation consists of learned patterns of behavior anchored in
cooperation, the basic mechanism of social integration. Like Mikhailov-

skii, Kareev was much stronger in criticizing Darwinian sociology built on the notion of conflict than in spelling out the natural and cultural attributes and mechanisms of cooperation.[41]

Despite all the anti-Darwinian arguments on sociological grounds, Kareev's philosophy of history reflected the structural principles of Darwin's evolutionism. This was made particularly clear in Kareev's noted effort to refute N. Ia. Danilevskii's philosophy of history, built upon two anti-Darwinian principles: denial of the idea of human history as a unilinear and universal evolutionary process, and rejection of the notion of progress. In *Russia and Europe* Danilevskii advanced a cyclical philosophy of history—or theory of cyclical evolution—which emphasized formal (morphological) similarities and substantive (normative) differences between the "historical-cultural types," the highest points of social and cultural activity in different historical eras. Every successive historical-cultural type selected a particular cluster of cultural material, or values, for emphasis and elaboration: the Hebrew type concentrated on monotheistic religion, the Greek type on art, the Roman type on law and government, and the western European type on "positive science." A true representative of classical evolutionism, a child of Darwinian tradition, Kareev emphasized the universally uniform standards and indicators of social and cultural progress. He believed in absolute and universal norms for measuring progress. Danilevskii, an early relativist, contended that every cultural-historical type had—and should be judged by—its own standards. In the final analysis, Kareev wanted to establish the cultural links that connected Russia with the universal values of humanity; Danilevskii was interested primarily in Russia's cultural uniqueness.

If all populist theorists were united in rejecting, or undervaluing, the evolutionary role of the struggle for existence, the reasons must be sought in the unique features of populist ideology, which made the *obshchina* communalism, built on cooperation and social solidarity, the most desirable model for the Russian society of the future. In making their positions known, each theorist responded in a unique way. Mikhailovskii ruled out the struggle for existence without subjecting it to a careful and orderly analysis; Lavrov lost track of it in an expansive but vague discussion of cooperation and solidarity; and Kareev relied on logical deductions to render it a sociological notion of secondary importance. Despite the broad sweep of their criticism of the strategic positions of Darwinian evolutionary biology, the three populist writers were united in recognizing the gigantic proportions of Darwin's contributions to sci-

entific methodology, the cultural elevation of science, and historical orientation in the study of the living world.

Anarchism: Petr Kropotkin

Classical anarchism, with Petr Aleksandrovich Kropotkin as one of
its most erudite champions, had two Russian roots. From nihilism it inherited a strong faith in "rational egotism"—a moral stance based on
the philosophy of individualism—and the belief in scientific knowledge
as the only sure weapon in the uncompromising war against ignorance
and superstition, man's major enemies. From populism it inherited a
firm belief in cooperation as the cement that held human society together and guaranteed social and cultural progress. A total war on state
authority of every description gave it its most distinctive feature. Although Kropotkin spent his most productive years in western Europe,
his social philosophy was shaped primarily by Russian ideals and realities. He shared the basic beliefs of the rebellious intelligentsia who saw
the future of Russia in a system of social relations based on secular wisdom, ethical individualism, and primitive socialism. Martin A. Miller
said it all: "[Kropotkin's] return to his native land in 1917 (after many
unsuccessful efforts to do so) was the culmination of tendencies which
indicate that, in the last analysis, Kropotkin was a *Russian* anarchist."[42]
    After graduating from the Corps of Pages in St. Petersburg and serv-
ing one year as a *page de chambre* to Alexander II, Kropotkin opted to
join a Cossack regiment in East Siberia. Imbued with the scientific spirit
of the 1860s, he became deeply involved in the natural history of Siberia; his interests touched on physical geography, geology, ethnography,
and psychology. Siberia presented him with two extremes: the challenge
of unique and puzzling geography, and the challenge of the grievous
plight of downtrodden humanity. The Siberian experience gave him a
strong background for becoming not only a scientist but also the articulator of an ideology, which he later identified as anarchism, a rebellion
against injustices created by—and intrinsic to—the state. In both endeavors he acted as a scientist: as he saw them, natural history and anarchism were different manifestations of a world outlook that stressed
the cognitive primacy of science as a body of knowledge and a method
of inquiry. He defined anarchism as a social science based on the inductive method of natural science.
    Anarchism, Kropotkin emphasized, "is the inevitable result of that
natural-scientific, intellectual movement which began at the close of the

eighteenth century, was hampered for half a century by the reaction that set in throughout Europe after the French Revolution, and has been appearing again in full vigor since the end of the fifties."[43] It became fully developed only after the triumph of the natural science approach to human society.

As Kropotkin saw it, the "awakening of naturalism" during the 1860s owed a primary debt to Darwin. Darwin made biology a solid science by giving it an integrative and interpretive principle of universal magnitude. Moreover, Darwin's theory made it possible to reconstruct the "progressive evolution" not only of organisms as biological phenomena, but also of human society as a sociological entity. The concept of evolution emancipated both biology and sociology from the fetters of metaphysics. This concept attracted the attention of many pre-Darwinian scholars, but Darwin was the first to teach historians "to treat the facts of history in the same manner that naturalists examine the gradual development of the organs of a plant or that of a new species."[44] Darwin gave sociology both a law of social evolution and a scientific method—"the inductive-deductive method."

Despite his unalloyed admiration and respect for the enormous scope of Darwin's contribution to science and the scientific method, Kropotkin became widely known primarily for his relentless criticism of the struggle for existence, the conceptual pillar of Darwinian evolutionism. His fame rests on *Mutual Aid* (1902), a detailed and discursive compendium of zoological and ethnographic information in support of evolution as a progressive development of organic nature and human society based on cooperation and mutual aid, rather than on Darwinian competition and the struggle for existence.

In tracing the genesis of his concern with mutual aid as a primary mechanism of organic and social survival, Kropotkin pointed out three developments that made a particularly strong impression on him.

First, as a young natural historian working in Siberia he found it exceedingly difficult to document the struggle for existence as the prime mover of organic and social evolution. He noted many years later:

> I recollect myself the impression produced upon me by the animal world of Siberia when I explored the Vitim regions in the company of so accomplished a zoologist as my friend Polyakoff was. We both were under the fresh impression of the *Origin of Species,* but we vainly looked for the keen competition between animals of the same species which the reading of Darwin's work had prepared us to expect. . . . We saw plenty of adaptations for struggling, very often in common, against the adverse circumstances of climate, or

against various enemies . . . ; we witnessed numbers of facts of mutual support, especially during migrations of birds and ruminants; but even in the
Amur and Usuri regions, where animal life swarms in abundance, facts of
real competition and the struggle between higher animals of the same species
came very seldom under my notice, though I eagerly searched for them.[45]

Second, in 1879 Karl Kessler, former rector of St. Petersburg University, read a paper, subsequently published in the *Proceedings* of the St.
Petersburg Society of Naturalists, that emphasized cooperation and mutual aid as the major vehicle of the struggle for existence.[46] Kessler's
statements made a particularly strong impression on Kropotkin, not
only because they confirmed his Siberian experience, but also because
they expressed an idea that was widely shared and extensively documented by Russian naturalists. Kessler did not deny the struggle for
existence: all he wanted to show was that in this struggle it was "mutual
support" rather than conflict that occupied the decisive position.[47]

Third, in 1888 Thomas Huxley published his controversial lecture at
Oxford University entitled "The Struggle for Existence in Human Society," in which he portrayed the evolution of moral principles as a social and cultural force contravening the cosmic principle of universal
struggle.[48] Kropotkin resented Huxley's claim that morality was not
built upon—but was directed against—man's natural endowments. In
an effort to refute Huxley's arguments Kropotkin published a series of
articles in the journal *Nineteenth Century*. These articles made up *Mutual Aid,* published in 1902 and quickly translated into many languages.
Kropotkin stated his thesis simply and categorically: ethical principles
rest on man's natural endowments—they go along with, rather than
against, human nature. Sociality and mutual aid are built upon the
natural—instinctive—base of man's behavior.

Greta Jones has given a precise description of Kropotkin's thinking:

> In his book *Mutual Aid* (1902), Kropotkin took the argument back to
> Darwin's *Descent of Man*. Kropotkin argued that Darwin had shown that
> the greatest evolutionary advantage enjoyed by a species was "sociability."
> Hence the instinct of altruism gradually replaced that of self-interest. This
> was achieved without the diminution of natural forces—as Huxley sug-
> gested. On the contrary, it was nature itself which brought about sociability.
> Nor was the intervention of social institutions necessary to mitigate struggle
> and move evolution onto a higher path as Ritchie and Huxley believed. Far
> from this being the case, it was the intervention of social institutions which
> disrupted the harmony of nature.[49]

In his effort to displace the Darwinian struggle for existence as a "law
of evolution," Kropotkin fought with the zeal and passion of a true

crusader. *Mutual Aid* became a bible for anti-Darwinists often involved in a relentless campaign against the "materialism" of natural science. In all this, however, Kropotkin retained his admiration for Darwin's unmatched contribution to the scientific foundations of biology. Despite his heavy emphasis on the "instinct of harmony" and "sociality," Kropotkin did not ignore the evolutionary role of competition and struggle. He went so far as to assert that both cooperation and competition are rooted in man's instinctive endowment, but that only cooperation ensures "the progressive evolution" of human society. In his view, Darwin used the term "struggle" in a figurative rather than in a literal sense.[50]

Kropotkin recognized competition, or conflict, as a functional component of social life: whatever he undertakes, man always encounters natural or social barriers that he must overcome. Struggle is an ever-present component of social relations, but it has an ethical value only when it is directed against institutionalized barriers, such as the bureaucratic structure of the state, that stand in the way of man's creative power. To create harmonious social relations man must destroy the forces inimical to the instinctive base of sociality.[51] Kropotkin did not reject the struggle for existence as an evolutionary mechanism; what he rejected was Darwin's choice of conflict, rather than of cooperation, as the primary mode of the struggle for existence.

Encouraged by the popularity of *Mutual Aid,* Kropotkin carried his research in two quite separate directions. First, he became involved in a systematic study of the ethical structure of human society, built upon biological endowments, and second, he became very much concerned with organic evolution as a natural foundation for social evolution. The latter effort led him to a close scrutiny of the modern state of the scientific study of heredity as a factor of evolution and a meeting ground for Lamarck's and Darwin's biological theories.

Martin A. Miller has pointed out the main drive behind Kropotkin's new preoccupation with the fundamental issues of ethics:

> In the midst of his researches on mutual aid in the Middle Ages, Kropotkin wrote Grave that he was simultaneously "tracing the ethical feelings of man." He realized the enormity of his subject, which entailed nothing else than an ethical history of modern society. Considering the failure of both religion and science to construct a social morality, Kropotkin saw it as his task to fill this gap. He wanted to pose an alternative to Huxley's "immoral cosmic process," to Spencer's "altruism and egoism," to "bourgeois morality" and to "the bankruptcy of science." . . . All of Kropotkin's major writings of this period, most of which were either researched or written in the 1890s, bring together these complex aspects of social transformation.[52]

Kropotkin worked primarily on linking social evolution to the instinctive base of "sociality" or "mutual support." A criticism of Auguste Comte's sociology helped him clarify his views on this matter. Comte, he wrote, did not recognize that the "moral sense" depends primarily on the biological nature of man and that both are results of "an extremely long" process of evolution. Consequently, Comte did not see that "the moral sense of man is nothing else but a further evolution of the mutual aid instincts," which emerged in animal societies long before the appearance of the first manlike creatures.[53] Kropotkin expressed full agreement with J. M. Guyau's assertion that "moral approval and disapproval were naturally prompted in man by instinctive justice."[54] "Guyau," he said, "understood that morality could not be built on egoism alone, as was the opinion of Epicurus, and later of the English utilitarians. He saw that . . . morality includes also the instinct of sociability."[55]

In *Ethics,* his last work, Kropotkin made it abundantly clear that he did not consider Huxley a true interpreter of Darwinian thought. He pointed out that Darwin, in contrast to Huxley, recognized the origin of morality in an instinctive—natural—endowment for sociality. Unlike Huxley—and Spencer—he sought the origin of morality outside "the theories of coercion, utilitarianism, and religion."[56] Kropotkin turned to *The Descent of Man,* which recognized the instinctive base of sociability—and cooperation—as the true source of morality. In this work, as Kropotkin now interpreted it, Darwin made conflict an important factor only in relations between different kinds—species or varieties—of organisms.

In elaborating the theoretical principles of ethics, Kropotkin admitted that all the germinal ideas of his basic orientation came from Darwin's writings. Darwin's ideas, he noted, reached far beyond biology. As early as 1837 Darwin entered in his notebook this prophetic remark: "My theory will lead to a new philosophy." The application of the idea of evolution to the entire living universe marked "a new era in philosophy." Subsequently, Darwin wrote a sketch of the development of moral sentiments, which marked the beginning of a new ethics. This sketch threw new light on the efficient causes of moral feelings: it placed the entire body of ethics on a scientific basis. Darwin's ethical ideas represented a new elaboration of the lines of thought bequeathed by Bacon. They assured Darwin, according to Kropotkin, of a place among such founders of ethical schools as Hobbes, Hume, and Kant.[57]

Darwin's ethics, as Kropotkin saw it, rested on four principles: (1) the foundations of all "moral feelings" are in "social instincts"; (2) all

animals could acquire a "moral sense" as soon as they reached the level
of intellectual development approximating that of man; (3) the development of mutual sympathy—anchored in the instinctive base of sociality—determines the criteria of "public approbation and disapproba-
tion"; and (4) habit plays an important role in strengthening "social
instinct and mutual sympathy." [58] In developing his principles, Darwin
examined "sociability in animals, their love of society, and the misery
which every one of them feels if it is left alone, their continual inter-
course, [and] their mutual warnings." [59]

In *Mutual Aid* Kropotkin criticized Darwin for his choice of competition and conflict over cooperation and mutual trust as a primary vehicle of the evolutionary process. "The Morality of Nature," published
in 1905, restored his loyalty to Darwin. Now he implied that his moral
philosophy was neither more nor less than an elaboration of Darwin's
sociological principles. In his opinion, none of Darwin's immediate followers had ventured to develop Darwin's ethical philosophy. He lamented the death of George Romanes, whose study of animal intelligence
had given him the background for a discussion of "animal ethics." [60]

Having refocused his attention, Kropotkin undertook the challenging task of bringing Darwin's theory in tune with modern developments
in biology. In this field, he made his task easy by dismissing all varieties
of neo-Lamarckism and neo-Darwinism—as well as much of Mendelian genetics—as mere aberrations in the development of modern biological thought. He operated on the convenient assumption that neo-
Lamarckians and neo-Darwinists were inspired by considerations lying
"outside the true domain of biology." [61] Lamarck, he said, had nothing in common with "the vitalist and other theories of the German
neo-Lamarckians, of whom Francé (a distinguished botanist) and Dr.
Adolph Wagner are prominent representatives." [62] Weismann's—or neo-
Darwinian—views, as Kropotkin saw them, suffered from compounded
inconsistencies. Weismann began by giving strong support to natural selection and by rejecting the inheritance of acquired characteristics; he
gradually adopted new hypotheses, which actually recognized the role
of inherited characteristics in the origin of new species. [63] Then he reread
Darwin's works to find out that the English naturalist, particularly in
The Variation, recognized the evolutionary role of the inheritance of acquired characteristics, without sacrificing any of the power of natural
selection. [64]

With neo-Darwinism and neo-Lamarckism eliminated from his
agenda, Kropotkin began to think of the possibility of effecting "a syn-

thetic view of evolution,"[65] which would integrate the main features of Darwin's and Lamarck's theories. The new synthesis assumed a recasting of Darwin's struggle for existence by making sociality, rather than conflict, its major feature. It also recognized the direct influence of environment on the evolution of plants and animals and the inheritance of acquired characteristics. Kropotkin's work on the new synthesis consisted mainly of showing how Darwin gradually modified his ideas to bring them closer to Lamarckian positions. Now he claimed that the exaggerated emphasis on the role of conflict in the struggle for existence came from Darwinists, not from Darwin.

At least some Darwinists found it necessary to rebuke Kropotkin's approach to evolutionary synthesis. E. Ray Lankester, a classic type of contemporary Darwinist, showed particular bitterness about Kropotkin's calculated effort to separate Darwinists from Darwin. In Kropotkin's perfunctory handling of evolutionary principles and dogmatic adherence to Lamarckian ideas, he saw the work of an amateur in biological theory, guided more by ideological sentiment than by scientific knowledge. He accused Kropotkin of flagrant abuses in citing authoritative sources, in misinterpreting Lamarck's "laws," and in substituting "reckless assertion for experiment."[66] Ludwig Plate, an eminent German supporter of Darwin's theory, noted that "despite some good observations," Kropotkin's *Mutual Aid* was "thoroughly uncritical," particularly in interpreting the instinctive base of cooperation.[67]

Theodosius Dobzhansky had noted that Kropotkin answered the call of those who were determined "to rescue Darwin, or at any rate the theory of evolution, from Social Darwinists."[68] Darwin had recognized the general fact that organisms could participate in the struggle for existence either by fighting or by cooperating with each other, but he was too preoccupied with the former to give sufficient attention to the latter. Kropotkin filled in the gap by placing the primary stress on cooperation as the mechanism of survival. This did not mean, however, that he had made a major contribution to evolutionary biology. In Dobzhansky's words:

> Kropotkin's arguments made only a ripple on the evolutionary thought of his day, partly because scientific vindication of callousness was more desired than that of compassion, and partly because of the uncritical character of some of the evidence with which he, a dilettante in biology, tried to bolster his case. Moreover, at the turn of the century and later the thinking of evolutionists was going in a different and . . . wrong direction.[69]

Peter J. Bowler has noted that without an assumption of the inheritance of acquired characteristics, Kropotkin's evolutionary theory "would degenerate into a form of Darwinism, in which successful groups eliminated those with a lesser degree of cooperation in the struggle for existence."[70] Lamarckism, rather than Darwinism, allowed for altruism to "be learned by all and then inherited as an instinct." For this reason, "it can be no coincidence that Kropotkin essentially began to write in favor of the inheritance of acquired characters."[71]

The French biologist Yves Delange and his associate Marie Goldsmith expressed a favorable attitude toward Kropotkin's shift of emphasis from the struggle for survival to mutual aid as the major mechanism of organic evolution. They concluded their widely read *Théories de l'évolution* with laudatory comments on Kropotkin's discussion of the primary role of association and solidarity in the evolutionary process. They were particularly impressed with the rich promise of extending Kropotkin's biological ideas to the scientific study of human society—of the animal roots of various manifestations of human life.[72] Delange and Goldsmith presented Kropotkin's theory as an expression of a strong tradition in Russian biological thought. They considered N. A. Severtsov, M. A. Menzbir, and A. F. Brandt the leading lights in this tradition.

In general, the biologists paid little attention to Kropotkin's dubious search for a synthesis of selected currents of evolutionary thought and to his effort to reaffirm his admiration for Darwin. They paid much more attention to *Mutual Aid*, which warned against one-sided—Malthusian—interpretations of the struggle for existence. In essence, however, *Mutual Aid* and the studies that followed can best be described as an effort to reinterpret evolutionary biology in the light of anarchist ideology. Anarchist ideology, in turn, was a variety of populist thought—a unique emphasis on the progressive affirmation of the wholeness and inner harmony of human personality. Kropotkin and Mikhailovskii contended that harmonious personality was the main goal of social evolution, but they saw different paths to the realization of this goal. Whereas Mikhailovskii would do away with excessive division of labor in society, Kropotkin emphasized the elimination of institutional clusters that stood in the way of a full realization of instinctive sociability, the natural base of mutual aid. With regard to Darwin, the two ideologues differed in one important respect: while Mikhailovskii did not go beyond the *Origin of Species* and its elaboration of mutual conflict as the main vehicle of the struggle for existence, Kropotkin ended his long writing career with a fresh look at *The Descent of Man* and its suggestive thoughts about the instinctive base of sociality and cooperation as a natural base of the moral code. In *Ethics* Kropotkin paid tribute to Darwin, not only for pointing out the supremacy of the instinctive base of altruism over the instinctive base of conflict, but also for making this supremacy the foundation of modern ethics. In the second half of the nineteenth century, he wrote, all experts in ethics had been under the spell of evolutionary thought. "Darwin's theory," he added, "had a tremendous and decisive influence upon the progress of modern realistic ethics, or at least on some of its divisions." [73]

In his emphasis on mutual aid as the main wheel of evolution and progress, Kropotkin was firmly linked with the ideology of populism, particularly as Mikhailovskii presented it. The two were united also in paying homage to Darwin's science as a triumph of reason and empirical wisdom over sacred thought safeguarded by mysticism and irrationalist metaphysics. Both recognized the gigantic scope of Darwin's contribution to historicism and inductionism as the pillars of modern scientific thought.

Maksim Kovalevskii, the most esteemed professional sociologist in Russia, offered a plan to integrate Kropotkin's ideas into the mainstream of modern sociological thought. Ready to recognize the struggle for existence as the prime fact of evolution, Kovalevskii criticized Darwinian extremists who considered "struggle" and "cooperation" mutually exclusive social processes. Relying on citations from Darwin, Russian naturalists, and Western sociologists, he gave full support to Kropotkin's conceptualization of cooperation and sociability as a special mode of the struggle for existence. [74] Whereas competition operates on an interspecific level, cooperation is the mainstay of intraspecific relations. In cooperation, as a variety of the struggle for existence, Kovalevskii, obviously leaning on Durkheim, saw the prime mover of social consciousness. He was careful to caution his readers to accept Darwin's view of the struggle for existence "in a larger and metaphorical sense." [75]

Iakov Aleksandrovich Novikov, the author of *La critique du darwinisme social*, gave a limited support to Kropotkin's sociological views. A prolific writer, he spent most of his time in Paris, where he was one of the founders of the International Sociological Institute and an active member of the Paris Sociological Society. He was first known as a defender of organismic theory in sociology and then as a noted student of the psychological foundations of human society. Unlike Kropotkin, he opposed socialism of any description. Widely read in the West, Novikov's writings attracted little attention in Russia. Encouraged by Norman Angell, George Nasmyth, an American scholar, published *Social Progress and the Darwinian Theory*, devoted primarily to an analysis of Novikov's evolutionary ideas and critique of Social Darwinism.[76]

Novikov saw the major contribution of Darwinism in its liberating biology from the reign of spirits and in its "emancipating the human mind from the shackles of theology." He called this feat "one of the most important developments in the history of our species."[77] An equally important contribution of Darwinism was in the recognition of struggle as the prime mover of the living world. In the doctrine of "the survival of the fittest" Novikov saw the reign of "justice" in the realm of life. These two contributions made Darwinism highly respected among "the most enlightened and most liberal persons of our epoch."[78]

As Novikov saw it, there was also a dark side to Darwinism, embodied in Social Darwinism. Making only passing remarks about the great contributions of Darwinism, he devoted most of *La critique* to exposing and dissecting the evils of Social Darwinism, particularly its recognition of warfare and of "collective homicide" as the real mechanism of social and cultural progress. While assembling massive arguments against warfare, Novikov recognized several kinds of struggle in human society, which he categorized as physiological, economic, political, and intellectual struggles. He arranged these categories on a scale of cultural progression: while the most primitive societies depend extensively on physiological struggle (centered on the procurement of the basic necessities of life), the most highly civilized societies are dominated by intellectual struggle. The most successful civilized societies are those that have the most efficient language, the most sublime literature, and the most universal philosophy.

Unlike Kropotkin, Novikov thought that Darwinism, despite its triumphs in biology, had no place in sociology. He was convinced that any effort to apply Darwin's theory to sociology led inevitably to Social Darwinism. In sociology, as he looked at it, Darwinism is "real poison" and a manifestation of "the eclipse of the human mind."[79] Kropotkin wanted to enrich sociology by returning to what he thought was pure Darwinism as presented in *The Descent of Man*. Novikov wanted to enrich sociology by keeping Darwinism out of it.

Marxism and Darwinism

In his eulogy at Marx's grave, Engels gave Marx credit for extending Darwin's theory to the study of the inner dynamics and change in human society. Engels paid a great tribute to Marx by elevating him to the heights of scholarship Darwin had pioneered.

From the very beginning, however, it was clear that Marxists would have little affinity with the struggle for existence as a key factor explaining social evolution. The struggle for existence, on the biological level, and class struggle, on the sociological level, are totally different notions. Darwin envisaged the struggle for existence as a universal mechanism of organic evolution, operating in every phase of the history of species and covering all species. Marx's class struggle is not universal: it does not apply to every phase of social evolution and to every society. Like populists and anarchists, the followers of Marx were dedicated to creating a society that would rule out every sort of social conflict. This has been one of the basic reasons for the persistent efforts by Marxists to discredit the Malthusian formula as a sociological category. Marxists have been consistently critical of Social Darwinists, as well as of all efforts to transpose biological laws to the social sciences. They have been categorical in dissociating themselves from Darwin, who was inclined to use biological arguments in explaining social phenomena.[80] In biology, Marx and Engels, according to Plekhanov, took the positions of Darwinism; in sociology, they took the positions of historical materialism.

No Russian Marxist theorist of the nineteenth and early twentieth centuries had undertaken a systematic study of Darwinism as a scientific theory and as an ideology. Georgii Valentinovich Plekhanov was more attracted to Darwin's principles than any other Russian Marxist; his discussion, however, was neither methodical nor particularly original. In his references to Darwin, Plekhanov was much more concerned with clarifying the basic principles of Marx's social theory than with the ongoing controversies centered on Darwin's theory of organic evolution and with the relationship of Darwinism to sociology. Plekhanov accepted the statement Engels had made at Marx's graveside; he cautioned, however, that the similarities between Darwin and Marx were not in the scientific laws they formulated but in the scientific spirit of their scholarly engagement. He wrote:

> Darwin succeeded in solving the problem of the origin of plant and animal species in the struggle for survival. Marx succeeded in solving the problem of the emergence of different types of social organization in the struggle

of men for their existence. Logically, Marx's investigation begins precisely where Darwin's ends. Animals and plants are under the influence of the physical environment. The physical environment acts on social man through the social relations that arise on the basis of the productive forces, which at first develop at various speeds, according to the characteristics of the physical environment. Darwin explains the origin of species, not by an inborn tendency of the animal organism toward development, as Lamarck had assumed, but by the adaptation of the organism to external conditions: not by the nature of the organism but by the influence of external nature. Marx explains the historical development of mankind, not by human nature, but by the qualities of social relations among men, which arise under the influence of social man on external nature. The spirit of research is absolutely the same in both thinkers. That is why one can say that Marxism is Darwinism in its application to social science (we know that chronologically this is not so, but that is unimportant).[81]

Showing a strong positivist inclination, Plekhanov noted that the mainstream of nineteenth-century intellectual development made it possible to give the philosophy of history a solid scientific footing. Marx's philosophy of the history of mankind, he added, was built on the firm foundation of social science in the same way that Darwin's "philosophy of the history of species" was built on the solid base of natural science. Marx and Darwin made complementary contributions to rendering the philosophy of history inseparable from science.[82]

Marx and Darwin, according to Plekhanov, adhered to a position of philosophical materialism and empirically grounded historicism. Plekhanov was ready to recognize, however, that the similarity between Marx and Darwin was more in the style of work than in the knowledge they produced and the theoretical orientation they adhered to. According to Plekhanov, both were true scientists, not because they conducted laboratory experiments, but because their generalizations were based on rigorous logic and were open to empirical verification. They applied the same scientific method to different universes of inquiry. In Plekhanov's words: "Just as Darwin has enriched biology by a strikingly simple and rigorously scientific theory of the origin of species, so the founders of scientific socialism have discovered the great principle of change in the species of social organization by studying the development of productive forces and of the struggle of these forces against the backward 'social conditions of production.'"[83]

In his discussion of ethical and aesthetic endowments, Plekhanov took the opportunity to compare the contributions of Marx and Darwin. Here again he acknowledged two common features of Darwin's evolutionary theory and Marx's historical materialism: both represented "solid

steps" forward in the development of science, and both added new substance and vigor to materialistic philosophy. The two theories, he added,
covered different levels of reality, each requiring a distinct approach.
Biological theory explained the natural capabilities for the development
of ethical behavior and aesthetic sentiment; sociological theory, in turn,
explained the level and direction of the actual development of morality
and the sense of beauty. Although Marx began to construct his sociological theory at the point where Darwin completed his biological theory, Plekhanov stood firmly against making sociological theory a mere
extension of biological theory. He claimed that Darwin did not always
adhere to this rule, but that his "transgressions" were relatively mild in
comparison with those of a group of Darwinists, typified by Haeckel,
who translated the instincts of prey animals into a universal principle of
"war of all against all" in human society.[84] Darwin's theory, in Plekhanov's view, was scientifically sound only insofar as it referred to the biological realm. The two theories complemented each other only insofar
as they covered the two sides of human reality.

Darwin, according to Plekhanov, made major contributions on two
fronts. His work made it unnecessary—and absurd—to resort to imaginary "innate tendencies" of organisms toward progress (as Erasmus
Darwin and Lamarck had believed) in order to account for the evolution of species. It made it also unnecessary—and absurd—to rely on
supernatural forces in order to understand social progress.[85]

Plekhanov could adhere to Engels's utterance at Marx's graveside
only by avoiding a discussion of the struggle for existence as the motive
force of organic and social evolution. He shifted the emphasis to a formal
similarity: to Darwin's and Marx's recognition of external causation as a
primary tool of scientific explanation. To go beyond these formal similarities would mean to risk succumbing to the lures of bourgeois sociology "with a very ugly content." The endorsement of Engels's eulogy did
not prevent Plekhanov from dismissing the Malthusian theory of population and Hobbesian sociology as sheer sophistry.

Plekhanov credited Darwin with making a decisive contribution to
the triumph of a historical orientation in biology. He wrote that so long
as biology adhered to a static view of nature, it relied on a metaphysical
style of thought. The same applied to French materialism, which repeatedly tried to relinquish the static philosophy of nature but without success. Darwin's theory, which ushered in modern biology, made a permanent break with the old view.[86]

Plekhanov went so far as to place the label "dialectical" on Darwin's orientation. He made it clear, however, that it was "dialectical" only to the extent that it recognized the study of change as the basic method for understanding living nature. In one important respect, Darwin's theory was not dialectical: by adhering strictly to the principle of *natura non facit saltum*, it drew attention only to gradual, quantitative change in given natural phenomena. It had no room for qualitative—or revolutionary—change. Darwinian sociologists, relying on the principle of *historia non facit saltum*, either ignored social revolutions or called them maladies.[87]

In 1908 Plekhanov praised the work of Hugo de Vries, whose mutation theory, challenging the Darwinian commitment to gradualism, gave support to the idea of the dialectics of nature.[88] He called de Vries's contribution an "epoch-making" discovery and welcomed the shift of emphasis in biology from "gradual changes" to "leaps" in the process of evolution. In the Darwinian emphasis on "gradual changes" he detected a source of unfavorable influence on the development of experimental biology. Plekhanov was happy to consider de Vries's mutations a confirmation of Engels's "dialectics of nature," based on the principle that the dialectics of human thought reflects the dialectics of the work of nature. In all this, he preferred to identify mutations as transformations of quantities into qualities. Plekhanov received help from A. M. Deborin, a young Marxist theorist attracted to dialectics as a backbone of the philosophy of science. Deborin noted that the closer the scientists came to recognizing the role of saltation in living nature the closer they came, consciously or unconsciously, to the standpoint of dialectical materialism.[89] De Vries received credit for providing another proof that the Marxists were correct in claiming that the theory of evolution, formulated in the spirit of Darwinian gradualism, did not give a true account of "the objective process of development in both nature and history."[90]

Plekhanov took note of Darwin's interpretation of the growth of technology as a continuation of natural history. He claimed that zoology, rather than history, gives the best picture of the emergence and early development of technology.[91] Zoological study leads to the recognition of the important role of natural environment in determining "the character of social environment."[92] Under the spell of Darwinian thought, Plekhanov placed stronger emphasis on the influence of external environment on the constitution and dynamics of human society than a typical Marxist theorist was inclined to do. Social progress, he said, is measured by the growth of mutual relations between "social man and geographical environment."

Plekhanov dealt extensively with various interpretations of the guiding principles of Darwin's theory, in both its biological and sociological aspects. Particularly notable were his comments on N. G. Chernyshevskii's critique of the Malthusian bias of Darwinism, on the Darwinian ideas built into Lewis H. Morgan's grand theory of cultural evolutionism, and on N. K. Mikhailovskii's comparisons of Darwin's and Marx's theoretical ideas. In all these comments Plekhanov made a concerted effort to illumine the basic premises of Marxist thought, to sharpen the methodological weapons of dialectical materialism, and to strengthen the Marxist position on the battlefield of clashing ideologies. In his comments he solicited help from Marx and Engels without fully surrendering the right to voice independent opinions on contemporary scientific and philosophical thought.

Plekhanov had the highest respect for Chernyshevskii, whom he considered the patriarch of the Russian revolutionary intelligentsia. He viewed the tsarist government's decision to exile Chernyshevskii to Siberia the most severe crime inflicted on "the intellectual development of Russia."[93] All the praise, however, did not prevent him from expressing dissatisfaction with "The Theory of the Origin of the Beneficial Nature of the Struggle for Life," which Chernyshevskii published in *Russian Thought* in 1888, after his return from exile. The Siberian experience, Plekhanov thought, had devastated Chernyshevskii's former "brilliance" and "depth of thought." The article on Darwin was "extremely weak" and created "a most painful impression." It was written by a man who was utterly shaken and broken.[94]

Plekhanov objected particularly to two of Chernyshevskii's criticisms of Darwin's theory. First, he thought that Chernyshevskii did not give an accurate interpretation of the struggle for existence, the key principle of Darwinian evolution. Plekhanov agreed with Chernyshevskii that Darwin exaggerated the role of the struggle for existence—and of natural selection—in the evolution of species. He thought, however, that Chernyshevskii erred in tying the struggle for existence exclusively to inadequate food supplies—and to Darwin's Malthusian bias. Darwin's definition of natural selection, as Plekhanov interpreted it, did not depend on food supplies alone. Had Chernyshevskii adopted a broader identification of natural selection, his criticism of Darwin's theory would have taken a different line of reasoning.[95] Second, Plekhanov thought that Chernyshevskii erred in criticizing Darwin for the growth of evolutionary sociology based on biological principles. He thought that Darwin should not be proclaimed responsible for the sins of Darwinian sociologists.[96]

Lewis H. Morgan, an expert on the Iroquois Indians and a leading pioneer in modern cultural anthropology, attracted Plekhanov's attention for two reasons. In the first place, Morgan made Darwin's grand theory of organic evolution a model for an equally grand theory of social and cultural evolution. He offered the first detailed picture of the unilinear evolution of human society and made a serious effort to legitimate the use of the comparative-historical method as the main tool of anthropological research. In the second place, Engels considered Morgan's theory an independent discovery of Marxism and wrote the *Origin of the Family, Private Property and the State* (1884) with the explicit purpose of making Morgan's ideas an integral part of Marxism. Plekhanov agreed with Engels that Morgan's "independent discovery" of the Marxist conception of history proved that "the time was ripe for it, that it simply had to be discovered."[97]

Plekhanov was particularly impressed with the fact that in *Ancient Society* Morgan emphasized the evolution of social institutions as a product of social adaptation to advancements in technology. In his view, Morgan—who had shown no interest in or awareness of Marxist social thought—supplied original material showing that the evolution of the means of production is the key cause of structural changes in human society.[98]

It was through Engels's *Origin of the Family*, translated into Russian in 1894, that the Russian reading public learned about Morgan's ideas.[99] Plekhanov noted that Morgan's "independent discovery" of the basic views of historical materialism represented a triumph for the theory of Marx and Engels.[100] He also observed that under Morgan's influence Marx modified his grand scheme of unilinear social evolution. Now, for example, Marx was ready to equate the social organization of production in classical antiquity with that of the Orient. The Marxian theory of social development, he thought, became much stronger after it incorporated Morgan's views on the structure and dynamics of clan organization and kinship systems.

Plekhanov argued that, at least in one important respect, Morgan was ahead of Darwin: whereas Darwin was imprecise and wavering in describing the causes and mechanisms of organic evolution, Morgan was eminently successful in pinpointing the basic causes and mechanisms of social evolution. Naturally, Marxist theorists were much more ready to accept Darwin's notion of gradual change in organic nature than Morgan's notion of gradual change in human society.

In honoring Morgan, Engels equated the importance of his work with that of Darwin and Marx. He thought that Morgan's discovery of maternal clan, a predecessor of the paternal clan of "modern" society, has the same significance for ancient history as Darwin's theory of evolution has for biology, and as Marx's theory of surplus value has for political economy.[101] The fact that his social theory came to Russia under Engels's auspices made Morgan both enemies and friends. The long article on Morgan in the Brockhaus-Efron *Encyclopedic Dictionary* managed to take a middle-of-the-road position: it stated that despite his oversimplifications and gaping omissions, Morgan was one of "the leading ethnologists and sociologists" of his time.[102]

Comparisons of Darwin and Marx came also from the populist critics of Marxism. During the 1890s, in a concentrated campaign against Marxism in general and Russian Marxism in particular, N. K. Mikhailovskii started a bitter feud with Plekhanov by stating that, as a social theory, Marxism did not find support in the world of scholarship. In answering the charges, Plekhanov pointed out the widely acknowledged role of Morgan in building the scientific foundations of evolutionary ethnology. "In North America," Plekhanov reported, "Morgan contributed to the creation of an entire school of ethnologists," and in Europe his most notable follower was H. Cunow, the author of studies on the kinship systems of Australian aborigines, the Inca empire, and the general economic attributes of the matriarchate.[103] Plekhanov was not able to detect a sure sign of Morgan's—and Engels's—influence on the development of cultural evolutionism in Russian ethnography. Undoubtedly, he selected Morgan's theory as a link between Marxist theory and the scholarly community because it represented, in his view, a unity of Darwin's evolutionary idea and a materialistic interpretation of history. It allied historical materialism with cultural evolutionism, strongly represented in the academic community. E. Terray is correct in stating that Morgan appealed to Marxist theorists not because he transferred Darwin's biological concepts to ethnology but because he made a direct contribution to "the materialist interpretation of history"—to the recognition of the evolution of "inventions" and of "arts of subsistence" as the wheels of social evolution.[104]

Mikhailovskii reacted angrily against the claim of M. A. Antonovich, in *Darwin and His Theory* (1896), that Darwin and Marx had common enemies: the supporters of "bourgeois exploitation and of every kind of economic and political injustice."[105] He pointed his finger at a large array of persons who, contrary to Antonovich's insinuation, thought that Darwin's theory supported "bourgeois competition and exploitation."[106] Not pressing the issue, he merely wanted Antonovich not to overlook the ties between Darwin's theory and Social Darwinism. Although of a liberal frame of mind, Antonovich was not a Marxist.

In an article published in the journal *Russian Wealth* in 1894, Mikhailovskii made a comparison of the theoretical work of Marx and Darwin. He arrived at a conclusion that did not flatter Marx:

> What is essentially the work of Darwin? A few generalizing ideas, most intimately interconnected, which crown a whole Mont Blanc of factual material. Where is the corresponding work of Marx? It does not exist . . . and not only is there no such work of Marx, but there is no such work in all Marxist literature, in spite of all its extensiveness and wide distribution. . . . The very foundations of economic materialism, repeated as axioms innumerable times, still remain unconnected among themselves and untested by facts, which particularly deserves attention in a theory which in principle relies upon material and tangible facts, and which arrogates to itself the title of being particularly "scientific." [107]

In answering Mikhailovskii's negative assessment of the theory of "economic materialism," Plekhanov followed a simple line of reasoning: he tried to show that Mikhailovskii's laudatory characterization of Darwin's work applied equally to Marx's theory. "The preceding history of social science and philosophy," he wrote, "had piled up a 'whole Mont Blanc' of *contradictions,* which urgently demanded solution. Marx *did precisely solve them with the help of a theory which, like Darwin's theory, consists of a 'few generalizing* ideas, most intimately connected among themselves.' When these ideas appeared, it turned out that, with their help, all the contradictions which threw previous thinkers into confusion could be resolved." [108]

Mikhailovskii made two additional observations on the relation of Marxism to Darwin's theory. In the first place, he stated that Darwin's theory, "the great scientific discovery of our century," owed no debt to the dialectical triad, and that the same was true for the discoveries that laid the foundations for thermodynamics and electromagnetic theory. For that matter, he hastened to add, neither was the discovery of "economic materialism," as "a special branch of scientific knowledge," indebted to the Hegelian triad. [109] In the second place, he noted that Engels fully ignored the question of the transferability of Darwin's theory to sociology. In his opinion, "economic materialism" did not offer a philosophical explanation of the ties between the natural sciences and the social sciences. [110] He noted that Engels's *Anti-Dühring* was full of praise for Darwin's theory, but that it made no reference to its usefulness in sociology.

Plekhanov kept up with current developments in biology that affected both the Darwinian evolutionary theory and the relationship of Darwin's scientific views to Marx's social philosophy. Some of his interpretations of new developments were snap judgments that did not help elucidate the philosophical contact between Marxism and Darwinism. He looked, for example, at Raoul Francé, a leader of psycho-Lamarckism, as a contributor to a scientific confirmation of modern materialism. This judgment contradicted the general—and correct—assessment of Francé's work as a complex system of metaphysical-idealistic constructions ornamented lavishly with metaphors borrowed from modern experimental biology. Influenced by Hans Driesch and Eduard Hartmann, Francé saw the future of evolutionary biology in a shifting of the emphasis from "external causality" to "internal teleology," and in the viewing of "internal teleology" as a result of self-generated "mental energy" operating on the cellular level. In the *Current State of Darwinian Questions* he attempted to ally the teleological bent and psychological underpinnings of neo-Lamarckism and vitalism with the mutationist view of the origin of species.[111] Perhaps Plekhanov was favorably impressed with Francé's stated intent not to reject Darwinism but to bring it in tune with contemporary developments in biology. He might have been swayed by Francé's favorable reference to Ivan Pavlov's contributions to the experimental methodology of the physiology of digestion.[112] He wrote:

> I may add that in present-day natural science, and especially among the neo-Lamarckians, there has been a fairly rapid spread of the theory of so-called *animism of matter,* i.e., that matter in general, and especially organized matter, possesses a degree of *sensibility.* This theory, which many regard as opposed to materialism,[113] is in fact, when properly understood, only a translation into the language of present-day natural science of Feuerbach's materialistic doctrine of the unity of being and thought, of object and subject. It may be confidently stated that, had they known of this theory, Marx and Engels would have been keenly interested in this trend in natural science that still waits for a detailed explanation.[114]

While ready to report an element of "materialism" and "naturalism" in one of the most forceful expressions of neovitalist thought, Plekhanov used every opportunity to emphasize the differences between Darwin's "naturalist materialism," fastened to a mechanistic philosophy of nature, and Marx's "historical materialism," steeped in dialectics that rejected a dogmatic acceptance of mechanical models. In carrying out this task he did not always show the requisite precision and consistency. Particularly in his effort to formulate a Marxist theory of aesthetics, he often found the naturalist models of Darwinian thought more appealing than the sociological method of historical materialism. As Andrzej Walicki has pointed out, he "conceived man not as the creator of his own nature and history, but merely as a product, a passive medium of objective processes, subject to the strict determinism of 'natural necessity.'"[115] The mechanistic lures of naturalism led him on occasion to digress from the Marxist view of human society as an entity sui generis, a uniquely structured reality dependent on self-generated causation.

Lenin showed little interest in Darwin. He wrote more about Haeckel than about Darwin. When he talked about Darwin his comments were perfunctory, never occupying a central position. He, too, likened Marx to Darwin on purely formal grounds: just as Darwin wrought a revolution in biology, so Marx wrought a revolution in social science. In Lenin's words:

> Just as Darwin put an end to the view that the species of animals and plants are . . . "created by God" and are immutable and was the first to place biology on a solid scientific foundation by establishing mutability and succession of species, so Marx put an end to the view of society as a mechanical aggregation of individuals . . . and was the first to put sociology on a scientific footing by establishing the concept of the economic formation of society as the sum total of the given relations of production and by showing that the development of these formations is a process of natural history.[116]

To Lenin, as to Plekhanov, Morgan represented a scientific orientation in ethnology which drew together Darwinian evolutionism and Marxian materialism. Morgan's explanation of the evolution of prehistoric societies, dominated by clan organization, filled an important gap in the general evolution of human society formulated by Marx and Engels. According to Lenin, Morgan broadened the perspectives of the Marxian philosophy of history by showing, on the basis of a vast amount of empirical data, that "material relations," rather than "ideological relations," unveiled the mysteries of the dynamics of tribal societies.[117] Lenin agreed with Engels that Morgan's analysis of the social organization of the North American Indians held "the key to all the great and hitherto unfathomable riddles of the earliest Greek, Roman, and German history."[118]

In exploring the unity of Marxism and Darwinism, Lenin and Plekhanov showed considerable interest in the sociologically significant ties between organic and social evolutions—particularly between "natural" and "artificial" technologies. As James Rogers has pointed out, Plekhanov was not averse to using biology as a source of explanations for sociological phenomena. For example, he talked about the genetic transmission of such a moral quality as social solidarity.[119] This, however, was a digression rather than a norm in Plekhanov's thinking; a good Marxist, he was generally opposed to a reduction of social behavior to biological explanations. Lenin was much more categorical in separating biological and sociological realities as subjects of scientific inquiry. In his view, every biological metaphor in sociology represented a dangerous concession to neopositivism. He chastised A. A. Bogdanov for relying on Darwinian "selection"—as well as on Ostwald's energeticist terminology—in his attempt to construct a general theory of society.[120]

Plekhanov differed from Lenin in yet another respect: he offered a detailed discussion of the influence of the environment on the evolution of individual societies. Flirting with geographical determinism, he was under the spell of the evolutionary theories of Lamarck and Darwin, both of which considered adaptation to the environment the propelling force of organic evolution. Plekhanov reasoned that the influence of the environment on the structure of society was indirect: changes in the environment cause changes in the natural and technological substratum of production, which in turn cause changes in "production relations," or social structure. He felt compelled to point out that Lamarck, contrary to the common interpretation of his position, counted environmental changes among indirect—rather than direct—influences on organic evolution. According to Plekhanov's citation from Lamarck's *Zoological Philosophy,* environmental changes induce new habits in specific groups of animals, and new habits, in turn, produce a novel division of labor among organs and a novel "structure of organs." [121]

F. Bersenev, an unheralded defender of historical materialism, was atypically direct and explicit in claiming close relations between Marx's and Darwin's views. He emphasized the complementarity—rather than the full identity—of the two theories. Darwin, he wrote, explained the mechanisms of organic evolution—the evolutionary role of the struggle for existence, and natural selection. Marx applied the same mechanism to man's mental and social life. By merging his ideas with those of Darwin he created a monistic world view, the cornerstone of dialectical materialism. Marx accepted all of Darwin's evolutionary principles, but he applied them to man's mental and social life and made "production relations" the real battlefield for the social struggle for existence and for social selection. Indeed, he established that social conditions are the most important factor of human evolution.[122]

The defenders of Marxist orthodoxy favored Darwinism generally on ideological rather than on scientific grounds. Darwin's theory of evolution supplied Marxists with arguments that could easily be marshaled and used against such archenemies of materialism as religious ethics, scriptural cosmogony, idealistic metaphysics, and subjective epistemology. It was for this reason that Marxist theorists made their admiration for Darwin strong and irrevocable. Their attitude toward Darwin's theory qua science was of a different order: they did not keep quiet about their basic disagreement with the principles of the evolutionary theory. They rejected the very heart of Darwinian science: the Malthusian premise of the struggle for existence, the gradualism and uniformitarianism of organic change, and the transferability of the underlying principles of organic evolution to human society. It was primarily because of the pronounced ambivalence in their attitude toward Darwin that they made no serious effort to produce a comprehensive study of the Marxist view of Darwinian thought.

A. A. Bogdanov represented a unique wing of Marxist revisionism. He acquired prominence by his assiduous work in overcoming what he considered the two major weaknesses of Marxist theorists: the pronounced lack of efforts to explore the Darwinian contributions to the understanding of natural and social evolution; and a flagrant disregard for the epistemological ideas of various schools of neopositivism. In *Knowledge from a Historical Standpoint* (1902) he made a brave effort to formulate an energeticist orientation in philosophy, inspired by Wilhelm Ostwald's philosophy of science, a modern substitute for both idealistic and materialistic ontologies. He presented the new orientation as a synthesis of two major and complementary approaches to natural and social realities: the static approach, emphasizing the phenomena of equilibrium and stability, and the historical approach, emphasizing the phenomena of change.[123] In his treatment of the static approach, Bogdanov was eager to find a middle ground between neopositivist and Marxist theories of knowledge; in elaborating the historical approach he pleaded for a blend of Marxist historicism and Darwin's evolutionary orientation.

Of particular interest to us is Bogdanov's criticism of Marxist theorists for their failure to explore the rich promise of Darwinian historicism. Selection and adaptation, the conceptual pillars of Darwin's theory, applied fully, in Bogdanov's opinion, to the organic universe and to human society. Asking himself the question whether there was an essential difference in the development and functioning of "social" and "biological" forms, he answered with a categorical no. In both cases, "the environment-induced changes in old forms produce a continuous line of new variations." Most variations are not adaptable to the environment, and selection eliminates them. "The surviving minority is a product of selection that operates in the same way in nature and in society." [124] "Regardless of the vantage point from which they are viewed, the social forms represent adaptations in the same sense and to the same degree as all biological forms." [125]

Bogdanov carried the analogy of the evolutionary aspect of "natural" and "social" forms a step further. He saw no reason why Haeckel's biogenetic law, which defines ontogeny as a brief recapitulation of phylogeny, could not be applied profitably to the study of the evolution of social forms. "The history and structure of every social formation—of every form of work and cognition—reflects the developmental path and the organizational principles of the entire society." [126] Unfortunately, Bogdanov had made no effort to provide a comprehensive and systematic analysis of the sociology of Haeckel's law, and of the points of contact between Darwinism and Marxism. Strongly suggested, but unelaborated, is also Bogdanov's view of psychology as a link between biology and sociology. "Sociality," he said, "is a higher result of the mental development of the animal kingdom. The mental life is the most complex form of biological development." [127] It holds the key to the understanding of both selection and adaptation. Bogdanov was willing to go much farther than either Darwin or Marx in recognizing the preeminent role of comparative psychology in the scientific study of the evolution of man as a species and as a social being.

The Russian theorists of Marxism, both pure and tempered, avoided a closer concern with the biological significance of Darwin's theory. The more firmly a writer adhered to Marxist orthodoxy, the more he avoided a direct confrontation with the most distinctive aspects of the scientific makeup of Darwin's evolutionary explorations. The Marxists expressed generally negative views on two social science extensions of Darwin's theory: social evolutionism, concerned with gradualism and universal stages in the development of human society, and Social Darwinism, whose representatives translated the struggle for existence and natural selection into general sociological laws. The cautious—and often negative—attitude toward Darwinism was in part an expression of uncertainty about the acceptability of individual propositions on scientific, philosophical, or ideological grounds. The Marxist theorists did not receive much encouragement from the writings of their teachers—Marx and Engels. At this time some of the key letters by the founders of Marxism were not yet in the public domain. Nor was Engels's *Dialectics of Nature*, containing references to Darwin, available in published form.

Conclusion

Russia can rightfully boast of a group of scientists who helped bolster Darwin's crucial arguments and give the idea of evolution a pivotal position in modern science, the philosophy of nature, and the world outlook. Aleksandr Kovalevskii and Il'ia Mechnikov made major contributions to the transformation of invertebrate embryology into an evolutionary discipline. They produced valuable empirical data and guarded generalizations suggesting challenging arguments in favor of the embryological and morphological unity of vertebrates and invertebrates and of the common origin of man and animals. Their basic embryological conclusions evoked lively and extensive comments in the West. Mechnikov also received wide acclaim for his phagocyte theory, the foundation of a special evolutionary orientation in comparative pathology. Vladimir Kovalevskii, Darwin's most intimate friend in Russia, must be counted among the most prominent pioneers in modern paleontology; George Gaylord Simpson called his study of the history of horses and related ungulates "one of the first and greatest triumphs of evolutionary paleontology."[1] Stephen Jay Gould has paid tribute to Kovalevskii by editing his collected papers, making them more readily accessible to the growing circle of interested readers.[2]

The list of contributors to Darwinian thought who have received international recognition must include also Aleksei Severtsov, who made an eminently successful effort to present—and to advance—an integrated approach to the general laws of evolutionary morphology. He belonged to the group of noted biologists who were firmly convinced that only a full elaboration, clarification, and integration of the general principles of organic evolution could create a basis for a grand synthesis of Darwinism and genetics. Richard Goldschmidt gave Severtsov credit for a full elaboration of the claim that "the general idea of evolution" required a linking of embryonic study with the facts of genetics.[3] His work represents a bold and logically rigorous effort, first, to elevate morphology to the position of a central discipline in evolutionary biology, and second, to coordinate embryological, paleontological, and physiological data as the building blocks of a unified body of biological theory.

Imperial Russia contributed not only widely heralded builders of the Darwinian theory, but also equally eminent critics. Karl von Baer, a distinguished member of the St. Petersburg Academy of Sciences and the most illustrious embryologist of the pre-Darwinian era, was one such critic. In opposition to Darwinism, he proposed an alternative theory of evolution that emphasized the purposiveness of vital processes, the duality of internal causation and geographical determinism, the combination of gradualism and heterogenesis, the secondary importance of the struggle for existence as a mechanism of evolution, and a retreat from the uniformitarian interpretation of organic transformation. His theory became one of the most popular sources of anti-Darwinian arguments depended on by philosophers, theologians, and conservative ideologues. N. Ia. Danilevskii's monumental critique of Darwinism owed a major debt to von Baer's rich reservoir of arguments against Darwin's ideas.[4] While criticizing Darwinism, von Baer advanced the broad outlines of a general theory of organic evolution whose influence on modern biological thought was second only to Darwin's.

During the 1870s to 1890s only a few Russian evolutionary biologists attracted more attention in the West than V. I. Shmankevich, the high school teacher who spent his weekends and summer vacations studying the role of changing environment in the transformation of lower crustaceans in the Odessa lagoons.[5] While the Lamarckians used his research data to bolster their own arguments, the anti-Lamarckians worked diligently on disproving his bold claims that the direct influence of the environment on the transformation of species was open to empirical study. George Russel Wallace, William Bateson, E. D. Cope, Y. Delage, L. Plate and many others took part in assessing the place of Shmankevich's findings in modern biological thought. Shmankevich did not criticize Darwin; he merely tried to show that Lamarck was correct.

S. I. Korzhinskii relied on his theory of heterogenesis to launch a frontal attack on every fundamental principle of Darwin's theory. He even recommended that the term "evolution" be scrapped as a misleading biological notion. Immediately translated into German, his papers on heterogenesis attracted instantaneous and widespread attention in the West. Hugo de Vries counted Korzhinskii among the most distinguished precursors of the mutation theory. Delage and Goldsmith noted that among de Vries's precursors Korzhinskii was the only one worth recording.[6] During the first two decades of the twentieth century he was one of the Russian experts in transformist biology most frequently referred to in Western studies. Rádl and Vernon Kellogg assigned him a place of honor among the pioneers of modern biological studies of variation and heredity.

No Russian critic of the Darwinian notion of the struggle for existence had attracted more international attention than P. A. Kropotkin. A grand summation of a strong Russian tradition in Darwinian criticism, Kropotkin's *Mutual Aid* was originally published in English and was quickly translated into French, German, Russian, and several other languages. In one respect, Kropotkin's "scientific" writings occupied an anomalous position in the Darwinian literature. Widely read and heralded as a major critic of the Darwinian struggle for existence, he was lavish in his praise of Darwin's contributions to the making of biology into an exact science, safely removed from metaphysical interference. His belated and hastily designed effort to integrate Darwinism into a general Lamarckian view of organic transformation did not evoke much appreciative comment.

Darwinian ideas arrived in Russia through many channels and types of intellectual endeavor. Four groups played a particularly important role in the interpretation of Darwin's scientific legacy: scientists, philosophers, theologians, and ideologues. The ideologues are singled out as a distinct group, not because they alone looked at Darwinism through an ideological prism, but because they made ideology their central concern.

Russian scientists covered every aspect of the elaborate structure of Darwinian thought. They alone represented the full spectrum of attitudes and critical stances toward the scientific, sociological, and general intellectual merits of the new theory of evolution. Members of the scientific community viewed Darwinism not only as an ordered system of scientific thought but also as a unique expression of philosophical ideas and ideological statements. There were also scientists—the zoologist A. A. Tikhomirov, for example—who showed no hesitation in blending scientific and theological arguments.

In the vast enterprise of evolutionary inquiry three interests received particularly strong emphasis: the evolution of individual species, the general evolutionary aspect of embryonic growth, and the evolutionary role of the environment. The last-named interest fertilized and gave new direction to biogeography (zoogeography and phytogeography) and ecology. Although their studies did not attract much attention in the West, Russian biogeographers and ecologists made lasting contributions to the study of organic evolution. In carrying out their tasks they depended heavily on Dawin's fertile ideas on the evolutionary aspects of migration and isolation, and of multiform manifestations of the struggle for existence and natural selection. Biogeographers and ecologists laid the foundations for a considerable national commitment to the scientific study of plant and animal "associations" and "communities."

As might have been expected, many scientists chose not to declare their views on Darwin. After all, Darwin was a controversial scientist, a source of ideas contradicting the autocratic value system. Many scientists made no comment, for the simple reason that they were not in the habit of passing judgment on scientific developments outside their narrow specialties. Those who declared their views—a formidable group—represented every position between the extremes of Darwinian orthodoxy, on the one hand, and total rejection, on the other.

Orthodox Darwinists assumed that Darwin's ideas made up the total core of the evolutionary theory. M. A. Menzbir and A. N. Severtsov recognized other theories only insofar as they amplified and bolstered the Darwinian core—insofar as they were recognized only as subsidiaries to Darwin's evolutionary principles. Kliment Timiriazev went even further: he accused the pioneers of modern experimental biology of sacrificing the best interests of science to the needs of conservative ideological causes. Orthodox Darwinists were a shrinking group. They were deeply engaged in defending the mechanistic foundations of Darwin's science.

At the beginning of the twentieth century, or somewhat earlier, many leading Darwinists showed unmistakable signs of readiness to retreat from orthodox theoretical positions. They contended that the ideas of the new genetics should be recognized as independent components of the core of the evolutionary theory. The time had come, they thought, to take a closer look at the theoretical and methodological challenges of the experimental branches of neo-Lamarckism. Represented by Mechnikov, Kholodkovskii, and Shimkevich, this group worked on making the Darwinian ideas the unifying force in a new synthesis of evolutionary principles. Its members were not engaged in experimental research in genetics. They showed little inclination to enter the rapidly spreading debate centered on the crisis in the mechanistic underpinnings of biology, in general, and Darwinism, in particular.

A small community, Russian geneticists were not united in their attitudes toward Darwinian thought. One group advocated a full integration of the relevant claims of the new experimental-biological theories of heredity and mutations into the Darwinian evolutionary theory. It advocated an evolutionary schema in which natural selection occupied the key position. N. K. Kol'tsov, the chief spokesman for this group, envisaged a new evolutionary theory that synchronized the "universal law" of natural selection with Mendel's and de Vries's biological principles. The second group, represented by Regel' and Filipchenko, endorsed the same synthesis with one major exception: it assigned natural selection a secondary place in the dynamics of evolution.

Both groups of geneticists went along with the defenders of Darwinian orthodoxy and with "unorthodox" Darwinists in recognizing Darwin's immense contribution to the triumph of the evolutionary view in modern biology. They, too, recognized Darwin's genius and unlimited vision. They agreed with the defenders of orthodoxy in viewing Darwin's empirical method and inductive logic as the method and the logic of true science. They saw Darwin as a dedicated guardian of the ethos of science—of the moral principles that guide scientists in their professional work. Darwin, they noted, showed remarkable consistency in defending science as the pivotal part of secular culture: to him, scientific truth was neither unchallengeable nor impervious to the inexorable processes of erosion. He created a new world outlook that could best be described as scientific optimism, dominated by the growing volume of positive knowledge and by the expanding power of man's rational endowments.

Mention should be made of an influential group of scientists who were not biologists but who held Darwin in high esteem and readily admitted their intellectual debt to the grand structure of his evolutionary thought. These scientists did not concern themselves with the building blocks or basic principles—such as the struggle for existence and natural selection—that accounted for the Darwinian explanation of the mechanisms of the evolutionary process. Their basic task was to carry a generalized evolutionary view to as many sciences as possible—and in so doing they helped make Darwin's contributions to science more visible and impressive. Vladimir Vernadskii, for example, made a serious effort to establish a genetic, or evolutionary, mineralogy: he studied paragenesis, the sequences in the formation of minerals located in the same area. Following the English chemist William Crookes, L. A. Chugaev argued in favor of an evolutionary chemistry concerned with the development of chemical elements from a simple primordial substance. N. A. Morozov wrote a long essay on the evolution of heavenly bodies, and N. A. Umov wrote one on the evolution of atoms as building blocks of the universe. Many scientists did not try to place their research into an evolutionary framework, but this did not prevent them from saluting the idea of evolution as the chief contributor to the triumph of historical orientation in nineteenth-century science.

Another group of scientists relied on the idea of evolution to create new scientific disciplines. The geologist V. V. Dokuchaev, for example, created soil science as a clearly demarcated discipline bringing together the methods and relevant substance of geology, geography, physics, chemistry, and biology. Evolutionary principles provided the glue that gave unity and meaning to his work on the genesis, history, classification, and distribution of soil. He described soil as a special product of the interaction of organic and inorganic nature. In his view, soil is a product of the laws of nature which determine both its temporal sequences and its spatial configuration. Dokuchaev disagreed with individual aspects of Darwin's intricate system of scientific thought; he thought, for example, that "the famous Darwin" exaggerated the role of worms in the formation of soil.[7] All this did not deter him from making evolution the chief integrating principle of the new science. His theoretical stance combined two principles that received a clear and most dramatic expression in Darwin's theory: the principle of the cosmic unity of nature, and the principle of the universality of evolution.

Transposed to history and the social sciences, the study of the origins of natural objects became the study of the origins of structural components of human society and culture. The historians and allied scholars found themselves working on the origins of such institutional bulwarks of Russian social history as serfdom, estate system, autocracy, *obshchina*, and the unique relationship of the state to society at large. The evolutionists in sociology asked even broader questions: they reflected the strong concern of Western ethnologists and sociologists with the origins of religion, art, family, clan organization, and property as universal aspects of human society and culture. Maksim Kovalevskii created a new discipline—"genetic sociology" or "social embryology"—dealing exclusively with the origins of universal institutional complexes.[8]

Diametrically opposite to the views of the defenders of orthodoxy was the exceedingly small group of biologists involved in a total and unrelenting war on Darwinism. A. A. Tikhomirov, the best-known and most determined representative of this group, considered both the idea of organic evolution and Darwin's specific theory as direct assaults on the facts of science and on the cardinal values of Russian culture. An active and outspoken defender of autocratic values, Tikhomirov placed his anti-Darwinian campaign within an ideological framework: he selected and defended only those facts of science that, in his opinion, supported a holy war on the "materialism" and "anti-Christian" implications of Darwinian thought. Although he rejected evolutionism in general, he directed his main arguments against Darwin's interpretation of the anthropoid origin of man. During the last twenty-five years of tsarist reign, Tikhomirov may well have been the only university scientist to wage a systematic and sustained war on Darwin's theory.

At the turn of the century it became quite clear that criticism of the mechanistic matrix of Darwin's philosophy came from liberal biologists no less than from conservative biologists. I. P. Borodin and A. S. Famintsyn, both members of the St. Petersburg Academy of Sciences, expressed sharp criticism of the reign of mechanism in biology—including evolutionary biology—but they also signed petitions against government efforts to curb academic autonomy and student activities.

What did the critics mean by the reign of mechanistic views in biology? They meant specifically and exclusively a primary dependence of biologists on the methodological and conceptual tools of physics and chemistry, which was rapidly expanding during the 1890s. Little did they know that the time was fast approaching when physics and chemistry also would be deeply involved in an effort to emancipate themselves from the exclusive reign of mechanistic views. In this case, however, the expressions "mechanistic views" and "mechanical models" designated something quite different: a guiding philosophy dominated by Newton's clockwork picture of the universe, Laplacian determinism, and Kantian notions of space and time as pure and a priori categories of cognition. All Russian biologists involved in the criticism of the increasing dependence of biology on physicochemical modes of scientific explanation favored a strong emphasis on the psychological approach to vital processes. They operated on the erroneous assumption that every psychological approach was automatically antimechanistic. In their opinion, the physicochemical analysis should continue to be depended on, but only to a limited measure. Darwin erred, they thought, in attaching only a secondary importance to the psychological aspect of evolution. In defending their newly discovered psychological orientation, most critics of the reign of mechanistic views in biology experienced a noticeable influence of revived vitalism.

Members of the scientific community, particularly biologists, depended primarily on popular periodicals—the so-called thick journals—to present their general scientific and philosophical interpretations of Darwin's theory. *Fatherland Notes, Russian Herald, Russian Thought,* the *Herald of Europe, God's World,* and *Russian Wealth,* in particular, served as main vehicles for the diffusion, elaboration, and generalized assessment of Darwinian evolutionism. Evolution in general and Darwin's theory in particular drew the attention of such popular scientific journals as *Science Review,* the *Word of Science,* the *Herald of Knowledge,* and *Nature.*[9] These journals contributed to a more thorough understanding of the new historical view of nature and of the enormous scientific, ethical, philosophical, and ideological challenges that came in its wake. All contributed to the triumphant ascent of Darwin's theory to the summit of modern thought. Very few comprehensive surveys and general appraisals of Darwin's theory appeared in professional scientific journals, which were few in number and generally unappealing to the lay public.

The main branches of biology were not equally involved in the process of absorbing and advancing Darwin's theoretical principles. As in Germany, botanists showed much less interest in the new evolutionary idea than zoologists. At the Eleventh Congress of Russian Naturalists and Physicians, held in St. Petersburg in 1902, close to 40 percent of the papers presented at the session of the zoological section dealt with evolutionary themes; only 10 percent of the papers presented at the botanical session dealt with evolution. The leading botanists showed much more interest in systematics and various branches of plant physiology than in the questions of species transformation. As a generic group, botanists were also the main critics of the growing dependence of biological research on the tools and explanatory principles of physics and chemistry. They were the first group in Russia to echo Gustav Bunge's antimechanistic pronouncements.

Russia lagged behind the Western nations, particularly the United States, Great Britain, and Germany, in developing an institutional base for—and a strong interest in—experimental work in genetics and related disciplines dealing with heredity and its role in the evolution of species. No doubt this phenomenon has many explanations. Two are most general and most obvious.

First, Russian biological thought was heavily steeped in the Lamarckian tradition. The strength of Russian Lamarckism was not in an elaborate body of integrated theory but in a widespread belief in the external environment as a direct and primary causal factor in the mechanism of transformation. The strong Lamarckian bias could not but hamper the reception of an idea favoring autogenesis. A typical Russian evolutionary biologist accepted Darwin's conciliatory statements on the Lamarckian type of environmentalism.

Second, vested academic interests worked against a rapid and efficacious development of experimental research in genetics. Moscow University and St. Petersburg University, the country's leading institutions of higher education, were dominated by professors who established their reputations as defenders and popularizers of Darwinian ideas. Although most of these professors recognized the scientific merit of the new theories, they were strongly inclined to place them, as subsidiary knowledge, within the framework of Darwinism. They had written almost all the best-selling university textbooks in zoology, which, without exception, helped them safeguard the superiority of Darwinian thought. Under these conditions, it was most unlikely for research in genetics to acquire strong institutional support. When, for example, N. K. Kol'tsov tried in 1908 to initiate an experimental study of heredity in the zoological laboratory of Moscow University, he promptly lost his position as a research associate. The beginnings of genetic research took place in Shaniavskii University, the Moscow Institute of Agriculture, and the Bureau of Applied Botany, all short in research tradition and laboratory equipment.

The philosophers, the second major channel for the flow of Darwinian ideas to Russia, did not represent a united body: they were divided into distinct groups, each with minimum interaction with other groups. Three of these groups stood out because of the clarity of their orientations: university professors of philosophy, university professors who did not teach philosophy but who dealt extensively with the epistemological crisis in science, and free-lance writers on philosophical themes.

The university professors of philosophy were a homogeneous group: all were committed to idealistic metaphysics, all considered mysticism a vital vehicle of philosophical exploration and communication, and all were deeply involved in an effort to show the superiority of the "absolute" knowledge of metaphysics over the "relative" knowledge of science. All considered the rapidly eroding preeminence of mechanistic principles the sure sign of an irretrievable decline of the dominance of science. That not a single university professor of philosophy sided with Darwin on any primary point of evolutionary theory or on any issue of philosophical import was indeed a unique feature of the Russian confrontation with Darwin's scientific legacy. Why they did not wage a more direct and sustained attack on Darwinism is a puzzling question. Perhaps, they did not want to invite open conflict with university biologists, the pivotal force of Russian Darwinism. Or perhaps they preferred to pass on this duty to theology professors concerned with the philosophical problems of scientific knowledge.

The second group of philosophers consisted of natural and social scientists who wrote on epistemological themes and whose collective effort gave Russia the first national literature in the philosophy of science. Unlike the philosophy professors, the spokesmen for this orientation interpreted the erosion of mechanistic views, not as a decline in the intellectual power of science, but as the onset of a new revolution in scientific thought. The professional metaphysicists interpreted the current crisis in science as the beginning of a new era in which scientific knowledge would play only a secondary role. The scientist-philosophers, by contrast, saw the crisis as a harbinger of the ascent of science to new heights of intellectual achievement. They were firm, however, in refusing to interpret the rapid widening of scientific horizons as a decline in religious and philosophical outlooks. Like the leading metaphysicists, they rejected the nihilist effort to link science with materialistic philosophy and the positivist denigration of nonscientific modes of inquiry.

Most scientist-philosophers did not reject the idea of evolution, nor did they challenge the greatness of Darwin's scientific stature. They protested against the views of Darwinism as a body of theory stated in mechanistic metaphors and built exclusively upon the axioms of external determinism. The sociologist B. A. Kistiakovskii questioned the exclusive reign of causality in the mechanism of scientific explanation. The physicist A. I. Bachinskii, one of the harsher critics of Darwinism, challenged the view of organic evolution as a specific extension of the Newtonian idea of the continuity of motion. On epistemological grounds, the histologist I. F. Ognev could not accept the alliance of Darwinism with the belief in the superiority of physicochemical analysis in biology. V. P. Karpov, another histologist, did not subject Darwin's theory to direct criticism, but he worked hard on dissolving it in a grand system of organicist philosophy. Prompted in no small measure by current advances in experimental biology and philosophical criticism, these sci-

entists went against the Darwinian orthodoxy by demanding an unqualified acceptance of saltatory changes and of the role of purposiveness in the process of the transformation of species. Most of them
turned to Western neopositivists and neo-Kantians—to Mach, Avenarius, Poincaré, Duhem, and Rickert—for a new philosophical orientation that opposed both idealistic metaphysics and mechanistic
philosophy.

The third group of philosophers included persons who published
popular accounts of current developments in science, but who did not
hold academic positions, did not engage in scientific research, and were
not counted as professional philosophers. Their attitude toward Darwin's ideas was generally favorable. They published philosophical papers
exclusively in liberal journals aimed at reaching a wide cross section of
the educated public. V. V. Lunkevich, one such writer, worked primarily
on the theoretical complexities of new currents in biology and their relationship to Darwinism. Most of his key writings dealt with the philosophical reaction to the growing crisis in biology. More than any other
writer, with or without academic affiliation, he contributed to a systematic diffusion of new biological thought in Russia and to a discussion of
the major issues of the current dilemmas in the philosophy of science.
He not only analyzed Ernst Mach's epistemological discussion of physical reality but also showed the relevance of this discussion to the study
of biological problems. His aim was to show that the crisis in modern
physics was at the same time a crisis in biology and in philosophy.

In all his excursions into the philosophical intricacies of the full spectrum of the modern theories of life, Lunkevich was guided by the sole
purpose of building the theoretical background for an assessment of the
current state of Darwin's ideas. He accepted the prediction of Claude
Bernard and Ernst Mach that the future of biology lay somewhere between mechanism and vitalism, and between causality and teleology.
His conclusion was simple and forthright: all new developments in the
evolutionary branches of biology contained elements of truth, but they
did not challenge the supremacy of Darwinian thought. Although Darwinism had much room for improvement, it continued to be the quintessential component of evolutionary biology. On both philosophical
and scientific grounds, the future of evolutionary biology was in Darwinism, and the future of Darwinism was in a perspective untrammeled
by the iron rigidity of mechanistic conventions. In the rise of the mutation theory he saw a crucial sign of developments in the right direction.
He did not hesitate, however, to view de Vries's contribution as an elaboration, rather than as a negation, of Darwinian thought.[10]

Arguing mainly on philosophical grounds, Lunkevich saw the future of evolutionary biology in a cross-fertilization of Darwinism and Lamarckism. A typical representative of Russian biological thought, he made no secret of considering Darwinism a more basic and comprehensive theory of evolution than Lamarckism. He rejected the German psycho-Lamarckism, which recognized the need for an alliance of Darwinism and Lamarckism but which considered Lamarck's internal stimulus—a psychological factor—the moving force of evolution. Nor did he favor the group of French Lamarckians, represented most energetically by Félix Le Dantec, A. Giard, and E. Perrier, who thought that Lamarck's theory was more basic than Darwin's because it invited a deeper concern with the origins of life in the world of chemistry.[11]

M. M. Filippov's two-volume *Philosophy of Reality* represented the first effort in Russia to lay the groundwork for a philosophy fastened to scientific knowledge. Philosophy, as he saw it, studies the scientific world outlook. It concentrates on major scientific concepts, particularly those that contribute to the unity of the sciences, to the epistemological clarification of scientific knowledge, and to a scientific theory of moral principles.[12] He saw evolution as the basic unifying principle of the biological and social sciences and as a cornerstone of modern philosophy. Although Darwin, in his view, did not give a completely satisfactory explanation of organic evolution—and although he vastly exaggerated the role of natural selection as a mechanism of evolution—he was much closer to the realities of the transformation of species than any other biologist or group of biologists.[13]

The theologians, the third major channel for the flow of Darwinian ideas to Russia, moved slowly and purposefully. Not until the 1880s did their anti-Darwinian offensive become a concerted and carefully designed effort. They had a common target for attack: the Darwinian thesis of the anthropoid origin of man. They were also united by a repeatedly stated assertion that Darwinism was one of the most striking manifestations of "natural science materialism." All agreed that there was no way to build a bridge between Darwinism and religious beliefs.[14] Rigid anti-Darwinism was the most distinguishing feature of the Russian theological criticism of biological evolutionism. When they chose to "defend" Darwinism, as in the case of V. Kudriavtsev-Platonov, Russian theologians did so only for tactical reasons, never allowing for substantive concessions.

In their war on Darwinism, theologians depended on several modes of argumentation. At first they rejected Darwin's theory on the grounds that it contradicted the scriptural explanation of the creation of the living world. Relying mainly on translated or paraphrased Western works, this literature created modest interest in natural theology, an effort to present a synthetic history of living forms that fully agreed with biblical explanations.[15] This effort did not produce the expected results, mainly because it took place at a time when interest in natural theology in the West was in sharp decline.

Particularly after the publication of *The Descent of Man* (1871), Darwin's theory became a target of moralistic criticism. The critics based their anti-Darwinian arguments on the assumption that moral sentiments, representing the indissoluble ties between man and divine authority, were the most distinctive features of the human species. They claimed that Darwin violated the dignity and the freedom of man by seeking the origin of moral rules in animal behavior. Relying on the most popular, and simplest, lines of argument, they emphasized the divine origin of the moral code, which made ethical norms impervious to evolutionary processes and eliminated the need for the struggle for existence as a key factor in social and moral dynamics. Other theological writers preferred to criticize Darwin on logical grounds. These critics pointed out unsupported generalizations, ambivalent abstractions, non sequiturs, contradictory conclusions from the same empirical data, and many other types of flaws in reasoning.

A solid group of theological critics relied on philosophical arguments. These critics were active from the very beginning of the Darwinian era but became particularly strong after 1880. All were guided by the same idea: Darwinism was a product, a most notable expression, and the main propelling force of the philosophical materialism that dominated nineteenth-century science. These writers did not agree in identifying the main western European schools of materialistic philosophy. Nikanor, Bishop of Kherson and Odessa, represented the most extreme case; his list of the architects of materialism included Kant, Schelling, Hegel, Vogt, Schopenhauer, "present-day monists," and "evolutionists."[16] All theological writers considered empiricism, positivism, and utilitarianism specific forms of materialistic thought. Relying heavily on Eduard Hartmann's metaphysical criticism and Vladimir Solov'ev's eloquent and masterful defense of metaphysics, they had little use for contemporary epistemology, even in its "idealistic" version. The materialistic basis of Darwin's theory, as they saw it, consisted primarily of a categorical denial of transcendental causation, teleology, the supernatural origin of morality, and scriptural creationism.

Particularly after 1890, the most notable group of theological critics of Darwinism relied primarily on the arguments advanced by the scientific community. They studied current developments in biology for the purpose of marshaling scientific arguments against evolutionary thought as Darwin had perceived it. The true Russian pioneer in this line of endeavor was S. S. Glagolev, professor of Christian apologetics at Moscow Theological Academy, who borrowed anti-Darwinian arguments from three generations of scientists: the generation of Wigand, von Baer, and Kölliker; the generation of Eimer and the persevering workers in "developmental mechanics" and "experimental morphology"; and the generation of Korzhinskii, de Vries, and the rapidly growing numbers of scholars interested in the rich promise of Mendelian genetics. His published work showed reasonable familiarity with current developments in paleontology, anthropology, and ethnography.

The primary significance of theological criticism was in expressions of disagreement with a specific interpretation of the harmony of nature—the cosmic unity. Theologians criticized the scientific world outlook refracted through the prism of Darwinian thought. They depended on science in combatting Darwinism because scientific facts were ready at hand and were rapidly growing in volume and weight. As a target of attack, Darwin's science interested them only insofar as it challenged their cosmic outlook and moral philosophy. One must agree with Karl Popper that every metaphysical and religious criticism of a scientific idea is always sociological, never scientific. Scientific or not, theological criticism helped make the Darwinian debate richer in pertinent ramifications and broader in philosophical compass. It was one of the notable manifestations of a major crisis in Darwinian thought.

Theological criticism passed through two general phases of development. During the first phase, which lasted until the end of the 1880s, the theological scholars were resolute and categorical in rejecting both Darwin's ideas and the evolutionary theory in general. During the second phase, which peaked during the first decade of the twentieth century, the theological scholars continued to be firm in their anti-Darwinian attitude, but they did not reject the idea of evolution in general. In the current contributions of experimental biology and in various neo-Lamarckian schools they detected promising material for a theory of evolution reconcilable with creationism. Confrontation with science in general and Darwinism in particular helped the Russian church scholars advance a new discipline which they named "theoretical theology." This discipline made Russian theologians active participants in the grand debate centered on the ongoing crisis in scientific thought.

Next to scientists, philosophers, and theologians, ideologues served as an important source of comments on the reigning aspects of Darwin's theory. The ideologues of conservatism—Pogodin, Danilevskii, Strakhov, Rozanov, and Pobedonostsev—rejected Darwinism as a branch of modern materialism and as a direct assault on Russia's sacred culture, the values clustered around the institutional complex of autocracy. As the ideologues of autocracy saw it, Western materialism included a wide variety of intellectual traditions interpreted as terrestrial and nature-bound in vision. The conservative ideologues—all enemies of Darwinism—counted the philosophical traditions of Bacon, Locke, Kant, Hegel, Mill, Comte, and Spencer among the chief sources of materialist impurities. The ideological spokesmen for radical movements demanding basic changes in the structure of Russian society considered Darwinism a great victory for the spirit of science—for critical thought, rationalism, and secularism. It is possible, wrote the nihilist D. I. Pisarev, that Darwin's discovery "has no match in the annals of science."[17] The populist N. K. Mikhailovskii noted that the future would show that Darwinism had produced "broader and more fruitful generalizations" than any other scientific theory. Darwin's theory, he added, belonged to the "eternal truths" of basic sciences.[18] The anarchist P. A. Kropotkin observed that Darwin's theory of the origin of species was "an established fact" and a new key to a better understanding of "the life of physical matter, the life of organisms, and the life and evolution of societies."[19] In the view of the Marxist G. V. Plekhanov, Marxism is Darwinism applied to the study of human society.

Behind fundamental differences between the two sides of the ideological spectrum was a major point of agreement: both sides expressed strong skepticism about the struggle for existence and natural selection as prime movers of organic evolution. A large majority of ideologues regarded cooperation rather than competition as a basic mechanism of transformation. Some recognized competition but showed a clear tendency to interpret it as a special case of cooperation or mutual aid. Some, Mikhailovskii for example, considered competition the law of organic nature, and cooperation the law of human society. Kropotkin, by contrast, considered cooperation the basic law of both nature and society. Widespread conflict in modern society, he observed, is rooted not in human nature but in social dynamics.

The ideological criticism of Darwin's interpretation of the struggle for existence accounted for the weak development of Social Darwinism in Russia. V. A. Zaitsev, who made an isolated and rather rash effort to justify political, cultural, and racial inequality on the basis of Darwinian principles, was attacked from all sides.[20] And so was Clemence Royer, the French translator of the *Origin of Species,* who leaned on Darwin to justify social inequality. Later on, the theories of L. Gumplowitz, Schäffle, and other German Social Darwinists found no appreciative audience in Russia. All leading Russian sociologists from N. K. Mikhailovskii and P. L. Lavrov to M. M. Kovalevskii and B. A. Kistiakovskii gave ample expression to a consistent opposition to Social Darwinism.

Two Russian scholars with strong ideological commitments were among the most widely recognized pioneers in the war against Social Darwinism. P. A. Kropotkin made the attack on Social Darwinism the central theme of his *Mutual Aid,* one of the most widely read and discussed works in modern sociology. Ia. A. Novikov produced a comprehensive, but rather eclectic, rebuttal of modern efforts to use Darwin's struggle for existence in interpreting warfare as a source of social progress. In his words, Social Darwinism is "a doctrine that considers collective homicide the cause of human progress."[21] He assembled anthropological, historical, economic, political, juridical, and intellectual arguments against Spencer's treatment of the struggle for existence as the instrument of social evolution, identified as social progress.

Opposition to the Darwinian struggle for existence on the ideological front found only a feeble reflection in scientific circles. This became eminently clear during the 1890s. I. I. Mechnikov, one of the most respected leaders in Russian biology, now had no doubt about the primary role of the struggle for existence in the transformation process. In his theory of inflammation he made the struggle for existence a pillar of comparative pathology, a new evolutionary discipline. Darwin's theory, he wrote, "demonstrated that only the characteristics which are advantageous to the organism survive in the struggle for existence while those that are harmful to the individual are readily eliminated by natural selection."[22] A. N. Beketov was busy trying to show the full compatibility of Christian morality and evolution propelled by two forces: the struggle of organisms with the physical environment, and the competition (*sostiazanie*) among individual organisms or species for food and space.[23] The novelist Leo Tolstoy found enough light in Beketov's tangled reasoning to scoff at his effort to prove the existence of close ties between Darwinian struggle for existence and Christian morality.[24]

A. S. Famintsyn and V. A. Vagner, noted for their effort to advance an empirically oriented psychological approach to organic evolution, recognized the struggle for existence as a primary factor in the transformation of species. Famintsyn considered the sense organs "the most powerful tool of organisms in the struggle for existence." Writing in a more general vein, Vagner considered all the general principles of Darwin's theory impervious to the ravages of time. He regarded the struggle for existence as the pivotal notion of Darwin's biology.

In 1896 K. A. Timiriazev, the most determined and consistent defender of Darwinian orthodoxy, surprised his friends and foes alike with the published statement that the struggle for existence was an unfortunate expression and that Darwin's theory would have fared much better without it. Timiriazev noted that he "could give an entire course on Darwinism without making the error of mentioning the struggle for existence."[25] He expressed dissatisfaction with this label, not because he did not agree with the meaning Darwin gave it, but because it was widely abused by all kinds of interpreters. Misinterpretation, he wrote, was particularly pronounced in humanistic studies. At no time did Timiriazev indicate the least inclination to disagree with Darwin's firm allegiance to Malthus's law.

During the 1890s the monumental *Encyclopedic Dictionary* published in forty-one volumes by Brockhaus and Efron devoted a long article to the struggle for existence as a basic law of living nature.[26] The article on species as a biological category was cast completely in a Darwinian mold.[27] The item on transformism stated that Darwin's notion of natural selection was responsible for a full collapse of the stationary view of living nature. Darwin and his followers gathered so much evidence in support of "the hypothesis of selection" that it had become a guiding idea of modern biology. "The entire embryology, entire comparative anatomy, and entire paleontology of the post-Darwinian era have worked to fulfill his legacy, and are still following the paths he had opened."[28] The editors of the biological section of the encyclopedia found it unnecessary to assign any space to mutual aid as a special topic. The same decade saw a proliferation of university textbooks in zoology, the main avenues for the presentation of the achievements of evolutionary theory to students at all institutions of higher education. All these textbooks recognized natural selection and the struggle for existence as the main mechanisms in the evolution of species.

Particularly during the early years of the twentieth century, more Russian biologists were inclined to take Darwin's struggle for existence for granted than ever before. Menzbir and Shimkevich, Knipovich and Komarov, Kholodkovskii and Kol'tsov were among the leading biologists who recognized both natural selection as the chief mechanism of evolution and the struggle for existence as the main propelling force of natural selection. As a biological label, the struggle for existence, they thought, needed only a more precise definition and a more comprehensive application. Knipovich, for example, thought that it should have a "broader meaning," covering the relations of organisms with the surrounding world, and a "narrower meaning" referring to "the struggle among organisms," and that both should be subject to further elaboration and clarification.[29] A. N. Severtsov, the leading Russian contributor to the codification of the general theory of organic evolution, relied on mathematical notation to express the dependence of the duration of animal growth on the intensity of the struggle for existence.[30] He also pointed out the need for a systematic study of specific features of the struggle for existence during different phases in the growth of organisms belonging to the same species and occupying common territory.[31] N. V. Tsinger took a novel and most innovative course: he completed a meticulously documented empirical study of the role of the struggle for existence and natural selection in the evolution of false flax (*Camelina linicola*).[32] R. E. Regel', a leading pioneer in the rapidly developing field of genetics, thought it possible to find a functional place for the struggle for existence and natural selection in the grand evolutionary schema dominated by Mendelian theory.[33] K. A. Timiriazev continued to remind his readers that Malthus's law applied not only to man but to the entire organic world as well.[34]

A typical ideologue—particularly if he was a representative of the populist or anarchist intelligentsia—recognized both the struggle for existence and mutual aid as distinctive features of plant and animal life. He was determined, however, to establish and defend the primacy of mutual aid. A typical biologist followed the opposite line of reasoning: he either ignored mutual aid altogether or treated it as one of the many ramifications of the struggle for existence.

The ideologues referred to mutual aid as a basic principle of social integration. They had no vision of mutual aid as a mechanism of evolution, either organic or social. The biologists referred to the struggle for existence strictly as a mechanism in the transformation of species—as a moving force in the origin of species.

In Russia, no less than in the West, Darwin found many unfriendly critics. On many occasions he became the target of attacks for elaborations and extrapolations that he did not originate but that came from various Darwinists. Despite the multiple streams of criticism, his theory found strong support in Russian society. It became a powerful source of ideas that changed the course of science, the pulse of philosophy, and the meaning of culture. It gave man a renewed faith in the empirical anchorage of his most profound wisdom. It marked a victory for human reason, inductive method, empirical knowledge, and natural causation. It opened wide realms of nature and culture to scientific study and made a major contribution to the victory of a cosmic view detached from the fanciful world of metaphysical constructions and mystical escapes. In Russia Darwin had friends and enemies. All of them called him a "great man."[35]

Abbreviations

BdCh. Biblioteka dlia chteniia
BMOIP Biulleten' Moskovskogo Obshchestva ispytatelei prirody
Brockhaus-Efron *Entsiklopedicheskii slovar'*. 41 vols. St. Petersburg: F. A. Brockhaus and I. A. Efron, 1890–1901
BS-PU Biograficheskii slovar' professorov i prepodavatelei Imperatorskogo S. Peterburgskogo universiteta, 1869–1894. 2 vols. St. Petersburg, 1896–98
CCD The Correspondence of Charles Darwin. The Archives of the American Philosophical Society
EiG Estestvoznanie i geografiia
Granat-*Ents. Entsiklopedicheskii slovar'*. 53 vols. Moscow: Granat Institute, n.d.–1937 (7th ed.)
IIAN Izvestiia Imperatorskoi Akademii nauk, St. Petersburg
ITEU Istoriia i teoriia evoliutsionnogo ucheniia
JHB Journal of the History of Biology
LRN: biol. Liudi russkoi nauki: biologiia, meditsina, sel'skokhoziastvennye nauki. I. V. Kuznetsov, ed. Moscow, 1963
LRN: geol. Liudi russkoi nauki: geologiia, geografiia. I. V. Kuznetsov, ed. Moscow, 1962
MAI Mémoires de l'Académie impériale des sciences de St-Pétersbourg
MB Mir bozhii
MBS-AN Materialy dlia biograficheskogo slovaria deistvitel'nykh chlenov Imperatorskoi Akademii nauk. 2 vols. Petrograd, 1915–17
NiT Nauka i tekhnika
NN Nauchnoe nasledstvo
NO Nauchnoe obozrenie
NS Nauchnoe slovo
Obr. Obrazovanie
OZ Otechestvennye zapiski

Pr. *Priroda*
RB *Russkoe bogatstvo*
RM *Russkaia mysl'*
RV *Russkii vestnik*
TBPB *Trudy Biuro po prikladnoi botanike*
TIIE *Trudy Instituta istorii estestvoznaniia*
TIIE-T *Trudy Instituta istorii estestvoznaniia i tekhniki*
TSPO *Trudy S. Peterburgskogo obshchestva ispytatelei prirody*
VAN *Vestnik Akademii nauk SSSR*
VE *Vestnik Evropy*
VFiP *Voprosy filosofii i psikhologii*
VIE-T *Voprosy istorii estestvoznaniia i tekhniki*
VZ *Vestnik znaniia*
ZAN *Zapiski Imperatorskoi S. Peterburgskoi Akademii nauk*
ZAN:F-MO *Zapiski Imperatorskoi S. Peterburgskoi Akademii nauk po Fiziko-matematicheskom otdelenii*
ZhMNP *Zhurnal Ministerstva narodnogo prosveshcheniia*
ZhOB *Zhurnal obshchei biologii*
ZWZ *Zeitung für wissenschaftliche Zoologie*

Notes

Introduction

1. Ernst Mayr, "The Concept of Finality," p. 112.
2. M. M. Kovalevskii, "Darvinizm v sotsiologii," p. 153.
3. J. B. Bury, "Darwinism and History," p. 535.
4. M. J. S. Hodge, "The Development," p. 57.
5. J. Monod, *Chance*, p. 24.
6. J. Royce, *The Spirit*, p. 286.
7. Charles Peirce, *Selected Writings*, p. 263.
8. W. Bateson, "Heredity and Variation," p. 85.
9. J. Royce, *The Spirit*, p. 287.

Chapter One

1. Herzen, *Polnoe sobranie sochinenii*, vol. 12, p. 106.
2. Nestor Kotliarevskii, "Ocherki," p. 275.
3. I. I. Mechnikov, *Akademicheskoe sobranie sochinenii*, vol. 14, p. 10.
4. A. P. Shchapov, *Sochineniia*, vol. 3, p. 111.
5. P. L. Lavrov, *Zadachi pozitivizma*, p. 5. For a perceptive analysis of the positivist basis of Lavrov's philosophy, see H.-G. Noetzel, *Petr L. Lavrovs Vorstellungen*, ch. 3.
6. N. Kareev, "P. L. Lavrov kak sotsiolog," pp. 94–95. For Lavrov's views on science as the foundation of philosophy, see P. L. Lavrov, *Filosofiia i sotsiologiia*, vol. 1, p. 621.
7. I. I. Mechnikov, *Etiudy optimizma*, p. 5; A. Vucinich, *Social Thought*, pp. 8, 115–16.
8. A. Stadlin, "Istoricheskaia teoriia," p. 258.
9. [E. Solov'ev] Andreevich, *Opyt*, p. 277.

10. M. A. Antonovich, "Teoriia proiskhozhdeniia," p. 65. For a brief review of the early development of positivism in Russia, see S. P. Ranskii, *Sotsiologiia*, pp. 101–2.

11. A. P. Shchapov, *Sotsial'no-pedagogicheskie usloviia*, p. 268.

12. Ibid., p. 278; Claude Bernard, *An Introduction*, p. 68.

13. Bernard, *An Introduction*, p. 38.

14. N. N. Strakhov, *Filosofskie ocherki*, p. 139. Strakhov's article on Bernard, presented in this collection, was originally published in 1867.

15. I. M. Kaufman, *Russkie entsiklopedii*, vol. 1, p. 30.

16. Vasetskii and Mikulinskii, eds., *Izbrannye proizvedeniia*, p. 117.

17. Mikulinskii and Iushkovich, eds., *Razvitie estestvoznaniia*, p. 233.

18. B. E. Raikov, *Grigorii Efimovich Shchurovskii*, pp. 23–36.

19. There is a vast literature on the Russian forerunners of Darwin. The most comprehensive study is Raikov, *Russkie biologi-evoliutsionisty*, 4 vols., which goes extensively into the early Darwinian period. For a brief and pertinent survey, see S. R. Mikulinskii and V. I. Nazarov, "Razvitie idei evoliutsii," pp. 274–83.

20. Charles Darwin, *The Origin*, p. 21.

21. Ibid., pp. 434–35.

22. For biological data on Karl von Baer, see L. Ia. Bliakher, "Karl Maksimovich Ber." The Darwin–von Baer relations and von Baer's views on transformism are discussed in Timothy Lenoir, *The Strategy of Life*, chapter 6; Jane Oppenheimer, *Essays*, pp. 224–55; and B. E. Raikov, *Karl Ber*, pp. 393–412. For a thoughtful review of Raikov's book, see Jane Oppenheimer, "Review."

23. K. von Baer, "Ueber Papuas," p. 342; Georg Seidlitz, *Darwin'sche Theorie*, 1871, pp. 186–88.

24. For details, see Georg Seidlitz, *Die Darwin'sche Theorie* (1871), pp. 186–88.

25. L. Ia. Bliakher, "Karl Maksimovich Ber," p. 63.

26. I. I. Mechnikov, *Izbrannye proizvedeniia*, p. 73.

27. G. S. Vasetskii and S. R. Mikulinskii, eds., *Izbrannye proizvedeniia*, p. 214.

28. S. R. Mikulinskii, "Vzgliady K. M. Bera," pp. 358–60. Von Baer's early allegiance to a limited idea of organic evolution and to a "causal-historical orientation" in biology has been analyzed and documented in T. Lenoir, "Kant," pp. 100–12. This orientation, according to Lenoir, led von Baer to the discovery of the mammalian egg.

29. S. R. Mikulinskii, *Razvitie*, p. 403. For a brief description of Rul'e's pre-Darwinian evolutionism, see F. M. Scudo and M. Acanfora, "Darwin," pp. 736–38. For an excellent description of Rul'e's ties with Lamarckian tradition, see D. R. Weiner, "The Roots of 'Michurinism,'" pp. 245–48.

30. K. F. Rul'e, *Izbrannye biologicheskie proizvedeniia*, p. 242.

31. S. R. Mikulinskii, *Razvitie*, p. 405.

32. Borzenkov, *Istoricheskii ocherk*, p. 43. See also D. N. Anuchin, *O liudiakh russkoi nauki*, pp. 186–87.

33. D. N. Anuchin, *O liudiakh russkoi nauki*, p. 187.

34. A. N. Beketov, "Garmoniia," pp. 545–82. For comments, see S. L. Sobol', "Voznikovenie," pp. 45–47.

35. For details on A. N. Beketov's life and work, see Beketov, A. N., "Avtobiografiia."

36. *Protokoly VII S'ezda*, vol. 1, p. 26.

37. I. I. Mechnikov, *Akademicheskoe sobranie sochinenii*, vol. 14, p. 52.

38. Charles Lyell, "On the Occurrence of Works," p. 95; [N. N. Strakhov], "Poiavlenie cheloveka," pp. 5–6.

39. [Strakhov] "Poiavlenie cheloveka"; S. L. Sobol', "Iz istorii," vol. 14, p. 200.

40. S. L. Sobol', "Pervye soobshcheniia," pp. 130–35. See also N. N. Ipatova, "Vospriiatie sochinenii Ch. Darvina," pp. 130–31.

41. "Darvin i ego teoriia," no. 11, pp. 34–36.

42. For comment, see B. E. Raikov, "Iz istorii darvinizma," part 2, vol. 31, pp. 27–28.

43. "Darvin i ego teoriia," no. 11, pp. 31–32.

44. J. Schönemann, "Teoriia Darvina," p. 211.

45. Rachinskii, "Tsvety," p. 392.

46. M. A. Antonovich, "Teoriia proiskhozhdeniia."

47. B. V. Nikol'skii, *Nikolai Nikolaevich Strakhov*, pp. 22–24.

48. N. N. Strakhov, "Durnye priznaki," p. 170. See also G. M. Fridlender, *Dostoevskii*, pp. 218–22, and *Realizm Dostoevskogo*, pp. 157–62.

49. N. N. Strakhov, "Durnye priznaki," pp. 165–67.

50. N. N. Strakhov, *O metode*, pp. 178–82. (This book is a collection of essays published in various journals during the late 1850s and early 1860s.) See also Gerstein, *Nikolai Strakhov*, pp. 155–57.

51. P. A. Bibikov, *Kriticheskie etiudy*, p. 105.

52. Ibid., p. 118.

53. For more details on the evolution of Strakhov's views, see B. V. Nikol'skii, *Nikolai Nikolaevich Strakhov*, pp. 24–39.

54. N. N. Strakhov, *Iz istorii literaturnogo nigilizma*, pp. 123–34.

55. N. N. Strakhov, *Bor'ba*, vol. 2, pp. 250–80.

56. K. A. Timiriazev, *Kratkii ocherk*.

57. K. A. Timiriazev, "Kniga Darvina," no. 12, p. 861.

58. Charles Darwin, *The Origin*, p. 478.

59. K. A. Timiriazev, "Kniga Darvina," no. 8, p. 880.

60. James A. Rogers, "Charles Darwin and Russian Scientists," pp. 382–83.

61. Darwin's name had been known in the Russian scientific literature since 1839 when the *Agricultural Journal* published a summary of his monograph on the formation of mold. In 1846 a summary of Darwin's paper on the origin of coral islands was published by the *Mining Journal*. (For details, see B. E. Raikov, "Iz istorii darvinizma," part 2, pp. 17–26.)

62. Francis Darwin, ed., *The Life and Letters*, vol. 2, p. 256.

63. While the publication of Darwin's work was delayed by the printer's work on the plates of illustrative material, the publishers of the Russian translation used the illustrations prepared for Brehm's *The Life of Animals*. Darwin had no objection to this move. (CCD no. 5562.)

64. J. Schönemann, "Charlz Darvin," p. 241.

65. "Teoriia Darvina i iazykoznanie," pp. 262–63.

66. [P. L. Lavrov], "Antropologi," pp. 31–41.

67. Charles Darwin, *The Origin*, p. 444.

68. A. P. Shchapov, *Sotsial'no-pedagogicheskie usloviia*, p. 118.

69. Ibid., p. 256.

70. Ibid., p. 295.

71. I. P. Pavlov, *Polnoe sobranie sochinenii*, vol. 6, p. 441. See also K. N. Davydov, "A. O. Kovalevskii kak chelovek," p. 333. For comments on Pisarev's general views on the place of science in modern society, see A Skabichevskii, *Sochineniia*, vol. 1, pp. 143–80, and *Istoriia*, pp. 97–107; A. Coquart, *Dmitri Pisarev*, pp. 333–41; E. Solov'ev, *D. I. Pisarev*, pp. 192–220; and V. A. Tsibenko, *Mirovozzrenie*, pp. 53–65.

72. D. I. Pisarev, *Selected Philosophical, Social and Political Essays*, p. 329.

73. Ibid., p. 340.

74. Pisarev's citation is from Schleicher, *Darwin's Theory and Linguistics* (Pisarev, *Selected Essays*, p. 495.)

75. Ibid., p. 456.

76. Pisarev, *Izbrannye filosofskie i obshchestvenno-politicheskie stat'i*, pp. 292–94.

77. Strakhov, *Filosofskie ocherki*, pp. 188–200.

78. E. Rádl, *The History of Biological Theories*, pp. 88–89.

79. *Russkoe slovo*, 1864.

80. F. Kuznetsov, *Publitsisty*, pp. 193–94.

81. [Antonovich, M. A.], Postoronnyi satirik, "Literaturnye melochi," p. 163.

82. For additional information on Nozhin and Zaitsev, see James A. Rogers, "Darwinism, Scientism, and Nihilism," pp. 18–23, and "Russia," pp. 258–60; Charles Moser, *Antinihilism*, pp. 34–35; A. E. Gaisinovich, "Biologshestidesiatnik," p. 388.

83. P. N. Tkachev, *Izbrannye sochineniia*, vol. 5, p. 300.

84. B. E. Raikov, "Iz istorii darvinizma," part 2, p. 60.

85. F. Brandt, "Zametki," p. 215.

86. K. S. Veselovskii, "Otchet: 1869," p. 16.

87. For details on these elections, see G. A. Kniazev, "Izbranie Ch. Darvina," pp. 117–20.

88. V. P. Alekseev, "Evoliutsionnaia ideia," p. 231.

89. James A. Rogers, "The Reception of Darwin's *Origin*," p. 503.

90. Charles Darwin, *The Origin*, pp. 434–45.

91. For biographical information on A. O. Kovalevskii, see L. Ia. Bliakher, "Aleksandr Onufrievich Kovalevskii," pp. 157–72; K. N. Davydov, "A. O. Kovalevskii kak chelovek"; A. D. Nekrasov and N. M. Artemov, "Aleksandr Onufrievich Kovalevskii"; E. N. Mirzoian, "K voprosu o formirovanii evoliutsionnykh vzgliadov."

92. I. I. Mechnikov, *Stranitsy*, pp. 17–18.

93. L. Ia. Bliakher, *Istoriia embriologii v Rossii (s serediny XIX do serediny XX veka)*, pp. 17–18.

94. For additional comments, see K. N. Davydov, "A. O. Kovalevskii kak chelovek," pp. 342–43. Francis M. Balfour stated in 1881 that "our knowledge of the development of *Amphioxus* is mainly due to Kovalevskii" (*A Treatise*, vol. 2, p. 1).

95. A. O. Kovalevskii, *Izbrannye raboty,* pp. 41–78.

96. Ibid., p. 164.

97. Charles Darwin, *The Descent of Man,* pp. 175–76.

98. Ernst Haeckel, *The Evolution of Man,* vol. 1, pp. 441–42. E. Ray Lankester wrote in *Nature* in 1902: "Before Kowalevsky's work on the development of Amphioxus, carried out in 1864–65, and on Ascidia in 1866, zoologists were content to regard the cell-masses resulting from cell-divisions of the animal egg-cell as intricate heaps which no one could expect to analyze. Some way was made in the direction of their comprehension by the application to invertebrate embryos of the doctrine of cell-layers, but it was not until the avowed task of the embryologist became the definite tracing of the genesis of the cells of cell-layers one by one from pre-existing cells and finally from the first cell-division of the egg-cell that Kowalevsky's work bore its full fruit and thoroughgoing cellular embryology was established" ("Alexander Kowalevsky," p. 304).

99. V. V. Zalenskii, "O nauchnoi deiatel'nosti," pp. 653–54. See also V. M. Shimkevich, "Proiskhozhdenie pozvonochnykh," p. 616.

100. A. D. Nekrasov and N. M. Artemov, "Aleksandr Onufrievich Kovalevskii," p. 551.

101. Ibid., pp. 557–58.

102. For details on Mechnikov's life and work, see G. K. Khrushchov, "Il'ia Il'ich Mechnikov"; V. A. Dogel' and A. E. Gaisinovich, "Osnovnye cherty tvorchestva I. I. Mechnikova"; A. E. Gaisinovich, "Velikii russkii biolog I. I. Mechnikov"; R. I. Belkin, "Tvorcheskaia razrabotka"; and Zalkind, *Ilya Mechnikov.*

103. S. Zalkind, *Ilya Mechnikov,* pp. 52–53.

104. I. I. Mechnikov, *Izbrannye biologicheskie proizvedeniia,* pp. 652–72. Mechnikov also criticized Fritz Müller's use of the biogenetic law as a proof of the Darwinian conception of evolution, rather than as a hypothesis in need of empirical testing and verification. (Mechnikov, "Sovremennoe sostoianie nauki," pp. 160–64.)

105. L. L. Gel'fenbein, *Russkaia embriologiia,* p. 82.

106. V. A. Dogel', "Embriologicheskie raboty A. O. Kovalevskogo," p. 210.

107. For details, see S. Zalkind, *Ilya Mechnikov,* pp. 47–49.

108. A. E. Gaisinovich, "Velikii russkii biolog," p. 14.

109. I. I. Mechnikov, "Sovremennoe sostoianie," p. 168.

110. N. M. Kulagin, "I. I. Mechnikov," p. 703.

111. "Protokoly," *ZAN,* 1867, vol. 11, pp. 76–77.

112. "Pervoe prisuzhdenie," p. 192.

113. L. Ia. Bliakher, *Istoriia embriologii v Rossii (s serediny XIX do serediny XX veka),* p. 23.

114. For details, see E. L. Rudnitskaia, "Pis'ma A. O. Kovalevskogo," pp. 202–3.

115. I. I. Mechnikov, *Stranitsy,* pp. 19–20.

116. N. D. Nozhin, "Nasha nauka," no. 1, p. 21. See also N. K. Davydov, "A. O. Kovalevskii kak chelovek," pp. 332–33; Viktor Chernov, "Gde kliuch?" p. 114.

117. Nozhin, "Nasha nauka," no. 7, p. 175.

118. Raikov, *Russkie biologi-evoliutsionisty,* vol. 4, pp. 109–23.

119. Raikov, "Iz istorii darvinizma," part 1, p. 4.
120. I. Krasovskii, "Noveishaia teoriia," p. 19.
121. K. Lindemann, "Eshche somneniia v teorii Darvina," p. 939.
122. Ibid., p. 946.
123. Von Baer's embryological influence on Darwin's early work is analyzed in Jane Oppenheimer, *Essays*, pp. 248–55.
124. CCD, no. 2891. F. Darwin, ed., *Life and Letters*, vol. 2, p. 122: "I have stated the same ideas on the transformation of types or the origin of species as Mr. Darwin."
125. CCD, no. 2893.
126. K. von Baer, *Reden*, vol. 1, pp. 282–84. See also K. S. Veselovskii, "Otchet: 1861," p. 26.
127. V. P. Alekseev, "Evoliutsionnaia ideia," p. 221.
128. K. von Baer, "Mesto cheloveka v prirode."

Chapter Two

1. CCD, no. 9251.
2. M. Vladislavlev, "Review," pp. 199–200.
3. P. M-v, "Razvitie vyrazitel'nosti," pp. 25–26.
4. N. P. Vagner, "Chuvstva," p. 317.
5. "G. Kavelin kak psikholog," p. 18.
6. A. S. Famintsyn, "Darvin i ego znachenie," pp. 132–33.
7. Borzenkov, *Istoricheskii ocherk*, pp. 13–25.
8. Ibid., pp. 36–37. See also [Borzenkov], "Chteniia," p. 125.
9. L. O., "Charlz Darvin," p. 112.
10. [Anuchin] D. "Nauchnoe obozrenie," pp. 17–18.
11. L. Ia. Bliakher, *Istoriia embriologii v Rossii (s serediny XIX do serediny XX veka)*, p. 20; S. Zalkind, *Ilya Mechnikov*, pp. 47–48.
12. L. Ia. Bliakher, *Istoriia embriologii v Rossii (s serediny XIX veka do serediny XX veka)*, pp. 19–21, and "Aleksandr Onufrievich Kovalevskii," pp. 162–63.
13. A. O. Kovalevskii, *Izbrannye raboty*, p. 184.
14. Ibid., p. 210. See also Jane M. Oppenheimer, *Essays*, pp. 265–66.
15. L. Ia. Bliakher, *Istoriia embriologii v Rossii (s serediny XIX do serediny XX veka)*, pp. 19–21.
16. K. von Baer, "Entwickelt sich die Larve . . . ?"
17. For details on von Baer's criticism, see L. Ia. Bliakher, *Istoriia embriologii v Rossii (s serediny XIX veka)*, ch. 2.
18. G. A. Kniazev and B. E. Raikov, eds., *Pis'ma*, p. 103.
19. V. A. Dogel', "Embriologicheskie raboty A. O. Kovalevskogo," p. 210.
20. V. M. Shimkevich, "A. O. Kovalevskii. (Nekrolog)," p. 114.
21. L. Ia. Bliakher, "Nauchnye sviazi," pp. 96–117.
22. *Trudy III-go S'ezda*, pp. 55–56.
23. *Protokoly VII S'ezda*, vol. 1, p. 26.
24. CCD, nos. 5443, 7326.
25. CCD, no. 7442.

26. CCD, no. 7583.

27. For details on the phagocytic theory, see A. E. Gaisinovich, "100 let," pp. 12–22; and T. I. Ul'iankina, "I. I. Mechnikov," pp. 12–20.

28. I. I. Mechnikov, *Izbrannye biologicheskie proizvedeniia*, p. 24; A. Besredka, *Histoire d'une idée*, pp. 9–26.

29. Mechnikov, *Akademicheskoe sobranie sochinenii*, vol. 9, p. 72.

30. As cited in Nekrasov and Artemov, "Aleksandr Onufrievich Kovalevskii," p. 595.

31. B. N. Mazurmovich, "Materialy," pp. 319–22.

32. P. A. Novikov, "Zoologiia," pp. 312–13.

33. For details on Zalenskii's election, see Iu. I. Polianskii et al., eds., *Pis'ma A. O. Kovalevskogo*, p. 285.

34. For details on V. O. Kovalevskii's life and work, see D. Todes, "V. O. Kovalevskii"; L. Sh. Davitashvili, *V. O. Kovalevskii* and "Biografiia"; A. A. Borisiak, *V. O. Kovalevskii*; D. N. Anuchin, "V. O. Kovalevskii (Nekrolog)." For Kovalevskii's place in the development of evolutionary paleontology, see Iu. Ia. Polianskii, "Osnovnye puti"; L. Sh. Davitashvili, *Istoriia evoliutsionnoi paleontologii*, pp. 74–122. Kovalevskii's scientific papers have been republished under the title *The Complete Works of Vladimir Kovalevsky*, edited by S. J. Gould.

35. Davitashvili, *V. O. Kovalevskii*, pp. 152–54.

36. D. Anuchin, "Kovalevskii, Vladimir Onufrievich," BS-PU, vol. 1, p. 326.

37. Albert Gaudry, *Les enchaînements*, p. 44.

38. Davitashvili, "Vladimir Onufrievich Kovalevskii," *LRN: geol.*, p. 25.

39. John M. Smith, *The Theory of Evolution*, p. 240.

40. D. M. S. Watson, "The Evidence," p. 48. See also George Gaylord Simpson, *Horses*, pp. 87–88.

41. Davitashvili, *Istoriia evoliutsionnoi paleontologii*, p. 108.

42. H. F. Osborn, *The Age of Mammals*, p. 6.

43. For details, see Davitashvili, *V. O. Kovalevskii*, pp. 144–71.

44. CCD, no. 8914. The dedication reads in part: "It gives me great joy to dedicate this work to you, not because I consider my work particularly worthy of such a dedication, but because it gives me an opportunity to express my profound respect for you personally. From the very beginning of my scientific work you have been my best teacher and friend: you have paid full attention to my entire work, and during my prolonged stay in England, you have created favorable conditions for my research. Your intercession opened many collections and libraries for me which otherwise might have remained closed. Your name and your friendship were the best recommendations which opened all the doors for me." (V. O. Kovalevskii, *Sobranie nauchnykh trudov*, vol. 3, p. 99.)

45. V. O. Kovalevskii, *Sobranie nauchnykh trudov*, vol. 3, p. 99.

46. A. A. Borisiak, *Izbrannye trudy*, p. 256.

47. Daniel Todes, "V. O. Kovalevskii," pp. 141–42.

48. Davitashvili, "Biografiia V. O. Kovalevskogo," p. 96.

49. K. von Baer, *Reden*, vol. 2, p. 371.

50. CCD, no. 12963.

51. D. N. Anuchin, "Vladimir Onufrievich Kovalevskii. (Nekrolog)."

52. Davitashvili, "V. O. Kovalevskii, ego nauchnaia deiatel'nost'," pp. 306–7.

53. S. N. Nikitin, "Darvinizm," no. 9, pp. 233–34. See also I. Ia. Spasskii, "Razvitie," p. 33.

54. [Anuchin, D. N.], "Nauchnyi obzor: antropologiia," p. 171.

55. For relevant biographical data, see M. G. Levin, *Ocherki po istorii antropologii*, pp. 106–37.

56. V. A. Alekseev, *Osnovy darvinizma*, p. 355.

57. For biographical data, see "Anuchin," *MBS-AN*, vol. 1, pp. 1–14.

58. [Anuchin, D. N.] "Nauchnyi obzor," p. 180.

59. K. S. Veselovskii, "Otchet: 1878," p. 20.

60. I. M. Sechenov, *Selected Physiological and Psychological Works*, p. 154.

61. Sechenov's views on Spencer's evolutionism are discussed in M. G. Iaroshevskii, *Ivan Mikhailovich Sechenov*, pp. 317–21.

62. M. T. Ghiselin, *The Triumph*, p. 211.

63. For an analysis of Sechenov's "physicochemical orientation," see N. A. Grigor'ian, "O pervoi fiziologicheskoi shkole," pp. 144–45.

64. K. A. Timiriazev, "Istoricheskii metod," 1892, no. 10, p. 163.

65. [Borzenkov], "Chteniia," pp. 123–24.

66. Borzenkov, *Istoricheskii ocherk*, p. 43.

67. [Borzenkov], "Chteniia," p. 33.

68. V. Sergeevich, "Issledovanie," pp. 204–5.

69. I. I. Mechnikov, *Akademicheskoe sobranie sochinenii*, vol. 4, pp. 155–327.

70. Mechnikov, "Bor'ba," p. 29.

71. Ibid., pp. 30–47.

72. Mechnikov, "Antropologiia," pp. 191–95.

73. Mechnikov, *Izbrannye biologicheskie proizvedeniia*," pp. 210–11.

74. N. N. Banina, "O nauchnom mirovozzrenii," pp. 264–65.

75. K. F. Kessler, "O zakone," p. 136; Banina, "O nauchnom mirovozzrenii," p. 265, and *K. F. Kessler*, pp. 127–29. See also A. F. Brandt, "Sozhitel'stvo," no. 6, pp. 98–100.

76. For more details on Borzenkov's work, see L. Ia. Bliakher, "I. A. Borzenkov."

77. K. A. Timiriazev, *Zhizn' rastenii*, 11th ed., pp. 313–14.

78. K. A. Timiriazev, *Sochineniia*, vol. 7, pp. 54–55.

79. Timiriazev, *Sochineniia*, vol. 7, p. 54.

80. Ibid., pp. 63–65.

81. Ibid., p. 64.

82. Ibid., p. 66.

83. Ibid., p. 66.

84. I. F. Shmal'gauzen, *O rastitel'nykh pomesiakh*, pp. 22, 25. See also A. E. Gaisinovich, "Pervoe izlozhenie," pp. 22–24.

85. For a brief biography of A. N. Beketov, see P. A. Baranov, "Andrei Nikolaevich Beketov," pp. 116–25. See also K. M. Zavadskii, *Razvitie*, pp. 156–60, 166–68.

86. For views on Beketov's interpretation of Darwin's theory, see P. A. Baranov, "A. N. Beketov," pp. 121–24.

87. A. N. Beketov, "Bor'ba," p. 582.

88. Ibid., pp. 592–93.

89. Ibid., p. 593; A. de Candolle, *Histoire*, ch. 5. De Candolle's analysis of the role of natural selection in the scientific community is examined in Korf, "Teoriia Darvina."

90. A. N. Beketov, "Darvinizm," p. 92.

91. Ibid., p. 110.

92. Ibid., pp. 103–4.

93. For more details on Beketov's views on Darwin's theory, see P. A. Baranov, "Aleksei Nikolaevich Beketov," pp. 122–24.

94. For a comprehensive summary of Shmankevich's studies, see his "Zur Kenntniss des Einflusses." See also his "Über das Verhältniss der *Artemia salina*." For contemporary reports on Shmankevich's studies, see [A. O. Kovalevskii] "Sitzungsberichte"; and Hoyer, "Protocolle." Kovalevskii and Hoyer summed up but did not comment on Shmankevich's work. For details on Shmankevich's ideas and their reception, see B. E. Raikov, "V. I. Shmankevich." Raikov's interpretation reflects a strong Lysenkoist bias. For a modern Soviet comment, see A. B. Georgievskii, "Osobennosti razvitiia," p. 55.

95. For more details, see Raikov, "V. I. Shmankevich," pp. 249–53.

96. William Bateson, *Materials*, pp. 96–101. See also August Weismann, *The Evolution Theory*, vol. 2, p. 272; and L. Plate, *Selektionsprinzip*, p. 167. For a more recent view, see Y. Conry, *Introduction*, pp. 307–9.

97. *Dnevnik XI-go S'ezda*, no. 5, pp. 198–99.

98. B. E. Raikov, "V. I. Shmankevich," p. 258.

99. D. R. Weiner, "Roots of 'Michurinism,'" pp. 259–60.

100. Charles Darwin, *The Descent of Man*, p. 5.

101. M. A. Menzbir, "Charlz Darvin i sovremennoe sostoianie evoliutsionnogo ucheniia," pp. 53–54.

102. *Protokoly VII S'ezda*, vol. 1, p. 26. See also N. A. Umov, "Po povodu sbornika," p. 1.

103. N. K. Mikhailovskii, *Polnoe sobranie sochinenii*, 4th ed., vol. 5, p. 626.

104. Ibid., p. 635.

105. A. Ren'iar, "Nauka," no. 1, p. 334.

106. Ibid., p. 335.

107. A. Ren'iar, "Nauka," no. 7, pp. 186–93.

108. L. Popov, "Charlz Robert Darvin," p. 102.

109. A. Moskvin, "Charlz Robert Darvin," p. 16.

110. Ibid., p. 11.

Chapter Three

1. L. Ia. Bliakher, "Karl Maksimovich Ber," pp. 64–68.

2. Karl von Baer, "Entwickelt sich die Larve . . . ?" p. 33; B. E. Raikov, *Karl Ber*, p. 400; T. Lenoir, *The Strategy of Life*, pp. 257–61.

3. D. L. Hull, *Darwin*, pp. 416–27.

4. K. von Baer, *Reden*, vol. 2, pp. 235–480.

5. Jane M. Oppenheimer, *Essays*, pp. 231–32.

6. Charles Darwin, *The Origin of Species*, p. 21.

7. Jane M. Oppenheimer, "An Embryological Enigma," p. 296.

8. K. von Baer, *Reden,* vol. 2, pp. 241–52.

9. For a brief discussion of these principles, see I. I. Mechnikov, *Akademicheskoe sobranie sochinenii,* vol. 4, pp. 197–200.

10. K. von Baer, *Reden,* vol. 2, p. 269.

11. Mechnikov, *Akademicheskoe sobranie sochinenii,* vol. 4, pp. 273–75.

12. Ibid., pp. 280–81.

13. K. von Baer, *Reden,* vol. 2, pp. 422–23.

14. Ibid., p. 473.

15. Raikov, *Karl Ber,* p. 412.

16. N. N. Strakhov, *Bor'ba,* vol. 2, p. 279.

17. As cited in L. Ia. Bliakher, "Karl Maksimovich Ber," p. 65.

18. Stephen Jay Gould, *Ontogeny,* p. 56. S. J. Holmes has expressed a different view: "The problem of evolution always remains for von Baer an open question. The evidence seemed to him sufficient to justify the view that species and even larger groups arise by transmutation. He appears to be finally convinced that he could not logically extend the doctrine to include whole classes such as birds and mammals. Denying creationism, realizing that the *generatio equivoca* of higher organisms had no scientific value, and strongly impressed with the indications of genetic connection among the higher animals, he still argued over this or that difficulty which precluded him from giving the doctrine his unreserved support." (S. J. Holmes, "K. E. von Baer's Perplexities," p. 14.)

19. T. Lenoir, *The Strategy of Life,* pp. 270–75.

20. L. Ia. Bliakher, "Karl Maksimovich Ber," p. 71.

21. For details on Seidlitz's analysis of Darwin's ideas, see L. Ia. Bliakher, "Georg Zeidlits," pp. 26–58.

22. Georg Seidlitz, *Die Darwin'sche Theorie,* pp. 286–333.

23. Seidlitz, *Beiträge,* pp. 39–170.

24. Seidlitz, *Beiträge,* p. 170. For a critical survey of the views on adaptation, see Seidlitz, *Beiträge,* pp. 123–25.

25. L. Ia. Bliakher, *Problema,* p. 89.

26. Seidlitz, *Die Darwin'sche Theorie,* p. 66; L. Ia. Bliakher, "Georg Zeidlits," p. 36; A. B. Georgievskii, "Osobennosti razvitiia," p. 52.

27. Darwin, *The Descent of Man,* p. 528.

28. M. Pogodin, *Prostaia rech',* p. 105.

29. Ibid., pp. 80–81.

30. N. N., "O razbore Darvinovoi sistemy."

31. Pogodin, *Prostaia rech',* p. 447. In a carefully documented study, S. L. Sobol' identified A. P. Bogdanov, University of Moscow professor, as the author of comments on Pogodin's paper. Sobol' provides interesting comments on the ideological underpinnings of the conflict between academic liberals, who defended Darwin's theory, and the conservative leaders of anti-Darwinism. (For details, see S. L. Sobol', "Iz istorii," vol. 14, pp. 210–16.)

32. A. P. Shchapov, *Sotsial'no-pedagogicheskie usloviia,* pp. 118, 256, 261, 279.

33. N. N. Strakhov, *Bor'ba,* vol. 2, pp. 259–70.

34. B. L. Modzalevskii, ed., *Perepiska,* p. 162.

35. N. N. Strakhov, *Bor'ba*, vol. 2, p. 266.
36. Ibid., pp. 252–54.
37. Ibid., p. 255.
38. Ibid., p. 259.
39. For comment by a contemporary writer, see B. L. Modzalevskii, ed., *Perepiska*, pp. 53–54.
40. I. Tsion, "Proiskhozhdenie cheloveka," p. 12.
41. Ibid.
42. Ibid., pp. 13–14.
43. N. Sergievskii, "O bibleiskoi istorii," no. 2, p. 187.
44. A. P. Lebedev, "Uchenie Darvina," no. 8, p. 509.
45. L. R. Kharakhorkin, "Iz istorii," p. 144.
46. V. S. Solov'ev, *Sobranie sochinenii*, vol. 1, p. 143.
47. Ibid.
48. Ibid., p. 192.
49. B. N. Chicherin, *Nauka i religiia*, pp. 492–93.
50. For critical comments, see N. N., "Religiia," pp. 220–21. Virchow's Munich speech has been analyzed in Kurt Bayertz, "Darwinism," pp. 299–303.
51. Hugh McLean, *Nikolai Leskov*, p. 218; F. Ia. Priima and N. I. Prutskov, eds., *Istoriia*, vol. 3, p. 297.
52. F. M. Dostoevsky, *Polnoe sobranie sochinenii*, vol. 11, 6th ed., p. 164.
53. E. Hartmann, *Wahrheit*, p. 177.
54. [Flerovskii], *Filosofiia*, pp. 4–8.
55. Ibid., pp. 196–97.
56. Ibid., p. 153.
57. Ibid., p. 177.
58. Ibid., pp. 2–3.
59. N. K. Mikhailovskii, *Polnoe sobranie sochinenii*, vol. 6, 2nd ed., pp. 278–79.
60. E. Hartmann, *Wahrheit*, p. 2.
61. F. A. Lange, *Geschichte*, vol. 2, pp. 216–20.
62. Ibid., p. 271.
63. Ibid., p. 294.
64. Ibid., p. 261.

Chapter Four

1. P. Miliukov, "Universitety," pp. 796–97; P. Miliukov, C. Seignobos, and L. Eisenmann, *History of Russia*, vol. 3, pp. 136–47.
2. K. Leont'ev, *Vostok*, vol. 1, pp. 300–312.
3. Ibid., p. 311.
4. Ibid.
5. T. A. Lukina, ed., *Kaspiiskaia ekspeditsiia*, pp. 25–26.
6. P. A. Sorokin, *Modern Historical and Social Philosophies*, p. 52.
7. N. Ia. Danilevskii, *Rossiia i Evropa*.
8. F. M. Dostoevsky, *Pis'ma*, vol. 2, p. 181.
9. N. Ia. Danilevskii, *Rossiia i Evropa*, pp. 146–47.

10. N. Ia. Danilevskii, *Darvinizm*, vol. 1, part 1, p. 25.

11. Ibid., p. 26.

12. Danilevskii, *Darvinizm*, vol. 1, part 1, pp. 11, 24.

13. Ibid., part 2, p. 19.

14. Ibid., part 1, p. 511.

15. Ibid., part 2, p. 479.

16. Ibid., part 2, p. 462.

17. Ibid., part 1, p. 7.

18. Ibid., p. 8.

19. Ibid., p. 19.

20. For comments on *Darvinizm*, see J. A. Rogers, "Russian Opposition," pp. 493–502. See also George L. Kline, "Darwinism," pp. 314–19.

21. P. Ia. Svetlov, "K voprose o darvinizme."

22. I. Chistovich, "Darvinizm," p. 136. For comments on other leading theological reviews of Danilevskii's *Darvinizm*, see George L. Kline, "Darwinism," pp. 315–16.

23. P. Ia. Svetlov, "K voprose o darvinizme," pp. 668–74.

24. Ibid., p. 673.

25. "Pravitel'stvennye rasporiazheniia," p. 49.

26. "Review of N. Ia. Danilevskii, *Darvinizm*," pp. 186–87.

27. [Popov, L. K.], El'pe, "Posmertnyi trud." See also B. E. Raikov, "Iz istorii darvinizma," part 1, p. 21.

28. P. Semenov, "O trude Nikolaia Iakovlevicha Danilevskogo," p. 785.

29. N. N. Strakhov, "Predislovie," p. 45.

30. N. N. Strakhov, *Bor'ba*, vol. 3, p. 342.

31. For comments on Strakhov's defense of Danilevskii, see Linda Gerstein, *Nikolai Strakhov*, pp. 159–62; George Kline, "Darwinism," pp. 316–17.

32. K. A. Timiriazev, *Sochineniia*, vol. 7, p. 325.

33. Ibid., p. 272.

34. Strakhov, *Bor'ba*, vol. 2, p. 344.

35. Timiriazev, *Sochineniia*, vol. 7, pp. 409–10.

36. Strakhov, *Bor'ba*, vol. 2, pp. 407–8.

37. Timiriazev, *Sochineniia*, vol. 7, pp. 289–90.

38. B. V. Nikol'skii, *Nikolai Nikolaevich Strakhov*, pp. 26–35.

39. Strakhov, *Bor'ba*, vol. 2, p. 344.

40. Ibid., p. 465.

41. P. Mokievskii, "Lavrov kak filosof," p. 68.

42. N. A. Kholodkovskii, *Biologicheskie ocherki*, pp. 73–74.

43. Ibid., p. 92.

44. Ibid., pp. 86–87.

45. Ibid., p. 96.

46. Ibid., p. 115.

47. Ibid.

48. A. S. Famintsyn, "N. Ia. Danilevskii," p. 643.

49. Ibid., p. 627.

50. Ibid., p. 624.

51. Ibid., p. 643.

52. Iu. A. Filipchenko, *Evoliutsionnaia ideia*, p. 111.

53. A. Borozdin, "Danilevskii, Nikolai Iakovlevich," p. 68. For a more balanced interpretation, see R. E. MacMaster, *Danilevsky,* pp. 162–73.

54. For comments, see A. Borozdin, "Danilevskii," pp. 68–69; V. S. Solov'ev, "Danilevskii," p. 77; Famintsyn, "N. Ia. Danilevskii," p. 642.

55. L. S. Berg, geographer and author of *Nomogenesis,* was the first Russian scientist to recognize Danilevskii as a serious contributor to modern biological theory. (See L. S. Berg, *Nomogenesis,* pp. 36, 60, 149, 322, 336, 377–78.) However, Berg cited only Danilevskii's "positive" contributions.

56. V. S. Solov'ev, "Danilevskii," p. 77.

57. N. V. Shelgunov, *Ocherki russkoi zhizni,* p. 486.

58. See particularly V. V. Rozanov, "Vopros o proiskhozhdeniiu organizmov," "Teoriia Charlza Darvina," and "Smena mirovozzrenii," republished in *Priroda i istoriia.*

59. Rozanov, *Priroda i istoriia,* p. 155.

60. Ibid., p. 150.

61. Rozanov, "Vopros," pp. 312–16. For comments on other theological reviews of Danilevskii's *Darvinizm,* see George Kline, "Darwinism," pp. 315–16.

62. P. F. Lesgaft, "Nasledstvennost'," no. 9, sect. 2, pp. 51–52, 57.

63. P. F. Lesgaft, "Nasledstvennost'," no. 9, sec. 2, pp. 59–60; no. 10, sec. 2, pp. 29, 42–43.

64. S. I. Korzhinskii, *Chto takoe zhizn'?*

65. Ibid., pp. 46–47.

66. [Chernyshevskii, N. G.], Staryi Transformist, "Proiskhozhdenie," pp. 79–114.

67. Ibid., p. 79.

68. Ibid., p. 113.

69. Ibid., p. 106.

70. Ibid.

71. Naturalist', "Strannoe napadenie," pp. 233–35. For a modern comment, see V. F. Evgrafov et al., eds., *Istoriia,* vol. 3, pp. 52–54.

72. M. E-dt, "Darvin," p. 135.

73. See particularly Rozanov, *O ponimanii.* In order to show the firmness of his commitment to the search for a new philosophy, Rozanov recognized no intellectual predecessors and made almost no reference to other philosophers. He was interested primarily in making science a corollary to idealistic metaphysics.

74. K. A. Timiriazev, *Sochineniia,* vol. 7, p. 28.

75. George Kennan, "A Visit," p. 264.

76. *L. N. Tolstoy o literature,* p. 232.

Chapter Five

1. A. de Quatrefages, *Les émules,* vol. 2, p. 288.

2. M. M. Filippov, "Darvinizm," no. 33, p. 1043.

3. N. M. Knipovich, "Lamarkizm," pp. 290–91.

4. "Review of Ch. Darvin, *Proiskhozhdenie,*" pp. 36–37.

5. V. A. Varsonof'eva, *Aleksei Petrovich Pavlov,* p. 234.

6. V. A. Vagner, "Granitsy," no. 8–9, p. 88.

7. K. A. Timiriazev, "Faktory," p. 62.

8. E. N. Mirzoian, *Razvitie*, pp. 96–116; A. Vucinich, *Science in Russian Culture: 1861–1917*, pp. 329–31, 406–9; V. L. Omelianskii, "Razvitie," pp. 128–29, 132–33.

9. A. A. Shcherbakova et al., *Istoriia botaniki*, p. 64.

10. I. A. Chemena, *Darvinizm*.

11. Ia. Kolubovskii, *Filosofskii ezhegodnik: 1893*, pp. 173–74.

12. Y. Conry, *L'introduction*, p. 312.

13. A. Giard, *Controverses*, pp. 135–37; Y. Conry, *L'introduction*, pp. 240–41.

14. Giard, *Controverses*, pp. 9–11.

15. Y. Delage and M. Goldsmith, *Les théories*, 1914, p. 252.

16. P. J. Bowler, *The Eclipse*, pp. 92–98.

17. L. Plate, *Selektionsprinzip*, p. 591. Plate gave credit to J. Nusbaum for introducing this label. (See J. Nusbaum, "Zur Beurteilung und Geschichte des Neolamarckismus," *Biologische Zentralblatt*, 1910, vol. 30, pp. 599–611.)

18. N. Iu. Zograf divided the neo-Lamarckians into "American" and "German" schools. The American school, headed by E. D. Cope, "operated strictly on scientific grounds and made very important contributions to the theory of evolution." The German neo-Lamarckians, by contrast, built less on Lamarck's science than on his natural philosophy, far removed from a solid scientific base. He considered A. Pauly and R. Francé the typical representatives of German neo-Lamarckism. (N. Iu. Zograf, "Noveishiia techeniia," pp. 184–85.)

19. A. N. Beketov, *Geografiia rastenii*, p. 5; A. N. Severtsov, *Evoliutsiia i psikhika*, p. 15.

20. A. R. Wallace, *Darwinism*, p. 428.

21. William Bateson, *Materials*, pp. 96–101.

22. V. M. Shimkevich, *Biologicheskie osnovy*, pp. 506–7.

23. V. Zalenskii, *Osnovnye nachala*, p. 109.

24. See particularly A. Pauly, *Darwinismus und Lamarckismus*, and R. Francé, *Der heutige Stand*.

25. L. Plate, *Selektionsprinzip*, pp. 592–93.

26. Y. Delage and M. Goldsmith, *Les théories*, 1914, pp. 279–80.

27. R. W. Burkhardt, Jr., "The Zoological Philosophy," pp. xxx–xxxii; L. J. Jordanova, *Lamarck*, pp. 102–6.

28. For a perceptive comment, see Kholodkovskii, *Biologicheskie ocherki*, 37–40.

29. R. Francé, *Der heutige Stand*, ch. 4; V. V. Lunkevich, "Staroe i novoe," no. 1, pp. 145–49.

30. R. Francé, *Der heutige Stand*, p. 117.

31. Ibid., pp. 8, 16.

32. V. A. Vagner, "Novye i starye puti," pp. 96–97.

33. V. V. Lunkevich, "Staroe i novoe," no. 1, pp. 150–51.

34. M. A. Menzbir, *Vvedenie*, p. 244.

35. M. M. Filippov, *Filosofiia*, vol. 2, ch. 9.

36. V. V. Lunkevich, "Staroe i novoe," no. 1, p. 136.

37. V. L. Komarov, "Vidoobrazovanie," p. 537.

38. M. M. Novikov, "Neolamarkizm," no. 6, p. 712.

39. Kholodkovskii, *Biologicheskie ocherki*, p. 193.

40. K. A. Timiriazev, "Charlz Darvin," pp. 25–26; K. M. Zavadskii, *Razvitie*, p. 300.

41. K. A. Timiriazev, *Sochineniia*, vol. 6, pp. 288–89.

42. A. N. Beketov, *Geografiia rastenii*, p. 5.

43. Ibid., pp. 20–21.

44. "Review of Lamark, *Filosofiia*," pp. 847–48.

45. V. I. Taliev, "Darvinizm," pp. 174–75.

46. Ernst Mayr, "The Triumph," p. 1261.

47. E. Nordenskiöld, *The History of Biology*, p. 565; Peter J. Bowler, *Evolution*, pp. 237–39.

48. M. A. Menzbir, "Glavneishie predstaviteli: IV. Avgust Veismann," p. 39.

49. M. G., "Voprosy obshchei biologii," p. 169.

50. Ibid., p. 160.

51. V. I. Taliev, "Biologicheskie idei," p. 274.

52. Kholodkovskii, *Biologicheskie ocherki*, pp. 195–96.

53. Actually, Wallace gave Weismann's theory a sympathetic airing but not an outright approval. See Wallace, *Darwinism*, pp. 437–39.

54. V. I. Taliev, "Biologicheskie idei," p. 275.

55. N. Iu. Zograf, "Noveishiia techeniia," p. 176.

56. M. S. Ganin, "Iadro kletki," no. 30, p. 937.

57. M. M. Filippov, "Eshche o Veismanne," p. 558.

58. For other comments, see V. L. Shimkevich, "Polemika," pp. 94–105.

59. V. A. Vagner, "Novye i starye puti," p. 97.

60. E. Hartmann, *Das Problem des Lebens*, p. 377.

61. George G. Simpson, *The Meaning of Evolution*, p. 247. For a brief survey of neovitalism in Germany, its real home, see Francé, *Der heutige Stand*, ch. 6.

62. V. B. Bekhterev, "Psikho-biologicheskie voprosy," no. 4, pp. 16–24.

63. A. Danilevskii, "Zhivoe veshchestvo," p. 290.

64. For a critical comment on this view, see M. Iu. Gol'dshtein, "O nereshennykh problemakh," no. 10, p. 33.

65. V. V. Lunkevich, *Nereshennye problemy*, p. 268.

66. A. Danilevskii, "Zhivoe veshchestvo," p. 336.

67. N. A. Kholodkovskii, *Biologicheskie ocherki*, pp. 172–73.

68. N. Vinogradov, "Biologicheskii mekhanizm," p. 416.

69. V. A. Fausek, ed., *Sushchnost' zhizni*.

70. Fausek, "Predislovie," pp. vii–viii.

71. V. V. Lunkevich, *Nereshennye problemy*, pp. 290–91.

72. M. A. Antonovich, *Izbrannye filosofskie sochineniia*, p. 316.

73. N. N. Strakhov, *Bor'ba*, vol. 3, p. 88.

74. A. V. Nemilov, "Analogii," pp. 47–48.

75. M. Fervorn, "Vitalizm," pp. 83–85.

76. N. K. Kol'tsov, "Review of Gans Drish, *Vitalizm*," pp. 1530–31.

77. A. N. Severtsov, "Evoliutsiia i embriologiia," p. 271.

78. S. M. Luk'ianov, "O predelakh tsitologicheskogo issledovaniia," pp. 673–78.

79. B. P. Stroganov, "Akademik A. S. Famintsyn," p. 30.

80. A. S. Famintsyn, "Darvin i ego znachenie."

81. A. S. Famintsyn, *Obmen veshchestva,* pp. ix–x.

82. V. V. Polevoi, "A. S. Famintsyn," p. 67.

83. Famintsyn, "Chto takoe lishainiki?" p. 281; L. N. Khakhina, "Kontseptsiia," pp. 180–81, "Teoreticheskie vzgliady," pp. 142–48, and "K istorii," pp. 63–68.

84. Khakhina, "Problema simbiogeneza," p. 424; K. M. Zavadskii, *Razvitie,* p. 173.

85. Famintsyn, "O psikhicheskoi zhizni" and "Blizhaishie zadachi."

86. Gustav Bunge, *Lehrbuch,* p. 5.

87. Bunge, *Lehrbuch,* p. 13. See also I. K. Brusilovskii, "Sovremennye teorii," no. 3, sec. 1, pp. 26–28.

88. For pertinent comments on Bunge's "vitalism," see M. Fervorn, "Vitalizm," pp. 78–80; N. Vinogradov, "Biologicheskii mekhanizm," pp. 425–26; E. Hartmann, *Das Problem des Lebens,* pp. 99–100; A. S. Famintsyn, "Sovremennoe estestvoznanie," no. 1, pp. 20–21; I. P. Borodin, "Protoplazma," pp. 20–21.

89. I. Tarkhanov, "Vitalizm," pp. 550–51.

90. Famintsyn, "Sovremennoe estestvoznanie," no. 1, p. 23.

91. Ibid., no. 1, pp. 11–23.

92. Ibid., no. 6, p. 173.

93. Ibid., no. 3, p. 181.

94. Lunkevich, *Nereshennye problemy,* pp. 315–16.

95. N. K. Mikhailovskii, *Polnoe sobranie sochinenii,* vol. 2, 4th ed., p. 354.

96. For critical comment, see A. A. Bogdanov, "Obmen i tekhnika," p. 337; and M. M. Filippov, "Psikhologiia," pp. 2013–18. See also S. L. Frank, "Psikhologicheskoe napravlenie," pp. 62–63, 110.

97. B. G. Safronov, *M. M. Kovalevskii,* p. 65.

98. Aleksandr Vvedenskii, *Filosofskie ocherki,* pp. 39–40. Vvedenskii made his observations in 1898.

99. In building the psychological foundations of the evolutionary theory, Famintsyn took note of current changes in the methodological orientation of psychology. "Before our eyes a rapid and thorough metamorphosis is taking place; from an almost completely speculative discipline psychology is being transformed into an experimental science, breaking the last ties with metaphysics. The first decisive steps in this direction were made some twenty years ago. Wundt in Germany and Charcot in France have placed psychology on a new path—Charcot by his research on the role of hypnosis in curing hysteria, and Wundt by founding a center for work in experimental psychology." (Famintsyn, "Sovremennoe estestvoznanie," no. 4, p. 201.) For a critical comment, see N. Vinogradov, "Review of A. Famintsyn, *Sovremennoe estestvoznanie,*" pp. 421–22. See also V. V. Lunkevich, "Interesnaia kniga."

100. Famintsyn's philosophical views are discussed in Iu. A. Urmantsev, "Ob opytno-filosofskom mirovozzrenii." See also N. Vinogradov, "Review of A. Famintsyn, *Sovremennoe estestvoznanie.*"

101. For a brief survey of Famintsyn's work on symbiogenesis, see L. N. Khakhina, "Eksperimental'nye istoki."

102. Famintsyn, "Sovremennoe estestvoznanie," no. 6, p. 170.

103. The theoretical aspects of zoopsychology are discussed in V. A. Vagner, "Voprosy zoopsikhologii," *Voprosy zoopsikhologii,* and "Novye i starye puti."

104. V. A. Vagner, "Izuchenie dushevnoi zhizni," pp. 517–18.

105. For another assessment of Romanes's views, see M. M. Filippov, *Filosofiia,* vol. 2, pp. 1060, 1071–74.

106. G. J. Romanes, *Mental Evolution,* p. 160.

107. Darwin, *The Descent of Man,* p. 83.

108. V. A. Vagner, *Biologicheskie osnovaniia.*

109. For a perceptive analysis, see A. V. Petrovskii, *Voprosy istorii,* pp. 54–62.

110. V. A. Vagner, "Gorodskaia lastochka," p. 44.

111. For pertinent general assessments of Vagner's contributions to zoopsychology, see V. I. Strel'chenko, "V. A. Vagner"; K. E. Fabri, "V. A. Vagner"; and A. V. Petrovskii, *Voprosy istorii,* pp. 54–62.

112. M. T. Ghiselin, *The Triumph,* p. 187.

113. Richard W. Burkhardt, Jr., "Darwin," pp. 361–62.

114. V. A. Vagner, "Voprosy zoopsikhologii."

115. E. Haeckel, "Charles Darwin," p. 141.

116. Referring to Haeckel's *Natural History of Creation,* Darwin noted: "If this work had appeared before my eyes before my essay [on the descent of man] had been written I should probably never have completed it. Almost all the conclusions at which I have arrived I find confirmed by this naturalist, whose knowledge on many points is much fuller than mine" (*The Descent of Man,* p. 19).

117. V. D. Vol'fson, "Sovremennaia biologiia," no. 3, p. 65.

118. I. Tsion, "Proiskhozhdenie."

119. [Ia. A. Borzenkov], "Chteniia," p. 142.

120. Ibid., p. 143.

121. For a typical comment, see A. Genkel', "Ernst Gekkel'," pp. 97–98.

122. N. M. Knipovich, "Gekkel'," p. 264.

123. N. N. Strakhov, *Bor'ba,* vol. 3, pp. 81–82.

124. Georg Uschmann, "Haeckel's Biological Materialism," p. 118.

125. E. Haeckel, "Gekkel'," p. 872.

126. CCD: no. 5840.

127. A. Genkel', "Ernst Gekkel'," p. 93.

128. M. A. Menzbir, "Glavneishie predstaviteli: III. Ernst Gekkel'," p. 5.

129. O. Kiulpe, "Sovremennaia filosofiia," p. 31.

130. O. D. Khvol'son, *Gegel'.*

131. P. Kudriavtsev, *Absoliutizm,* p. 212.

132. V. I. Lenin, *Materializm,* p. 362.

133. A. Genkel', "Ernst Gekkel'," p. 90.

134. N. K. Kol'tsov, "Ernst Gekkel'," p. 206.

135. V. M. Shimkevich, "A. O. Kovalevskii. (Nekrolog)," p. 108.

136. V. L. Komarov, "Vidoobrazovanie," p. 527.

137. A. N. Severtsov, *Sobranie sochinenii,* vol. 3, p. 13.

Chapter Six

1. K. A. Timiriazev, *Nauka i demokratiia*, p. 120.
2. Timiriazev, *Sochineniia*, vol. 8, p. 63. See also E. Broda, "Darwin and Boltzmann," p. 61.
3. Timiriazev, *Sochineniia*, vol. 5, pp. 32–33. See also V. L. Komarov, "Zhizn' i tvorchestvo," pp. 50–51.
4. Timiriazev, *Sochineniia*, vol. 8, p. 114.
5. Ibid., p. 125.
6. Timiriazev, *Sochineniia*, vol. 5, pp. 168–69.
7. Ibid., vol. 6, p. 247. See also L. Ia. Bliakher, *Problema nasledovaniia*, pp. 27–28.
8. Timiriazev, *Sochineniia*, vol. 6, p. 288.
9. Ibid., vol. 5, pp. 174–77.
10. N. A. Maksimov, "Ocherk," p. 31.
11. Timiriazev, *Sochineniia*, vol. 8, pp. 122–23.
12. Ibid., vol. 6, p. 263.
13. Timiriazev, "Charlz Darvin," in M. M. Kovalevskii et al., *Pamiati*, p. 34.
14. Timiriazev, *Sochineniia*, vol. 6, p. 264.
15. Ibid., p. 265.
16. Timiriazev, "Charlz Darvin," in M. M. Kovalevskii et al., *Pamiati*, p. 34. For more details on Timiriazev's war on Mendelism, see A. E. Gaisinovich, "Contradictory Appraisal," pp. 257–74.
17. Timiriazev, *Sochineniia*, vol. 8, pp. 116–17.
18. M. Berthelot, *Science et morale*, p. 2; Timiriazev, *Sochineniia*, vol. 8, p. 306.
19. M. M. Filippov, "Darvinizm," no. 32, p. 993.
20. Ibid., p. 994. For a direct attack on Timiriazev's "materialistic" philosophy, see N. M. Solov'ev, *Nauchnyi ateizm*, pp. 68–93.
21. M. M. Filippov, "Darvinizm," no. 33, p. 1043.
22. Filippov's Lamarckian leanings are discussed in L. Ia. Bliakher, *Problema nasledovaniia*, pp. 44–45.
23. Timiriazev, *Izbrannye sochineniia*, vol. 3, pp. 478–79.
24. Timiriazev, *Sochineniia*, vol. 5, pp. 78–106.
25. Ibid., p. 31.
26. Ibid., p. 426.
27. Timiriazev, "Charlz Darvin," p. 36. For a description of Weldon's research, see V. L. Kellogg, *Darwinism Today*, pp. 157–62.
28. M. A. Menzbir, "Charlz Darvin i sovremennoe sostoianie evoliutsionnogo ucheniia," p. 66. For biographical data on Menzbir, see N. Ia. Rosina, *Mikhail Aleksandrovich Menzbir*; L. S. Tsetlin, "Akademik M. A. Menzbir"; I. I. Puzanov, "M. A. Menzbir kak zoogeograf"; S. I. Ognev, "Mikhail Aleksandrovich Menzbir"; and B. S. Matveev, "Mikhail Aleksandrovich Menzbir."
29. Menzbir, "Korennoi vopros," pp. 112–15.
30. P. P. Sushkin, "K 60-letnemu iubileiu," p. 668.

31. Menzbir, *Vvedenie,* ch. 10.
32. Menzbir, "Istoricheskii ocherk," no. 11, sec. 1, p. 83. For comments, see S. L. Sobol', "Bor'ba M. A. Menzbira," pp. 35–36.
33. Charles Darwin, *The Variation,* vol. 1, p. 9.
34. Menzbir, "Glavneishie predstaviteli: III. Ernst Gekkel'," pp. 4–5.
35. Menzbir, "Istoricheskii ocherk," no. 11, sec. 1, pp. 84–85.
36. Menzbir, "P. A. Kropotkin," pp. 102–5.
37. Menzbir, "Sovremennye zadachi," pp. 172–73.
38. Ibid., p. 172.
39. Menzbir, "Glavneishie predstaviteli: IV. Avgust Veismann," p. 55.
40. Menzbir, "Opyt teorii nasledstvennosti," p. 219.
41. Ibid., pp. 229–30.
42. Menzbir, "Mnimyi krizis," p. 201.
43. Ibid., p. 198.
44. Ibid., p. 199.
45. Ibid., pp. 200–1.
46. Menzbir, *Za Darvina,* p. 106.
47. Menzbir, "Ocherk uspekhov biologii," pp. 91–92.
48. Menzbir, "Istoricheskii ocherk," no. 11, sec. 1, pp. 94–95.
49. Menzbir, "Ocherk uspekhov biologii," p. 91.
50. Menzbir, "Evoliutsionnoe uchenie," p. 656.
51. Menzbir, "Istoricheskii ocherk," no. 10, sec. 1, p. 154.
52. N. Ia. Rosina, *Mikhail Aleksandrovich Menzbir,* p. 101.
53. Menzbir, "Glavneishie predstaviteli: I. Al'fred Uolles," pp. 65, 74–75.
54. Menzbir, "Glavneishie predstaviteli: II. G. D. Romanes," pp. 46–50.
55. For details on Severtsov's life and work, see L. B. Severtsova, *Aleksei Nikolaevich Severtsov;* B. S. Matveev, "Aleksei Nikolaevich Severtsov"; and B. S. Matveev and A. N. Druzhinin, "The Life and Work." See also E. A. Veselov, *A. N. Severtsov.*
56. L. F. Panteleev, *Iz vospominanii proshlogo,* vol. 1, p. 164.
57. As cited in R. L. Zolotnitskaia, "Nikolai Alekseevich Severtsov," p. 58.
58. N. A. Severtsov, *Puteshestviia,* p. 171.
59. Zolotnitskaia, "Nikolai Alekseevich Severtsov," p. 56.
60. L. B. Severtsova, *Aleksei Nikolaevich Severtsov,* ch. 8.
61. Matveev and Druzhinin, "The Life and Work," p. 51.
62. L. B. Severtsova, *Aleksei Nikolaevich Severtsov,* p. 256.
63. For a comprehensive and systematic study of Severtsov's contributions after the October Revolution, see Mark Adams, "Severtsov and Schmalhausen."
64. A. N. Severtsov, *Sobranie sochinenii,* vol. 3, pp. 13–14.
65. For comments on the theory of phylembryogenesis, see S. J. Gould, *Ontogeny,* pp. 216–20.
66. L. Ia. Bliakher, *Ocherk istorii morfologii zhivotnykh,* pp. 160–61; E. N. Mirzoian, "Zakonomernosti," pp. 381–82.
67. A. N. Severtsov, *Sobranie sochinenii,* vol. 3, p. 13.
68. E. A. Veselov, *A. N. Severtsov,* p. 187.
69. Zavadskii, *Razvitie,* p. 302.

70. B. Rensch, *Evolution*, p. 241.

71. A. N. Severtsov, *Sobranie sochinenii*, vol. 3, pp. 13–14; Matveev and Druzhinin, "The Life and Work," p. 44.

72. A. N. Severtsov, *Sobranie sochinenii*, vol. 3, pp. 341–42.

73. E. N. Mirzoian, "Istoriko-nauchnyi analiz," p. 141.

74. Sewertzoff, *Morphologische Gesetzmässigkeiten*, pp. 198ff; Ilse Jahn et al., eds., *Geschichte der Biologie*, p. 408; Matveev and Druzhinin, "The Life and Work," p. 48; E. N. Mirzoian, "Istoriko-nauchnyi analiz," p. 141.

75. A. N. Severtsov, *Sobranie sochinenii*, vol. 3, p. 14.

76. Ernst Mayr, *Populations*, p. 362.

77. B. S. Matveev, "Aleksei Nikolaevich Severtsov," p. 336.

78. I. I. Shmal'gauzen, "Nauchnaia deiatel'nost'," p. 61.

79. A. N. Severtsov, *Sobranie sochinenii*, vol. 3, p. 279; V. I. Polianskii and Iu. I. Polianskii, eds., *Istoriia evoliutsionnykh uchenii*, pp. 306–7.

80. A. N. Severtsov, *Sobranie sochinenii*, vol. 3, p. 322.

81. I. I. Shmal'gauzen, "Novoe v sovremennom darvinizme," p. 44.

82. B. S. Matveev, "Progress i regress," p. 72.

83. A. N. Severtsov, *Sobranie sochinenii*, vol. 3, p. 326.

84. Stephen Jay Gould has given a brief description of the modern interpretation of cenogenesis: "Haeckel's definition of cenogenesis had been much broader, but de Beer restricts it to embryonic and larval adaptations that have no effect upon adult organization. Severtzov . . . had introduced this restriction by arguing (quite justly) that many of Haeckel's cenogeneses are primarily adaptations of the adult stage, even though they may affect early stages of ontogeny as well. He wanted to make a clear distinction between truly larval adaptations and phylogenetic elaborations of the adult that begin early in ontogeny. Sewertsov's restriction has spread via de Beer through almost all literature in English on the subject" (S. J. Gould, *Ontogeny*, p. 223).

85. For comments on the progressive role of "regression," see L. B. Severtsova, *Aleksei Nikolaevich Severtsov*, pp. 302–3.

86. A. N. Severtsov, *Sobranie sochinenii*, vol. 3, pp. 328–29. For an analytical comment, see L. Sh. Davitashvili, *Uchenie*, pp. 11–17. See also B. S. Matveev, "Progress i regress," p. 72.

87. A. N. Severtsov, *Sobranie sochinenii*, vol. 3, p. 281.

88. Ibid., p. 276.

89. Ibid., pp. 317–19.

90. Ibid., p. 280.

91. For pertinent comments, see L. Ia. Bliakher, *Problema nasledovaniia*, pp. 156–62.

92. K. M. Zavadskii and E. I. Kolchinskii, *Evoliutsiia evoliutsii*, p. 137.

93. A. N. Severtsov, *Sobranie sochinenii*, vol. 3, p. 279.

94. Ibid., pp. 31–32.

95. Ibid., pp. 14–15.

96. See, for example, A. N. Severtsov, *Evoliutsiia i psikhika*, pp. 14–15.

97. A. N. Severtsov, *Sobranie sochinenii*, vol. 3, p. 16.

98. Ibid., p. 16–17.

99. Ibid., p. 17.

100. Mark Adams, "Severtsov and Schmalhausen," p. 216. See also E. N. Mirzoian, "Printsip istorizma," p. 70, and "Istoriko-nauchnyi analiz," p. 143.

101. A. N. Severtsov, *Sobranie sochinenii*, vol. 3, p. 523.

102. A. N. Severtsov, *Morfologicheskie zakonomernosti*, p. 76.

103. L. Ia. Bliakher, "A. N. Severtsov i neolamarkizm," p. 112.

104. A. N. Severtsov, *Sobranie sochinenii*, vol. 3, p. 218.

105. N. V. Tsinger, "O zasoriaiushchikh posevy l'na vidakh *Camelina*." For comments, see N. Navashin, "Tsinger, Nikolai Vasil'evich," pp. 432–33; K. A. Timiriazev, "Otbor estestvennyi," pp. 721–41; I. I. Shmal'gauzen, *Puti zakonomernosti*, p. 15; V. L. Komarov, "Vidoobrazovanie," p. 539.

106. I. P. Pavlov, *Dvadtsatiletnii opyt*, pp. 14–15.

107. Ibid., p. 40.

108. M. Kreps et al., eds., *Neopublikovannye i maloizvestnye materialy*, p. 24.

109. Pavlov, *Dvadtsatiletnii opyt*, p. 150.

110. Ibid., p. 207.

111. As cited in P. K. Anokhin, *Izbrannye trudy*, p. 126.

112. V. Polynin, *Prorok*, pp. 120–21; N. K. Kol'tsov, "Eksperimental'naia biologiia," pp. 40–41.

113. A. E. Gaisinovich, "The Origins," pp. 37–38.

114. W. H. Gantt, "Pavlov and Darwin," p. 236.

115. A. I. Bachinskii, "Zamechatel'nyi russkii uchenyi," p. 33.

116. N. Umov, "Po povodu sbornika," p. 5.

117. Umov, "Fiziko-mekhanicheskaia model'," pp. 689–90. See also N. K. Kol'tsov, "Organizatsiia kletki," pp. 211–12.

118. P. P. Lazarev, "Zakony fiziki," pp. 181–82.

119. M. M. Kovalevskii, *Ocherki proiskhozhdeniia i razvitiia*, p. 18.

120. M. M. Kovalevskii, "Darvinizm v sotsiologii," p. 153.

121. Ibid., p. 154.

122. P. G. Vinogradov, "O progresse," p. 282.

123. Ibid., p. 284.

124. John Dewey, *The Influence of Darwin*, pp. 8–19.

125. H. Höffding, *A History*, vol. 2, p. 436.

126. M. M. Filippov, *Filosofiia*, vol. 1, pp. x–xii.

127. Ibid., vol. 2, p. 837.

Chapter Seven

1. L. Plate, ed., *Ultramontane Weltanschauung*, p. 13.

2. S. S. Glagolev, "Novoe miroponimanie," pp. 37–41.

3. See particularly I. A. Chemena, *Darvinizm*; S. S. Glagolev, *O proiskhozhdenii*.

4. For comments on Glagolev's anti-Darwinism, see George L. Kline, "Darwinism," pp. 311–12.

5. Glagolev, *O proiskhozhdenii*, pp. 275–76.

6. See, for example, Glagolev, "Vzgliad Vasmanna."

7. For comments on Wasmann's evolutionary ideas, see V. V. Lunkevich, "Staroe i novoe," no. 1, pp. 154–55.

8. Erich Wasmann, *Die moderne Biologie*, pp. 197–98.

9. Wasmann, *Die moderne Biologie*, p. 304.

10. Lunkevich, "Staroe i novoe," no. 1, pp. 154–55.

11. Glagolev, "Vzgliad Vasmanna," May, p. 1.

12. Glagolev, "Vzgliad Vasmanna," Oct., p. 264.

13. Glagolev, "Mendelizm," Dec., pp. 727–36. See also N. Rumiantsev, "Darvinizm," 1895, no. 1, p. 2.

14. Glagolev, "Mendelizm," Oct., p. 502.

15. Glagolev, "Mendelizm," Oct., p. 502.

16. P. N. Kapterev, "Teleologiia neolamarkistov," Jan., pp. 88–89.

17. Ibid., May, pp. 115–18.

18. Ibid., pp. 136–39.

19. See, for example, Kapterev, "Teleologiia," March, pp. 576–91.

20. Kapterev, "Teleologiia."

21. V. D. Kudriavtsev-Platonov, *Sochineniia*, vol. 3, part 1, pp. 181–201.

22. Ibid., part 3, p. 60.

23. Ibid., p. 62.

24. O. D. Khvol'son, "Pozitivnaia filosofiia," p. 57.

25. V. I. Vernadskii, *Izbrannye trudy*, p. 51. The rapidly spreading interest in the philosophical foundations of science was anticipated by V. V. Bobynin, an eminent historian of Russian mathematics, who stated in 1886 that "in the philosophy of science, science achieves a higher consciousness—a consciousness of itself; therefore, the philosophy of science may be defined as a self-consciousness of science" (V. V. Bobynin, *Filosofskoe, nauchnoe i pedagogicheskoe znachenie*, p. 5).

26. As cited in S. M. Luk'ianov, "O predelakh tsitologicheskogo issledovaniia," p. 678.

27. For comments on the relations between science, philosophy, and theology, see V. F. Ern, "Priroda nauchnoi mysli," Jan., pp. 154–59.

28. B. S. Bychkovskii, *Sovremennaia filosofiia*, vol. 1, p. 30.

29. A. I. Bachinskii, "Pis'ma po filosofii: Pis'mo II," p. 97.

30. Bachinskii, "Pis'ma: II," pp. 107–8.

31. *L. N. Tolstoy o literature*, p. 557.

32. N. K. Mikhailovskii, "Literatura," 1898, no. 2, sec. 2, pp. 159–60.

33. B. A. Kistiakovskii, *Sotsial'nye nauki*, pp. 144–50.

34. S. N. Bulgakov, "Osnovnye problemy," pp. 43–46.

35. N. K. Kol'tsov, "Review of *Novye idei*," p. 123.

36. S. L. Frank, *Filosofiia i zhizn'*, pp. 164–217.

37. Ibid., pp. 167–68.

38. Ibid., p. 209.

39. Ibid., p. 208.

40. P. I. Novgorodtsev, *Ob obshchestvennom ideale*, vol. 1, p. 54.

41. Novgorodtsev, "Nravstvennyi idealizm," p. 279.

42. N. A. Berdiaev, *Sub'ektivizm*, p. 143.

43. Berdiaev, "Eticheskaia problema," pp. 103–4.

44. L. Tolstoy, "Protivorechiia," pp. 293–95.

45. B. N. Chicherin, *Polozhitel'naia filosofiia*, Supplement, p. 3.

46. Chicherin, *Polozhitel'naia filosofiia*, p. 9.

47. Ibid., p. 17.

48. Ibid., p. 317.

49. S. Bulgakov, *Ot marksizma k idealizmu*, pp. 63–64.

50. N. M. Solov'ev, *Nauchnyi ateizm*, p. 87.

51. Aleksandr Vvedenskii, *Filosofskie ocherki*, p. 40.

52. R. E. Regel', "Selektsiia," p. 469; Iu. A. Filipchenko, *Izmenchivost'*, pp. 53–54, 71–72.

53. A. A. Tikhomirov, *Sud'ba darvinizma*, p. iii.

54. Ibid., p. 57. See also his *Osnovnoi vopros*, pp. 74–75.

55. Tikhomirov, "Nashe universitetskoe delo."

56. Tikhomirov, *Samoobman v nauke*.

57. Ibid., p. 7.

58. Ibid., p. 8. In *Polozhenie cheloveka* Tikhomirov concentrated on a bitter attack on I. I. Mechnikov's defense of the idea of the anthropoid origin of man, pp. 3–5, 46–51.

59. V. N. and I. I. Liubimenko, "Ivan Parfen'evich Borodin," p. 26.

60. S. I. Korzhinskii, "Geterogenezis i evoliutsiia," *IIAN*, pp. 262–64.

61. For a critical appraisal, see V. I. Taliev, "Biologicheskie idei," pp. 271–77.

62. R. Francé, *Der heutige Stand*, p. 123.

63. V. G. Alekseev, *N. V. Bugaev*, p. 27.

64. N. V. Bugaev, "Matematika," pp. 42–43.

65. V. G. Alekseev, *Matematika*, pp. 10–11; P. A. Nekrasov, *Moskovskaia filosofsko-matematicheskaia shkola*, pp. 59–66.

66. Ibid., p. 107.

67. A. I. Bachinskii, "Pis'ma: II," pp. 97–98.

68. Ibid., pp. 106–7.

69. P. A. Nekrasov, *Moskovskaia filosofsko-matematicheskaia shkola*, p. 157.

70. [V. G. Alekseev] W. Alexejeff, "N. W. Bugajew und die idealistische Probleme," p. 362. For a summary of Bugaev's general orientation, see N. F. Utkina, *Pozitivizm*, pp. 144–45.

71. L. M. Lopatin, "Filosofskoe mirovozzrenie," p. 192.

72. L. M. Lopatin, "K voprosu o matematicheskikh istinakh," p. 70.

73. P. Tikhomirov, "Matematicheskii proekt," p. 362.

74. Ibid.

75. Felix Klein, the noted student of nineteenth-century developments in mathematics, made the observation that "the difference between the discrete magnitude of arithmetic and the continuous magnitude of geometry has always had a prominent place in history and in philosophical speculation," but "in recent times, the discrete magnitude, as conceptually the simplest, has come into the foreground." (F. Klein, *Elementary Mathematics*, p. 266.)

76. I. F. Ognev, "Vitalizm," pp. 728–29.

77. Ibid., p. 729.

78. I. F. Ognev, "Rech'," pp. 62–63.
79. V. Karpov, "Vitalizm," "Naturfilosofiia Aristotelia," and *Osnovnye cherty.*
80. Karpov, "Vitalizm," no. 3, p. 392.
81. Ibid., no. 4, p. 572.
82. Ibid., no. 4, p. 539.
83. Ibid., pp. 555–56.
84. Karpov, *Osnovnye cherty,* p. 63.
85. Karpov, "Vitalizm," no. 3, p. 392.
86. V. N. and I. I. Liubimenko, "Ivan Parfen'evich Borodin," pp. 26–27.

Chapter Eight

1. For biographical details, see E. N. Pavlovskii, "Nikolai Aleksandrovich Kholodkovskii," pp. 313–18.
2. N. A. Kholodkovskii, *Biologicheskie ocherki,* p. 158.
3. Ibid., p. 19.
4. E. N. Mirzoian, "Istoriko-nauchnyi analiz," pp. 125–27.
5. Kholodkovskii, *Biologicheskie ocherki,* p. 154.
6. Ibid., pp. 153–54; Mirzoian, "Istoriko-nauchnyi analiz," pp. 130–31.
7. Kholodkovskii, *Biologicheskie ocherki,* pp. 220–21.
8. Ibid., p. 138.
9. Ibid., p. 134.
10. Ibid, p. 155.
11. Ibid., pp. 157–58, 190–93.
12. Ibid., pp. 172–73.
13. I. I. Mechnikov, "Zadachi," pp. 761–62.
14. Ibid., p. 763.
15. Mechnikov, *O darvinizme,* p. 129.
16. Elie Metchnikoff, *Lectures,* p. 16; A. E. Gaisinovich, "100 let," p. 16.
17. Mechnikov, *Akademicheskoe sobranie sochinenii,* vol. 7, p. 169; Gaisinovich, "100 let," p. 20.
18. Mechnikov, *Akademicheskoe sobranie sochinenii,* vol. 9, p. 72.
19. "Notes," *Nature,* p. 386.
20. Mechnikov, *O darvinizme,* p. 222.
21. Mechnikov, "Darvinizm i meditsina."
22. Mechnikov, *Akademicheskoe sobranie sochinenii,* vol. 14, pp. 100–1.
23. E. Ray Lankester, "Elias Metchnikoff," p. 445.
24. Mechnikov, *Sorok let iskaniia,* p. 275.
25. Mechnikov, *Akademicheskoe sobranie sochinenii,* vol. 4, p. 401.
26. Ibid., vol. 4, pp. 398–99.
27. Ibid., vol. 14, p. 102.
28. I. M. Poliakov, "Razrabotka," p. 456. For details on the evolution of Mechnikov's attitude toward Darwin's theory, see Gaisinovich, "Problems of Variation," pp. 104–8.
29. For biographical details, see K. M. Deriugin, "Biografiia," D. M. Fedotov, "Vydaiushchiisia russkii zoolog-darvinist," and Iu. I. Polianskii, "Vydaiushchiisia russkii morfolog-darvinist."

30. V. Shimkevich, "Evoliutsionnaia ideia," p. 183.
31. Shimkevich, "Spenser i Veismann," no. 16, pp. 496–97.
32. Shimkevich, "Polemika," pp. 94–105.
33. Shimkevich, Biologicheskie osnovy, 3rd ed., pp. 308–11.
34. Ibid., pp. 446–47; K. M. Zavadskii, Razvitie, p. 276.
35. N. Ia. Rosina, Mikhail Aleksandrovich Menzbir, p. 154.
36. V. L. Shimkevich, "Embrional'nye plasti," p. 49.
37. Shimkevich, "Polemika," pp. 101–2.
38. Shimkevich, "Evoliutsionnaia ideia," pp. 167–84.
39. T. A. Lukina, Boris Evgen'evich Raikov, p. 30.
40. For more details, see E. M. Lavrenko, "Pamiati," pp. 1334–35.
41. V. A. Taliev, "Biologicheskie idei," pp. 270–71.
42. Taliev, "Biologicheskie idei," p. 271.
43. Ibid., p. 278.
44. For a general survey of Taliev's evolutionary ideas, see D. V. Lebedev, "Evoliutsionnye vzgliady."
45. Taliev, "O tselesoobraznosti."
46. Taliev, "Darvinizm," p. 174.
47. Taliev, Opyt, p. 6.
48. D. Ostrianin, Bor'ba, p. 144.
49. Taliev, Opyt, p. 259.
50. Ibid., p. 5.
51. V. K., "Review," p. 1451.
52. V. L. Komarov, "Review," p. 218.
53. R. E. Regel', "K voprosu o vidoobrazovanii." See also K. M. Zavadskii, Vid, p. 95.
54. Regel', "K voprosu o vidoobrazovanii," p. 168.
55. V. M. Shimkevich, "Nasledstvennost'," pp. 353–62.
56. Major types of criticism directed at Darwin's theory at the end of the nineteenth century—before Hugo de Vries published his first study on mutations and Gregor Mendel's theory of heredity was rediscovered—are discussed in Vladimir Vagner, "Novye i starye puti." Vagner dealt exclusively with developments in the West. He devoted particular attention to the strengths and weaknesses of embryological studies of evolution, implications of the principle of "change in one direction" (Eimer), the role of "associations" as an evolutionary factor, and recent interpretations of the struggle for existence. He lamented the lack of effort at a synthesis of various theories that would give evolution a broader theoretical and experimental base. Embryology, he thought, cannot explain the "character of relations between freely associated animals," but the embryological and sociopsychological approaches can elucidate different aspects of a more general notion of evolution.
57. E. Shul'ts, "Irratsional'noe v biologii," pp. 788–92.
58. S. I. Korzhinskii, "Geterogenezis," IIAN. For an abbreviated version, see Korzhinskii, "Geterogenezis," ZAN. For a biographical sketch of Korzhinskii, see "Sergei Ivanovich Korzhinskii."
59. For a summary of Korzhinskii's theoretical propositions, see Iu. A. Filipchenko, Evoliutsionnaia ideia, pp. 161–63. See also G. D. Berdyshev and V. N. Splivinskii, Pervyi sibirskii professor, chs. 5 and 6; L. Plate, Selektionsprin-

zip, pp. 169, 502; E. Delage and M. Goldsmith, *Les théories*, 1911, pp. 319–20; V. L. Kellogg, *Darwinism Today*, pp. 91–93.

60. M. A. Menzbir, "Mnimyi krizis," p. 201.

61. Hugo de Vries, *Die Mutationslehre*, vol. 1, p. 50.

62. Dobzhansky, *Genetics and the Evolutionary Process*, p. 42.

63. V. A. Vagner, ed., *Novye idei: Nasledstvennost'*.

64. Lunkevich, "Staroe i novoe," no. 6, p. 111; V. L. Komarov, "Vidoobrazovanie," p. 530.

65. O. V. Baranetskii, "Vydaiushchiisia iavleniia," p. 2.

66. Ibid., p. 62.

67. Ibid., p. 58.

68. A. S. Serebrovskii, "Sovremennoe sostoianie," pp. 1244–47.

69. Korzhinskii, "Geterogenezis," *ZAN*, p. 1.

70. I. P. Borodin, "Ocherki," no. 12, pp. 257–67.

71. Ibid., p. 268.

72. Ibid., p. 274.

73. Shimkevich, *Biologicheskie osnovy*, pp. 308–11.

74. T. H. Morgan, *Experimental Zoology*, p. 232.

75. For biographical data, see K. A. Fliaksberger, "Robert Eduardovich Regel'," pp. 3–15.

76. R. E. Regel', "Selektsiia," pp. 493–96.

77. Ibid., p. 511.

78. Ibid., p. 507. See also T. M. Aver'ianova, "Evoliutsonnye vzgliady," pp. 147–48.

79. P. Mishchenko, "Review," p. 63.

80. A review in *Priroda* noted that the translation of Punnett's book provided the first detailed discussion of Mendel's theory available in the Russian language. ("Review of *Mendelizm* by R. K. Punnet," pp. 389–91.)

81. S. S. Glagolev, "Mendelizm."

82. E. Baur, "Vvedenie."

83. N. I. Vavilov, "Immunity," p. 62.

84. See, particularly, R. E. Regel', "K voprosu o vidoobrazovanii."

85. E. A. Bogdanov, *Mendelizm*, p. 620.

86. Ibid., p. 622.

87. N. K. Kol'tsov, "Review of E. A. Bogdanov, *Mendelizm*," p. 1361; A. E. Gaisinovich, "Vospriiatie mendelizma," pp. 49–50.

88. Kol'tsov, "Review of E. A. Bogdanov, *Mendelizm*," p. 1391.

89. In 1921 Kol'tsov played a major role in the founding of the Laboratory of Genetics at the Institute of Experimental Biology for the purpose of giving Darwinism a base in genetics. This research led to S. S. Chetverikov's pioneering contribution to population genetics, a solid step toward a synthetic theory of evolution. (For a detailed study of the early history of this laboratory and of Chetverikov's work, see V. V. Babkov, *Moskovskaia shkola*.)

90. Iu. A. Filipchenko, *Izmenchivost'*, pp. 53–68.

91. Ibid., pp. 88–90.

92. Filipchenko, *Nasledstrennost'*, p. 5.

93. Filipchenko, "Review," p. 281.

94. Ibid.

95. Filipchenko, "Obzor . . . za 1915 god," pp. 731–34, and "Obzor . . za 1916 god," pp. 397–99.

96. D. Ivanovskii, "Eksperimental'nyi metod," p. 32.

97. Ibid., p. 39. S. L. Sobol', writing under the spell of Lysenkoism, claimed that Ivanovskii placed the primary emphasis on hereditary transmission of acquired characteristics, rather than on mutations. (Sobol', "Evoliutsionnye vozzreniia," p. 51.)

98. G. Allen, *Life Sciences*, p. 57.

99. P. J. Bowler, *Evolution*, p. 292.

100. V. L. Kellogg, *Darwinism Today*, p. 5.

101. Hugo de Vries, "Variation," p. 84.

102. W. Bateson, "Heredity," p. 87.

103. M. V. Arnol'di, "Sovremennoe sostoianie," p. 40.

104. "Darvin," *Bol'shaia entsiklopediia*, vol. 8, p. 138.

105. Ibid., p. 133.

Chapter Nine

1. M. M. Kovalevskii et al., *Pamiati Darvina*.

2. I. I. Mechnikov, "Darvinizm i meditsina," p. 116.

3. K. A. Timiriazev, "Charlz Darvin," in M. M. Kovalevskii et al., *Pamiati Darvina*, p. 8.

4. Timiriazev, "Sezon," p. 367.

5. Ibid., p. 368.

6. Timiriazev, "Charlz Darvin," p. 38.

7. N. A. Umov, "Evoliutsiia," pp. 36–38.

8. D. N. Anuchin, "Stoletie."

9. "Chestvovanie pamiati," p. 92.

10. V. A. Vagner, "Vliianie Darvina," p. 8.

11. Ibid., p. 15.

12. V. I. Taliev, "Darvinizm," pp. 174–75.

13. A. N. Severtsov, *Sobranie sochinenii*, vol. 3, pp. 14–15.

14. Represented at the Cambridge celebration were also Moscow University (K. A. Timiriazev), St. Petersburg University (V. M. Shimkevich), and Iur'ev (Dorpat) University (A. I. Iarotskii and N. I. Kuznetsov). (A. Verner, "Iubilei," p. 92.)

15. K. V. Manoilenko and L. N. Khakhina, "Iz istorii," p. 311.

16. R. Gertvig, "Pamiati Darvina."

17. [E. Haeckel,] Ernst Gekkel', "Mirosozertsanie Darvina," p. 67.

18. R. Francé, "Darvin," p. 18.

19. *Vestnik znaniia*, 1909, no. 2, pp. 168–72, 209–50. (Darwin commemorative issue.)

20. V. Bitner, "Charlz Darvin," p. 171.

21. Jeffrey Brooks, *When Russia Learned to Read*, p. 334.

22. K. A. Timiriazev, "Kembridzh i Darvin," no. 11, p. 257.

23. Ibid., pp. 255–56.

24. Timiriazev, *Nauka*, p. 113.

25. A. Bers, "Darvinizm," p. 119.
26. Ibid., p. 116.
27. S. S. Glagolev, "K voprosu o proiskhozhdenii," p. 452.
28. Glagolev, "K voprosu o proiskhozhdenii," p. 450.
29. V. I. Taliev, *Charlz Darvin*, p. 42; K. M. Zavadskii, *Razvitie*, pp. 303–4.
30. Zavadskii, *Razvitie*, p. 303.
31. S. L. Sobol', "Ob izdanii Charlza Darvina," p. 223.
32. K. A. Timiriazev, "Piatidesiatiletnii iubilei."
33. S. Chulok, "K iubileiu," no. 4, sec. 1, pp. 61–63.
34. V. V. Lunkevich, *Osnovy zhizni*, vol. 2, p. 526.
35. Ibid., p. 441.
36. Lunkevich, "Staroe i novoe," no. 6, p. 119.
37. I. Iasinskii, "Darvin," p. 39.
38. V. A. Vagner, "Lamark," no. 8, pp. 14–18.
39. Lunkevich, "Staroe i novoe," no. 1, p. 144.
40. P. F. Lesgaft, "Pamiati Zhana Lamarka," p. 2. For biographical data, see L. Slonimskii, "Kniga o P. F. Lesgafte."
41. Timiriazev, "God itogov," no. 11, p. 325; N. Iu. Zograf, "Noveishiia techeniia," p. 194.
42. Ia. P. Shchelkanovtsev, *Kratkii kurs*, ch. VIII.
43. Shchelkanovtsev, *Kratkii kurs*, p. 89.
44. K. M. Deriugin, "Proiskhozhdenie," p. 619.
45. V. L. Komarov, "Vidoobrazovanie," p. 525.
46. Ibid., p. 532.
47. N. M. Knipovich, "Vzaimootnoshenie organizmov," p. 416.
48. D. N. Anuchin, "Proiskhozhdenie cheloveka i ego iskopaemye predki," p. 704.
49. [Anuchin] D. A., "Proiskhozhdenie cheloveka," 1898, p. 380, and "Pitekantropos," p. 735.
50. Anuchin, "Proiskhozhdenie cheloveka i ego iskopaemye predki," pp. 771–84.
51. B. L. Astaurov and P. F. Rokitskii, *Nikolai Konstantinovich Kol'tsov*, ch. 1; V. Polynin, *Prorok*, ch. 3.
52. N. K. Kol'tsov, "Organizatsiia kletki," pp. 211–12.
53. M. M. Zavadovskii, "Osnovnye etapy," p. 54.
54. Kol'tsov, "Vzgliady Lotsi," p. 1253.
55. Ibid., p. 1264.
56. Komarov, "Vidoobrazovanie," p. 540.

Chapter Ten

1. N. K. Mikhailovskii, "Literatura," 1895, no. 1, sec. 2, p. 145.
2. N. D. Nozhin, "Nasha nauka," no. 7, p. 175.
3. For additional comments on Nozhin's influence on Mikhailovskii, see E. E. Kolosov, *Ocherki*, pp. 157–65.
4. S. N. Iuzhakov, "Sotsiologicheskaia doktrina," pp. 354–55.
5. For details, see James A. Rogers, "The Russian Populists' Response,"

pp. 458–60; Kolosov, *Ocherki,* ch. 3; E. S. Vilenskaia, *N. K. Mikhailovskii,* ch. 2; S. P. Ranskii, *Sotsiologiia,* pp. 41–49; A. I. Krasnosel'skii, *Mirovozzrenie,* pp. 14–32.

6. Mikhailovskii, *Polnoe sobranie sochinenii,* vol. 1, 5th ed., p. 459.

7. Ibid., p. 458.

8. Lunkevich, *N. K. Mikhailovskii,* pp. 48–49.

9. Mikhailovskii, *Polnoe sobranie sochinenii,* vol. 1, 5th ed., p. 459.

10. Mikhailovskii, *Sochineniia,* vol. 1, pp. 37, 291.

11. Mikhailovskii, *Polnoe sobranie sochinenii,* vol. 1, 5th ed., p. 150.

12. Ibid., p. 536.

13. Mikhailovskii, *Sochineniia,* vol. 1, pp. 914–15.

14. Mikhailovskii, *Sochineniia,* vol. 3, pp. 284–86.

15. For a general comment, see A. Walicki, *A History of Russian Thought,* pp. 222–25.

16. According to A. Walicki, "The Russian peasant, argued Mikhailovskii, lives a life that is primitive but full; he is economically self-sufficient, therefore independent, 'well-rounded' and 'whole'; he satisfies all his needs by his own work, making use of all his capacities, being a tiller and a fisherman, a shepherd, and an artist in one person. The absence of underdevelopment of 'complex co-operation' makes the Russian peasant mutually independent, while 'simple co-operation' (i.e., a cooperation in which men are involved as 'whole beings') unites them in a moral solidarity based upon mutual sympathy and understand-ing" (A. Walicki, *Russian Social Thought,* p. 29).

17. Mikhailovskii, *Polnoe sobranie sochinenii,* vol. 1, 5th ed., p. 284.

18. Ivanov-Razumnik, *Istoriia,* vol. 2, p. 122.

19. P. L. Lavrov, *Opyt istorii mysli,* vol. 1, part 1, p. 121.

20. V. V. Bogatov, "P. L. Lavrov," p. 120.

21. Lavrov, *Filosofiia,* vol. 2, pp. 36, 300.

22. Ibid., vol. 1, p. 208.

23. For more details, see Bogatov, *Filosofiia,* pp. 97–99.

24. Lavrov, *Zadachi pozitivizma,* p. 17.

25. For a systematic and thorough study of Lavrov's theory of personality, see N. I. Kareev, *Teoriia lichnosti.*

26. Lavrov, *Formula progressa,* pp. 41–43.

27. Lavrov, *Opyt istorii mysli,* vol. 1, part 1, p. 197.

28. Ibid., part 2, p. 1427.

29. Lavrov, *Izbrannye sochineniia,* vol. 2, pp. 101–2.

30. Lavrov, *Formula progressa,* pp. 16–17.

31. A. Walicki, *The Controversy,* p. 55.

32. This study makes up vol. 1, part 2, of Lavrov's *Opyt istorii mysli.* See particularly pp. 1426–51.

33. For a critical review of Lavrov's "history of thought" as a "philosophy of history" fastened to cultural evolutionism, see A. A. Gizetti, "P. L. Lavrov," pp. 292–354.

34. Lavrov, *Opyt,* vol. 1, part 1, pp. 202–3.

35. Ibid., p. 121.

36. N. I. Kareev, *Vvedenie,* ch. 14.

37. Kareev, *Istoriko-filosofskie i sotsiologicheskie etiudy,* pp. 286–87.
38. Kareev, *Vvedenie,* p. 70.
39. For relevant comments, see Lavrov, *Formula progressa,* pp. 111–12.
40. Kareev, *Vvedenie,* pp. 72–74.
41. For Kareev's comments on Lavrov's views on solidarity, see Kareev, "Novyi istoriko-filosofskii trud," pp. 401–5.
42. M. A. Miller, *Kropotkin,* p. 255.
43. P. A. Kropotkin, *Revolutionary Pamphlets,* p. 192.
44. Kropotkin, *Modern Science,* p. 27.
45. Kropotkin, *Mutual Aid,* p. 9.
46. K. F. Kessler, "O zakone," pp. 128–36.
47. Kropotkin, *Mutual Aid,* p. 8. See also Y. Delage and M. Goldsmith, *Les théories,* 1911, p. 359.
48. Thomas H. Huxley, "The Struggle for Existence," pp. 329–41.
49. Greta Jones, *Social Darwinism,* p. 76.
50. M. M. Kovalevskii, "Darvinizm," pp. 136, 154.
51. N. Lebedev, "P. A. Kropotkin," pp. 98–99.
52. Miller, *Kropotkin,* p. 173.
53. Kropotkin, *Modern Science,* pp. 20–21.
54. Kropotkin, *Ethics,* pp. 325–26.
55. Ibid., p. 327.
56. Ibid., p. 48. For more details on this work, see James Hulse, *Revolutionists,* pp. 187–91.
57. Kropotkin, "The Morality of Nature," p. 407.
58. Ibid., p. 408.
59. Ibid., pp. 408–9.
60. Ibid., p. 418.
61. Kropotkin, "The Direct Action," p. 427.
62. He referred to R. H. Francé, *Der heutige Stand;* and Adolph Wagner, *Geschichte des Lamarckismus.*
63. Kropotkin, "The Direct Action," p. 424.
64. Kropotkin, "Inherited Variation," p. 833.
65. Kropotkin, "The Direct Action," p. 427.
66. Lankester, "Heredity," p. 491.
67. L. Plate, *Selektionsprinzip,* p. 117.
68. T. Dobzhansky, *Mankind Evolving,* p. 123.
69. Ibid., p. 134.
70. Peter J. Bowler, *Evolution,* pp. 215–16, *The Eclipse,* p. 55, and *Theories,* pp. 211–12.
71. Bowler, *The Eclipse,* p. 56.
72. Y. Delage and M. Goldsmith, *Les théories* 1920, pp. 359–64.
73. Kropotkin, *Ethics,* p. 279.
74. M. M. Kovalevskii, *Ocherki proiskhozhdeniia sem'i,* pp. 18–21.
75. Charles Darwin, *The Origin,* p. 75; M. M. Kovalevskii, *Ocherki,* p. 20.
76. G. Nasmyth, *Social Progress.*
77. J. Novicow, *La critique,* p. 10.
78. J. Novicow, *La critique,* p. 11.

79. Ibid., p. 389.
80. G. V. Plekhanov, *Izbrannye filosofskie proizvedeniia*, vol. 2, p. 678.
81. Ibid., vol. 1, pp. 690–91.
82. Plekhanov, *Filosofsko-literaturnoe nasledie*, vol. 3, p. 48.
83. Plekhanov, *Izbrannye filosofskie proizvedeniia*, vol. 1, p. 69.
84. Plekhanov, *Unaddressed Letters*, p. 19.
85. Plekhanov, *Izbrannye filosofskie proizvedeniia*, vol. 2, pp. 162–63.
86. Plekhanov, *Essays*, p. 172.
87. Ibid., p. 173.
88. Plekhanov, *Fundamental Problems*, pp. 46–47.
89. A. M. Deborin, *Filosofiia i politika*, p. 65.
90. Ibid., pp. 65–66.
91. Plekhanov, *Essays*, pp. 208–13.
92. Ibid., p. 215.
93. Plekhanov, *Selected Philosophical Works*, vol. 4, p. 156.
94. Ibid., p. 156.
95. Ibid., pp. 264–65.
96. Ibid., p. 266.
97. Plekhanov, *Fundamental Problems*, trans. Katzer, p. 80.
98. In an assessment of Morgan's anthropological theory, E. Terray states: "When Morgan thought diachronically, he thought in terms of evolution and used Darwinian language; when he thought synchronically he thought in terms of models and structures and his language prefigured that of Claude Lévi-Strauss. But when he overcame this ideological opposition between the diachronic and the synchronic, he set himself a worthwhile task and tried to think in terms of history." This stage in his thinking, according to Terray, brought Morgan close to Marxism. (E. Terray, *Marxism*, pp. 89–90.)
99. A Russian translation of Morgan's *Ancient Society* appeared in 1900.
100. Plekhanov, *Izbrannye filosofskie proizvedeniia*, vol. 1, p. 686.
101. F. Engels, *The Origin of the Family*, p. 16.
102. R., "Morgan, Lewis Henry," p. 835.
103. Plekhanov, *Izbrannye filosofskie proizvedeniia*, vol. 1, p. 496.
104. Terray, *Marxism*, pp. 23–25.
105. M. A. Antonovich, *Izbrannye filosofskie sochineniia*, p. 342.
106. N. K. Mikhailovskii, "Literatura," 1898, no. 2, sec. 2, p. 140.
107. Plekhanov, *The Development*, p. 254.
108. Ibid., p. 255.
109. Mikhailovskii, *Literaturnye vospominaniia*, vol. 2, p. 415.
110. Ibid., p. 433.
111. R. Francé, *Der heutige Stand*, chs. 4–7.
112. Ibid., p. 84.
113. See, for example, Francé, *Der heutige Stand*.
114. Plekhanov, *Fundamental Problems*, trans. Katzer, p. 45, and *Izbrannye filosofskie proizvedeniia*, vol. 5, p. 150; B. A. Chagin, *G. V. Plekhanov*, p. 167.
115. A. Walicki, *A History*, p. 424.
116. V. I. Lenin, *Polnoe sobranie sochinenii*, vol. 1, p. 139.
117. Ibid., p. 149.

118. Ibid., p. 146.
119. James A. Rogers, "Marxist and Russian Darwinism," pp. 207–8.
120. Lenin, *Materialism and Empirio-Criticism*, p. 340.
121. Plekhanov, *The Development*, pp. 217–18.
122. F. Bersenev, "Nechto o 'kriterii istiny'."
123. A. A. Bogdanov, *Poznanie*, p. 24.
124. Bogdanov, *Iz psikhologii obshchestva*, p. 55.
125. Ibid., p. 56.
126. [A. A. Bogdanov] N. Verner, "Filosofiia," p. 114.
127. A. A. Bogdanov, *Poznanie*, p. 25.

Conclusion

1. G. G. Simpson, *Tempo and Mode*, p. 103, and *Horses*, pp. 87–88.
2. [V. O. Kovalevskii], *The Complete Works*.
3. Richard Goldschmidt, *The Material Basis*, pp. 309–10; Simpson, *This View of Life*, p. 182.
4. N. Ia. Danilevskii, *Darvinizm*.
5. See particularly V. I. Shmankevich, "Zur Kenntniss des Einflusses der äusseren Lebensbedingungen."
6. Y. Delage and M. Goldsmith, *Les théories*, 1911, pp. 319–20.
7. V. V. Dokuchaev, *Russkii chernozem*, 2nd ed., pp. 510–11.
8. M. M. Kovalevskii, *Sotsiologiia*, vol. 1, p. 81; vol. 2, ch. 1.
9. The first series of *Priroda* was published in 1873–77; the second series has been published without interruption since 1912. (See N. V. Uspenskaia, "'Priroda.'")
10. V. V. Lunkevich, "Staroe i novoe," 1910, no. 6, p. 111.
11. Ibid., no. 1, pp. 134–35.
12. M. M. Filippov, *Filosofiia*, vol. 1, pp. x–xiii.
13. Ibid., vol. 2, p. 837.
14. V. D. Kudriavtsev-Platonov, *Sochineniia*, vol. 3, part 3, p. 60.
15. See, for example, N. Sergievskii, "O bibleiskoi istorii tvoreniia"; and G. Malevanskii, "O dukhovnoi prirode cheloveka."
16. [Nikanor], "Pouchenie na novyi god," p. 40.
17. D. I. Pisarev, *Izbrannye filosofskie i obshchestvenno-politicheskie stat'i*, p. 293.
18. E. E. Kolosov, *Ocherki*, p. 136.
19. P. A. Kropotkin, *Modern Science*, pp. 25–26.
20. V. Evgen'ev-Maksimov and G. Tinzengauzen, *Poslednie gody*, p. 302.
21. J. Novicow, *La critique*, p. 3. For a critical comment, see B. Naccache, *Marx*, p. 85.
22. E. Metchnikoff, *Lectures*, p. 4.
23. A. N. Beketov, "Nravstvennost'," pp. 58–63.
24. L. N. Tolstoy, "Protivorechiia," pp. 293–94.
25. K. A. Timiriazev, *Sochineniia*, vol. 7, p. 252.
26. V. Fausek, "Bor'ba za sushchestvovanie."
27. V. Fausek, "Vid," p. 244.

28. V. M. Shimkevich, "Transformizm," p. 722.
29. N. M. Knipovich, "Vzaimootnoshenie organizmov," p. 413.
30. A. N. Severtsov, *Sochineniia,* vol. 3, pp. 283–88.
31. Ibid., pp. 73–75.
32. N. V. Tsinger, "O zasoriaiushchikh posevy l'na vidakh."
33. R. E. Regel', "Selektsiia," p. 507.
34. K. A. Timiriazev, "Charlz Darvin," in M. M. Kovalevskii et al., *Pamiati Darvina,* p. 22.
35. For a brief statement on "the characteristics of the evolutionary theory in Russia," see K. M. Zavadskii and M. T. Ermolenko, "Evoliutsionnaia teoriia," pp. 505–9. See also James A. Rogers, "Reception," "Russian Opposition," and "Charles Darwin and Russian Scientists"; and A. Vucinich, *Science in Russian Culture, 1861–1917,* ch. 9.

Works Consulted

Adams, Mark B. "Severtsov and Schmalhausen: Russian Morphology and the Evolutionary Synthesis." In E. Mayr and W. B. Provine, eds., *The Evolutionary Synthesis,* pp. 193–225.

Alekseev, V. A. *Osnovy darvinizma.* Moscow, 1964.

Alekseev, V. G. *Matematika kak osnovanie kritiki nauchno-filosofskogo mirovozzreniia.* Iur'ev, 1903.

———. *N. V. Bugaev i problemy idealizma Moskovskoi matematicheskoi shkoly.* Iur'ev, 1905. *See also* Alexejeff, W.

[Alekseev, V. G.] Alexejeff, W. "N. W. Bugajev und die idealistische Probleme der Moskauer mathematischen Schule." *Vierteljahrsschrift für wissenschaftliche Philosophie und Soziologie.* Series 4 (1905), vol. 29, no. 3, pp. 335–67.

Alekseev, V. P. "Evoliutsionnaia ideia proiskhozhdeniia cheloveka v russkoi nauke do Darvina i proniknovenie v nee darvinizma." *Trudy Instituta etnografii,* 1971, vol. 95, pp. 213–33.

Allen, Garland. *Life Sciences in the Twentieth Century.* New York: Wiley, 1975.

Anokhin, P. K. *Izbrannye trudy: filosofskie aspekty teorii funktsional'noi sistemy.* Moscow, 1978.

Antonovich, M. A. *Charlz Darvin i ego teoriia.* St. Petersburg, 1896.

———. *Izbrannye filosofskie sochineniia.* Moscow, 1945.

———. "Teoriia proiskhozhdeniia vidov v tsarstve zhivotnykh." *Sovremennik,* 1864, no. 3, pp. 63–107.

[Antonovich, M. A.] "Charlz Darvin i ego teoriia." *RM,* 1893, no. 5, sec. 2, pp. 50–72; no. 6, sec. 2, pp. 1–23; no. 12, sec. 2, pp. 112–39; 1894, no. 2, sec. 2, pp. 1–27.

———. Postoronnii satirik, "Literaturnye melochi." *Sovremennik,* 1865, no. 1, sec. 2, pp. 157–71.

Anuchin, D. N. *O liudiakh russkoi nauki i kul'tury.* Moscow, 1950.

———. "Proiskhozhdenie cheloveka i ego iskopaemye predki." In M. M. Kovalevskii et al., eds., *Itogi nauki v teorii i praktike*, vol. 6, pp. 691–784.

———. "Stoletie so dnia rozhdeniia Ch. Darvina." *Russkie vedomosti*, 1909, no. 26 (Feb. 1), pp. 3–4.

———. "Vladimir Onufrievich Kovalevskii (Nekrolog)." *Rech' i otchet, chitanie v torzhestvennom sobranii Moskovskogo universiteta 12 ianvaria 1884 g.* Pp. 266–81.

———. "Kovalevskii, Vladimir Onufrievich." *BS-PU*, vol. 1 (St. Petersburg, 1896), pp. 324–36.

[Anuchin, D. N.] "Nauchnyi obzor: antropologiia i etnografiia." *RM*, 1884, no. 12, sec. 2, pp. 170–201.

[Anuchin,] D. "Nauchnoe obozrenie." *RM*, 1881, no. 11, sec. 13, pp. 17–38.

[Anuchin,] D. A. "Pitekantropos." Brockhaus-Efron, 1898, vol. 23A, pp. 734–35.

———. "Proiskhozhdenie cheloveka." Brockhaus-Efron, 1898, vol. 25, pp. 378–80.

Arnol'di, M. V. "Sovremennoe sostoianie voprosa o proiskhozhdenii vidov u rastenii." *NS*, 1904, no. 4, pp. 23–40.

Astaurov, B. L., and P. F. Rokitskii. *Nikolai Konstantinovich Kol'tsov.* Moscow, 1975.

Aver'ianova, T. M. "Evoliutsionnye vzgliady R. E. Regelia." *NiT*, 1973, vol. 8, part 2, pp. 146–49.

———. "K voprosu o istorii podgotovki sinteza genetiki i darvinizma (po rabotam R. E. Regelia)." In *Voprosy razvitiia evoliutsionnoi teorii v XX veke*, pp. 124–32. Leningrad, 1979.

Babkov, V. V. *Moskovskaia shkola evoliutsionnoi genetiki.* Moscow, 1985.

Bachinskii, A. I. "Dukh beskonechno malykh." In A. Bachinskii et al., *Sbornik po filosofii estestvoznaniia*, pp. 103–205.

——— "Pis'ma po filosofii estestvoznaniia. I. Chto takoe naturalisticheskii idealizm." *VFiP*, 1903, no. 70, pp. 806–21.

———. "Pis'ma po filosofii estestvoznaniia. II. O vozmozhnom vliianii matematicheskikh metodov na cherty nauchnogo miroponimaniia." *VFiP*, 1905, no. 76, pp. 86–110.

———. "Zamechatel'nyi russkii uchenyi: N. A. Umov." *RM*, 1916, no. 3, pp. 26–34.

Bachinskii, A. I., et al. *Sbornik po filosofii estestvoznaniia.* Moscow, 1906.

Baer, Karl von. "Entwickelt sich die Larve der einfachen Ascidien in der ersten Zeit nach dem Typus der Wirbelthiere?" *MAI*, series 7, 1873, vol. 19, no. 8, pp. 1–36.

———. "Mesto cheloveka v prirode ili kakoe polozhenie zanimaet chelovek v otnoshenii ko vsei ostal'noi prirode." *Naturalist*, 1865, nos. 2–5, 19–24; 1866, nos. 9, 18, 22–24; 1867, nos. 1–3.

———. *Reden gehalten in wissenschaftlichen Versammlungen und kleinere Aufsätze vermischten Inhalts.* St. Petersburg. Vol. 1, 1864; vol. 2, 1876; vol. 3, 1873.

———. "Ueber Papuas und Alfuren," *MAI*, series 6, 1859, vol. 8, pp. 269–346.

———. "Zum Streit über den Darwinismus," *Augsburger Allgemeine Zeitung*, 1873, no. 130, pp. 1986–88.

Balfour, Francis M. *A Treatise on Comparative Embryology.* 2 vols. London: Macmillan, 1881.

Banina, N. N. *K. F. Kessler i ego rol' v razvitii biologii v Rossii.* Moscow-Leningrad, 1962.

———. "O nauchnom mirovozzrenii K. F. Kesslera." *TIIE-T,* 1960, vol. 31, pp. 247–67.

Baranetskii, O. V. "Vydaiushchiisia iavleniia po noveishei literature o darvinizme." *Universitetskie izvestiia.* Kiev University, 1903, no. 1, pp. 1–62.

Baranov, P. A. "Andrei Nikolaevich Beketov." *LRN: biol,* pp. 116–25.

Bateson, W. "Heredity and Variation in Modern Light." In A. C. Seward, ed., *Darwin in Modern Science,* pp. 85–101.

———. *Materials for the Study of Variation Treated with Especial Regard to Discontinuity in the Origin of Species.* London: Macmillan, 1894.

Baur, E. *Vvedenie v eksperimental'noe izuchenie nasledstvennosti. TBPB,* Supplement, 1913.

Bayertz, Kurt. "Darwinism and Scientific Freedom." *Scientia,* 1983, vol. 118, pp. 297–307.

Beketov, A. N. "Avtobiografiia." In Vengerov, S. A., ed., *Kritiko-biograficheskii slovar' russkikh pisatelei i uchenykh,* 1891, vol. 2, pp. 353–63.

———. "Bor'ba za sushchestvovanie v organicheskom mire." *VE,* 1873, no. 10, pp. 558–93.

———. "Darvinizm s tochki zreniia obshchefizicheskikh nauk." *TSPO,* 1882, vol. 13, no. 1, pp. 92–110.

———. "Garmoniia v prirode." In Vasetskii, G. S., and S. R. Mikulinskii, *Izbrannye proizvedeniia . . . ,* pp. 545–82.

———. *Geografiia rastenii.* St. Petersburg, 1896.

———. "Nravstvennost' i estestvoznanie." *VFiP,* 1890, no. 5, pp. 7–67.

Beketov, N. N. *Rechi khimika.* St. Petersburg, 1908.

Bekhterev, V. M. "Psikho-biologicheskie voprosy." *NO,* 1902, no. 4, pp. 1–24; no. 5, pp. 1–25.

Belkin, R. I. "Tvorcheskaia razrabotka darvinizma I. I. Mechnikovym." In I. I. Mechnikov, *Izbrannye raboty po darvinizmu,* pp. 307–82.

Bendall, D. S., ed. *Evolution From Molecules to Men.* Cambridge: Cambridge University Press, 1985.

Berdiaev, N. A. "Eticheskaia problema v svete filosofskogo idealizma." In P. I. Novgorodtsev, ed., *Problemy idealizma,* pp. 91–136.

———. *Sub'ektivizm i individualizm v obshchestvennoi filosofii: kriticheskii etiud o N. K. Mikhailovskom.* St. Petersburg, 1901.

Berdyshev, G. D., and V. N. Splivinskii. *Pervyi sibirskii professor botaniki Korzhinskii.* Novosibirsk, 1961.

Berezin, N. "Po povodu novykh izdanii sochinenii Charlza Darvina." *Obr.,* 1896, no. 2, sec. 2, pp. 77–84.

Berg, L. S. *Nomogenesis or Evolution Determined by Law.* Translated from the Russian by J. N. Rostovtsov. Cambridge: M.I.T. Press, 1969.

———. *Trudy po teorii evoliutsii.* Leningrad, 1977.

Berlin, P. "O pervobytnom obshchestve." *NO,* 1900, no. 11, pp. 1946–2122.

Bernard, Claude. *An Introduction to the Study of Experimental Medicine.* New York: Collier, 1961.

Bers, A. "Darvinizm i khristianskaia nravstvennost'." *VE*, 1910, no. 5, pp. 109–19.

Bersenev, F. "Nechto o 'kriterii istiny'." *RM*, 1901, no. 7, sec. 2, pp. 123–43.

Berthelot, M. *Science et morale.* Paris, 1897.

Besredka, A. *Histoire d'une idée: L'oeuvre de E. Metchnikoff.* Paris, 1921.

Bibikov, P. A. *Kriticheskie etiudy.* St. Petersburg, 1865.

Bitner, V. "Charlz Darvin, 1809–1909." *VZ*, 1909, no. 2, pp. 168–72.

Bliakher, L. Ia. "A. N. Severtsov i neolamarkizm." *Iz istorii biologii*, 1970, no. 2, pp. 112–22.

———. "Aleksandr Onufrievich Kovalevskii." *LRN: biol*, pp. 157–72.

———. "Georg Zeidlits i ego kurs darvinizma." *Iz istorii biologii*, 1971, no. 3, pp. 3–58.

———. "Ia. A. Borzenkov—istorik biologii." *VIE-T*, 1982, no. 41, pp. 36–44.

———. *Istoriia embriologii v Rossii (s serediny XVIII do serediny XIX veka).* Moscow, 1955.

———. *Istoriia embriologii v Rossii (s serediny XIX do serediny XX veka): Bespozvochnye.* Moscow, 1959.

———. "Karl Maksimovich Ber." *LRN: biol*, pp. 56–72.

———. "Nauchnye sviazi A. O. Kovalevskogo i I. I. Mechnikova s zarubezhnimi zoologami." *TIIE-T*, 1959, vol. 23, pp. 93–143.

———. *Ocherk istorii morfologii zhivotnykh.* Moscow, 1962.

———. *Problema nasledovaniia priobretennykh priznakov.* Moscow, 1971.

———, ed. *Istoriia biologii s nachala XX veka do nashikh dnei.* Moscow, 1975.

Bliakher, L. Ia., et al., eds. *Istoriia estestvoznaniia v Rossii.* Vol. 3. Moscow, 1962.

Bobynin, V. V. *Filosofskoe, nauchnoe i pedagogicheskoe znachenie istorii matematiki.* Moscow, 1886.

Bogatov, V. V. *Filosofiia P. L. Lavrova.* Moscow, 1972.

———. "P. L. Lavrov i estestvoznanie. (K 150-letiiu so dnia rozhdeniia P. L. Lavrova)." *Filosofskie nauki*, 1973, no. 5, pp. 118–29.

Bogdanov, A. A. *Empiriomonizm.* Vol. 3. St. Petersburg, 1906.

———. *Iz psikhologii obshchestva.* 2nd ed. St. Petersburg, 1906.

———. "Obmen i tekhnika." In *Ocherki realisticheskogo mirovozzreniia*, 2nd ed. (St. Petersburg, 1905), pp. 279–342.

———. *Poznanie s istoricheskoi tochki zreniia.* Moscow, 1902.

———. "Psikhicheskii podbor. (Empiriomonizm v uchenii o psikhike)." *VFiP*, 1904, no. 73, pp. 335–78; no. 74, pp. 485–519.

[Bogdanov, A. A.] Verner, N. "Filosofiia sovremennogo estestvoispytatelia." In [Bogdanov, A.] N. Verner et al., *Ocherki . . . *, pp. 37–142.

[Bogdanov, A. A.] Verner, N., et al., *Ocherki filosofii kollektivizma.* Vol. 1. St. Petersburg, 1909.

Bogdanov, E. A. *Mendelizm ili teoriia skreshchivaniia. (Novoe napravlenie v izuchenii izmenchivosti).* Moscow, 1914.

Borisiak, A. A. *Izbrannye trudy.* Moscow, 1973.

———. *V. O. Kovalevskii: Ego zhizn' i nauchnye trudy.* Leningrad, 1926.

Borodin, I. P. "Protoplazma i vitalizm." *MB*, 1894, no. 5, pp. 1–28.

Borovoi, A., and N. Lebedev, eds. *Sbornik statei posviashchennyi pamiati P. A. Kropotkina.* Moscow, 1922.

Borozdin, A. "Danilevskii, Nikolai Iakovlevich." *Russkii biograficheskii slovar'.* Vol. 6 (St. Petersburg, 1905), pp. 67–72.

Borzenkov, Ia. A. *Istoricheskii ocherk napravlenii sushchestvovaiushchikh v zoologicheskikh naukakh v XIX stoletii.* Moscow, 1881.

[Borzenkov, Ia. A.] "Chteniia Ia. A. Borzenkova po sravnitel'noi anatomii." *Uchenye zapiski Imperatorskogo Moskovskogo universiteta: Otdel estestvennoistoricheskii,* 1884. Vol. 4.

Bowen, Francis. *Modern Philosophy From Bacon to Schopenhauer and Hartmann.* New York: Scribner, 1877.

Bowler, Peter J. *The Eclipse of Darwinism.* Baltimore: Johns Hopkins University Press, 1983.

————. *Evolution: The History of an Idea.* Berkeley and Los Angeles: University of California Press, 1984.

————. *Theories of Human Evolution.* Baltimore: Johns Hopkins University Press, 1986.

Brandt, A. F. "Sozhitel'stvo i vzaimnaia pomoshch'." *MB,* 1896, no. 5, pp. 1–23; no. 6, pp. 93–117.

Brandt, F. "Zametki o soderzhanii vtorogo i tret'ego otdelov moikh soobshchenii o morskikh korovakh (*Symbolae sirenologicae*)." *Trudy Pervogo s'ezda,* pp. 211–15.

Broda, E. "Darwin and Boltzmann." In E. Geissler and W. Scheler, eds., *Darwin Today,* pp. 61–70.

Brooks, Jeffrey. *When Russia Learned to Read.* Princeton, N.J.: Princeton University Press, 1985.

Brusilovskii, I. K. "Sovremennye teorii stroeniia zhivogo veshchestva," *RB,* 1898, no. 1, sec. 1, pp. 105–33; no. 2, sec. 1, pp. 53–83; no. 3, sec. 1, pp. 14–38.

Bugaev, N. V. "Matematika i nauchno-filosofskoe mirosozertsanie." *Dnevnik X-go S'ezda.* Pp. 36–52.

Bulgakov, S. N. "Chto daet sovremennom soznanii filosofiia Vl. Solov'eva." *VFiP,* 1903, no. 1, pp. 52–96.

————. "Osnovnye problemy teorii progressa." In P. O. Novgorodtsev, ed., *Problemy idealizma,* pp. 1–47.

————. *Ot marksizma k idealizmu.* St. Petersburg, 1903.

Bunge, Gustav. *Lehrbuch der physiologischen und pathologischen Chemie.* 2nd ed. Leipzig, 1889.

Burkhardt, Richard W., Jr. "Darwin on Animal Behavior and Evolution." In David Kohn, ed., *The Darwinian Heritage,* pp. 327–65.

————. "The Zoological Philosophy of J. B. Lamarck." In J. B. Lamarck, *Zoological Philosophy,* pp. xv–xxxix.

Bury, J. B. "Darwinism and History." In A. C. Seward, ed., *Darwin in Modern Science,* pp. 529–42.

Bychkovskii, B. S. *Sovremennaia filosofiia.* Vol. 1. St. Petersburg, 1911.

Chagin, B. A. *G. V. Plekhanov i ego rol' v razvitii marksistskoi filosofii.* Leningrad, 1963.

Chemena, I. A. *Darvinizm: nauchnoe issledovanie teorii Darvina o proiskhozh-denii cheloveka.* Odessa, 1892.

Chernov, Viktor. "Gde kliuch ponimaniiu N. K. Mikhailovskogo?" *Zavety,* 1913, no. 3, sec. 2, pp. 82–132.

[Chernyshevskii, N. G.] Staryi Transformist. "Proiskhozhdenie teorii blagotvor-nosti bor'by za zhizn'." *RM,* 1888, no. 9, sec. 2, pp. 79–114.

"Chestvovanie pamiati Ch. Darvina." *EiG,* 1909, no. 4, p. 92.

Chicherin, B. N. *Nauka i religiia.* 2nd ed. Moscow, 1901.

————. *Polozhitel'naia filosofiia i edinstvo nauki,* Moscow, 1892.

Chistovich, I., "*Darvinizm. Kriticheskoe issledovanie* N. Ia. Danilevskogo." *Vera i razum,* 1886, vol. 2, part 1, pp. 131–36.

Chugaev, L. "Evoliutsiia veshchestva v mertvoi i zhivoi prirode." *Obr.,* 1909, no. 2, pp. 69–84.

Chulok, S. "K iubileiu darvinizma." *RB,* 1909, no. 3, sec. 1, pp. 103–29; no. 4, sec. 1, pp. 45–65; no. 5, sec. 1, pp. 37–62.

Conry, Yvette. *L'introduction du darwinisme en France au XIXe siècle.* Paris, 1974.

Coquart, Armand. *Dmitri Pisarev (1840–1868) et l'idéologie du nihilisme russe.* Paris, 1946.

Danilevskii, Aleksandr. "Zhivoe veshchestvo." *VE,* 1896, no. 5, pp. 289–336.

Danilevskii, N. Ia. *Darvinizm: kriticheskoe issledovanie.* 2 vols. St. Petersburg, 1885–89.

————. *Rossiia i Evropa.* 3rd ed. St. Petersburg, 1888.

"Darvin." *Bol'shaia entsiklopediia.* Vol. 8, 1902, pp. 137–39.

"Darvin i ego teoriia obrazovaniia vidov." *BdCh.* 1861, no. 11, pp. 1–40; no. 12, pp. 1–36.

"Darvinizm." *Bol'shaia entsiklopediia.* Vol. 8, 1902, pp. 127–37.

Darwin, Charles. *The Descent of Man.* New York: Collier, 1900.

————. *The Expression of the Emotions in Man and Animals.* New York: Appleton, 1915.

————. *The Origin of Species.* Garden City, N.Y.: Doubleday, n.d.

————. *The Variation of Animals and Plants Under Domestication.* 2 vols. New York: D. Appleton, 1897.

Darwin, Francis, ed., *The Life and Letters of Charles Darwin.* 2 vols. New York: Basic Books, 1959.

Davitashvili, L. Sh. "Biografiia Vladimira Onufrievicha Kovalevskogo." In V. O. Kovalevskii, *Sobranie nauchnykh trudov,* vol. 1, pp. 7–155.

————. *Istoriia evoliutsionnoi paleontologii ot Darvina do nashikh dnei.* Moscow-Leningrad, 1948.

————. *Uchenie ob evoliutsionnom progresse. (Teoriia aromorfoza).* Tbilisi, 1972.

————. *V. O. Kovalevskii.* 2nd ed. Moscow, 1951.

————. "Vladimir Onufrievich Kovalevskii." *LRN: geol.,* pp. 23–32.

————. "V. O. Kovalevskii, ego nauchnaia deiatel'nost' i znachenie ego trudov po paleontologicheskoi istorii semeistva loshadinykh." Postscript to V. O. Kovalevskii, *Paleontologiia loshadei,* pp. 258–314.

Davydov, K. N. "A. O. Kovalevskii i ego rol' v sozdanii sravnitel'noi embrio-logii." *Pr.,* 1916, no. 4, pp. 579–98.

———. "A. O. Kovalevskii kak chelovek i kak uchenyi." *TIIE-T,* 1960, vol. 31, pp. 326–63.

Deborin, A. M. *Filosofiia i politika.* Moscow, 1961.

Delage, Yves, and M. Goldsmith. *Les théories de l'évolution.* Paris, 1911, 1914, 1920.

Deriugin, K. M. "Biografiia i nauchnye trudy V. M. Shimkevicha." *Trudy Leningradskogo Obshchestva estestvoispytatelei,* 1926, vol. 56, no. 2, pp. 5–24.

———. "Proiskhozhdenie razlichnykh grupp pozvonochnykh zhivotnykh." In M. M. Kovalevskii, et al., eds., *Itogi nauki v teorii i praktike,* vol. 6, pp. 619–89.

De Vries, Hugo. *Die Mutationslehre.* 2 vols. Leipzig, 1901–3.

———. "Variation." In A. C. Seward, ed., *Darwin in Modern Science,* pp. 66–84.

Dewey, John. *The Influence of Darwin on Philosophy and Other Essays in Contemporary Thought,* New York: Henry Holt, 1910.

Dnevnik X-go S'ezda russkikh estestvoispytatelei i vrachei v Kieve. L. L. Lund, ed. Vol. 2. Kiev, 1898.

Dnevnik XI-go S'ezda russkikh estestvoispytatelei i vrachei. No. 5. St. Petersburg, 1902.

Dnevnik XII-go S'ezda russkikh estestvoispytatelei i vrachei. Sec. 1. Moscow, 1910.

Dobzhansky, Theodosius. *Genetics and the Evolutionary Process.* New York: Columbia University Press, 1970.

———. *Genetics and the Origin of Species.* 3rd ed. New York: Columbia University Press, 1969.

———. *Mankind Evolving.* New Haven, Conn.: Yale University Press, 1962.

Dogel', V. A. "Embriologicheskie raboty A. O. Kovalevskogo v 60-kh—80-kh godakh XIX v." *NN,* 1948, vol. 1, pp. 206–18.

Dogel', V. A., and A. E. Gaisinovich. "Osnovnye cherty tvorchestva I. I. Mechnikova kak biologa." In I. I. Mechnikov, *Izbrannye biologicheskie proizvedeniia,* pp. 677–725.

Dokuchaev, V. V. *Russkii chernozem.* 2nd ed. Moscow, 1952.

Dostoevsky, F. M. *Pis'ma.* Edited by A. S. Dolinin, 4 vols. Moscow-Leningrad, 1928–59.

———. *Polnoe sobranie sochinenii.* 6th ed. Vol. 11. St. Petersburg, 1905.

E. "Novaia biologicheskaia teoriia," *Severnyi vestnik,* 1897, no. 10, sec. 1, pp. 187–92.

E-dt, M. "Darvin (Charlz-Robert)." Brockhaus-Efron, 1893, vol. 10, pp. 133–35.

Engels, F. *The Origin of the Family, Private Property and the State.* New York: International Publishers, 1942.

Ern, V. F. "Priroda nauchnoi mysli." *Bogoslovskii vestnik,* 1914, January, pp. 154–73; February, pp. 342–68.

Evgen'ev-Maksimov, V., and G. Tizengauzen. *Poslednie gody "Sovremennika," 1863–1866.* Leningrad, 1939.

Evgrafov, V. E., et al., eds., *Istoriia filosofii v SSSR.* Vol. 3. Moscow, 1968.

Fabri, K. E. "V. A. Vagner i sovremennaia zoopsikhologiia." *Voprosy psikhologii,* 1969, no. 9, pp. 100–107.

432   Works Consulted

Famintsyn, A. S. "Andrei Nikolaevich Beketov." *ZhMNP*, 1902, vol. 344, sec. 4, pp. 51–65.

———. "Blizhaishie zadachi biologii." *VE*, 1894, no. 5, pp. 132–53.

———. "Chto takoe lishainiki?" *Pr.*, 1918, April–June, pp. 266–88.

———. "Darvin i ego znachenie v biologii." *OZ*, 1874, no. 3, pp. 129–50.

———. "N. Ia. Danilevskii i darvinizm." *VE*, 1889, no. 2, pp. 616–43.

———. *Obmen veshchestva i prevrashchenie energii v rasteniiakh*. *ZAN*, Supplement, 1883, vol. 46.

———. "O psikhicheskoi zhizni." *VIII S'ezd: Obshchii otdel*, pp. 32–39.

———. "S. I. Korzhinskii. (Nekrolog)." *TSPO*, 1900, vol. 31, no. 4, pp. 357–64.

———. "Sovremennoe estestvoznanie i psikhologiia." *MB*, 1898, no. 1, pp. 1–29; no. 2, pp. 121–42; no. 3, pp. 167–99; no. 4, pp. 201–31; no. 5, pp. 209–40; no. 6, pp. 152–73; no. 7, pp. 92–113.

Fausek, V. A. "Bor'ba za sushchestvovanie." Brockhaus-Efron, 1891, vol. 4, pp. 458–60.

———. "Predislovie." In V. A. Fausek, ed., *Sushchnost' zhizni*, pp. v–viii.

———. "Vid." Brockhaus-Efron, 1892, vol. 6, pp. 240–48.

———, ed. *Sushchnost' zhizni*. St. Petersburg, 1903.

Fedotov, D. M. "Vydaiushchiisia russkii zoolog-darvinist. K stoletiiu so dnia rozhdeniia V. M. Shimkevicha." *Pr.*, 1958, no. 11, pp. 68–70.

Fervorn [Verworn], M. "Vitalizm." *Novoe slovo*, 1897, no. 8, pp. 67–85.

Filipchenko, Iu. A. *Evoliutsionnaia ideia v biologii*. Moscow, 1977.

———. *Izmenchivost' i evoliutsiia*. Petrograd, 1915.

———. *Nasledstvennost'*. Moscow, 1917.

———. "Obzor glavneishei literatury po genetike za 1915 god." *Pr.*, 1916, no. 5–6, pp. 731–34.

———. "Obzor glavneishei literatury po genetike za 1916 god." *Pr.*, 1917, no. 3, pp. 397–99.

———. "Review of I. Delazh and M. I. Gol'dsmit, *Teorii evoliutsii*." *Pr.*, 1917, no. 2, pp. 280–82.

Filippov, M. M. "Darvinizm na russkoi pochve." *NO*, 1894, no. 32, pp. 993–1005; no. 33, pp. 1025–43; no. 34, pp. 1057–71; no. 35, pp. 1089–1105.

———. "Eshche o Veismanne." *NO*, 1894, no. 18, pp. 555–62.

———. *Filosofiia deistvitel'nosti*. 2 vols. St. Petersburg, 1895–98.

———. "Psikhologiia v politicheskoi ekonomii." *NO*, 1900, no. 11, pp. 1998–2018.

———, ed. *Deviatnadtsatyi vek: obzor nauki, tekhniki i politicheskikh sobytii*. St. Petersburg, 1900.

[Flerovskii, N. N.] *Filosofiia bessoznatel'nogo, darvinizam i real'naia istina*. St. Petersburg, 1878.

Fliaksberger, K. A. "Robert Eduardovich Regel'." *TBPB*, 1922, vol. 12, no. 1, pp. 3–24.

Francé, Raoul. "Darvin—issledovatel'." *EiG*, 1909, no. 2, pp. 16–18.

———. *Der heutige Stand der Darwinischen Fragen*. Leipzig, 1907.

Frank, S. L. *Filosofiia i zhizn'*. St. Petersburg, 1910.

———. "Psikhologicheskoe napravlenie v teorii tsennosti." *RB*, 1898, no. 8, sec. 1, pp. 60–110.

Fridlender, G. M. *Dostoevskii i mirovaia literatura*. Moscow, 1979.

———. *Realizm Dostoevskogo*. Leningrad, 1964.

Gaisinovich, A. E. "Biolog-shestidesiatnik N. D. Nozhin i ego rol' v razvitii embriologii i darvinizma v Rossii." *ZhOB*, 1952, vol. 13, no. 5, pp. 377–92.

———. "Contradictory Appraisal by K. A. Timiriazev of Mendelian Principles and Its Subsequent Perception." *History of the Philosophy of Life Sciences*, 1985, vol. 7, pp. 257–86.

———. "The Origins of Soviet Genetics and the Struggle with Lamarckism, 1922–1929." Translated by Mark Adams. *JHB*, 1980, vol. 13, no. 1, pp. 1–51.

———. "Pervoe izlozhenie raboty G. Mendelia v Rossii (I. F. Shmal'gauzen, 1874)." *BMOIP*, 1965, vol. 70, no. 4, pp. 22–24.

———. "Problems of Variation and Heredity in Russian Biology in the Late Nineteenth Century." *JHB*, 1973, vol. 6, no. 1, pp. 97–127.

———. "Velikii russkii biolog I. I. Mechnikov." In I. I. Mechnikov, *Izbrannye proizvedeniia*, pp. 6–27.

———. "Vospriiatie mendelizma v Rossii i ego rol' v razvitii darvinizma." *Pr*, 1982, no. 9, pp. 42–52.

———. "100 let fagotsitarnoi teorii I. I. Mechnikova." *Pr.*, 1983, no. 8, pp. 12–22.

Gaissinovitch, A. E. *See* Gaisinovich, A. E.

Gamaleia, N. F. *Sobranie sochinenii*. Vol. 5. Moscow, 1953.

Ganin, M. S. "Iadro kletki kak organ nasledstvennosti." *NO*, 1894, no. 28, pp. 872–80; no. 29, pp. 897–905; no. 30, pp. 929–38.

Gantt, W. Horsley. "Pavlov and Darwin." In Sol Tax, ed., *Evolution After Darwin*, vol. 2, pp. 219–38.

Gaudry, Albert. *Les enchaînements du monde animal dans les temps géologiques*. Paris, 1878.

Geissler, E., and W. Scheler, eds. *Darwin Today: The 8th Kühlungsborn Colloquium on Philosophical and Ethical Problems of Biosciences*. Berlin, 1983.

Gel'fenbein, L. L. *Russkaia embriologiia vtoroi poloviny XIX veka*. Kharkov, 1956.

Genkel', A. "Ernst Gekkel'." *Obr.*, 1909, no. 3, pp. 89–100.

Georgievskii, A. B. "Osobennosti razvitiia evoliutsionnoi teorii v Rossii." In S. R. Mikulinskii and Iu. I. Polianskii, eds., *Razvitie evoliutsionnoi teorii*, pp. 43–61.

Gerstein, Linda. *Nikolai Strakhov*. Cambridge: Harvard University Press, 1971.

Gertvig, R. "Pamiati Darvina." *Obr.*, 1909, no. 4, pp. 64–82.

Ghiselin, Michael T. *The Triumph of the Darwinian Method*. Berkeley and Los Angeles: University of California Press, 1969.

Giard, Alfred. *Controverses transformistes*. Paris, 1904.

Gizetti, A. A. "P. L. Lavrov kak istorik mysli." In E. L. Radlov, ed., *P. L. Lavrov*, pp. 292–354.

Glagolev, S. S. "K voprosu o proiskhozhdenii cheloveka." *Bogoslovskii vestnik*, 1909, January–April, pp. 450–78.

———. "Mendelizm." *Bogoslovskii vestnik*, 1913, May, pp. 250–85; October, pp. 476–505; December, 725–37.

————. "Novoe miroponimanie." *Bogoslovskii vestnik,* 1911, no. 1, pp. 1–42.

————. "Novyi antidarvinisticheskii trud." *Bogoslovskii vestnik,* 1893, no. 8, pp. 478–90.

————. *O proiskhozhdenii i pervobytnom sostoianii roda chelovecheskogo.* Moscow, 1894.

————. "Vzgliad Vasmanna na proiskhozhdenie cheloveka." *Bogoslovskii vestnik,* 1911, April, pp. 621–42; May, pp. 1–16; July–August, pp. 417–50; October, pp. 233–64.

Glass, Bentley, et al., eds. *Forerunners of Darwin: 1745–1859.* Baltimore: Johns Hopkins University Press, 1968.

Glick, Thomas, ed. *The Comparative Reception of Darwinism.* Austin: University of Texas Press, 1974.

G., M. "Voprosy obshchei biologii." *RM,* 1896, no. 5, sec. 2, pp. 159–75.

Goldschmidt, Richard. *The Material Basis of Evolution.* New Haven, Conn.: Yale University Press, 1982.

Gol'dshtein, M. Iu. "O 'nereshennykh problemakh' biologii." *Obr.,* 1904, no. 9, pp. 59–93; no. 10, pp. 18–53.

Gould, Stephen Jay. *Ontogeny and Phylogeny.* Cambridge: Harvard University Press, 1977.

Grigor'ian, N. A. "O pervoi russkoi fiziologicheskoi shkole." *Iz istorii biologii,* 1975, no. 5, pp. 137–51.

Haeckel, Ernst. "Charles Darwin as an Anthropologist." In A. C. Seward, ed., *Darwin in Modern Science* pp. 137–51.

————. *The Evolution of Man,* 2 vols., New York: Appleton, 1879.

[Haeckel, Ernst] "Gekkel', Ernst." *Novyi entsiklopedicheskii slovar'.* St. Petersburg: Brockhaus and Efron, n.d., vol. 12, p. 872.

[Haeckel, Ernst] Gekkel', Ernst. "Mirosozertsanie Darvina i Lamarka." *EiG,* 1910, no. 5, pp. 49–67.

Hartmann, Eduard. *Das Problem des Lebens: Biologische Studien.* Bad Sachsa in Harz, 1906.

————. *Wahrheit und Irrtum im Darwinismus.* Berlin, 1875.

Hertwig, R. *See* Gertvig, R.

Herzen, A. I. *Polnoe sobranie sochinenii i pisem.* Vol. 12. Edited by M. K. Lemke. Petrograd, 1919.

Hodge, M. J. S. "The Development of Darwin's General Biological Theorizing." In D. S. Bendall, ed., *Evolution from Molecules to Men,* pp. 43–62.

Höffding, Harald. *A History of Modern Philosophy.* Translated by B. E. Meyer. Vol. 2. London: Macmillan, 1908.

Holmes, S. J. "K. E. von Baer's Perplexities Over Evolution." *ISIS,* 1947, vol. 37, no. 107–8, p. 7–14.

Hoyer. "Protocolle der Sitzungen der Section für Zoologie und vergleichende Anatomie der V. Versammlung russischer Naturforscher und Aerzte in Warschau in September, 1876." *ZWZ,* 1877, vol. 28, pp. 385–418.

Hull, David L. *Darwin and His Critics: The Reception of Darwin's Theory by the Scientific Community.* Chicago: University of Chicago Press, 1973.

Hulse, James W. *Revolutionists in London.* Oxford: Clarendon Press, 1970.

Huxley, Thomas H. "The Struggle for Existence in Human Society." In Petr Kropotkin, *Mutual Aid,* pp. 329–41.

Iachevskii, A. A., ed. *Iubileinyi sbornik posviashchennyi I. P. Borodinu.* Leningrad, 1927.

Iaroshevskii, N. G. *Ivan Mikhailovich Sechenov.* Leningrad, 1968.

Iasinskii, I. I. "Darvin i Lamark." *Novoe slovo,* 1909, no. 4, pp. 36–39.

Ioffe, A. F., et al., eds. *Nauka i tekhnika v SSSR: 1917–1927.* 3 vols. Moscow, 1927.

Ipatova, N. N. "Vospriiatie sochinenii Ch. Darvina v Rossii (vtoraia polovina XIX veka)." In *Istoriia russkogo chitatelia,* Leningrad, 1982, pp. 129–47.

Iuzhakov, S. N. "Sotsiologicheskaia doktrina N. K. Mikhailovskogo." In D. Mamin-Sibiriak, et al., *Na slavnom postu; literaturnyi sbornik posviashchennyi N. K. Mikhailovskomu,* 2nd ed., St. Petersburg, 1906, pp. 352–69.

———. *Sotsiologicheskie etiudy.* St. Petersburg, 1896.

Ivanov-Razumnik. *Istoriia russkoi obshchestvennoi mysli.* Vol. 2, 3rd ed. St. Petersburg, 1911.

Ivanovskii, D. "Eksperimental'nyi metod v voprosakh evoliutsii." *Izvestiia Varshavskogo universiteta,* 1908, no. 3, pp. 1–40.

Jahn, Ilse, et al., eds. *Geschichte der Biologie: Theorien, Methoden, Institutionen, Kurzbiographhien.* Jena, 1982 (2nd ed., 1985.)

Jepsen, Glenn L., George Gaylord Simpson, and Ernst Mayr, eds. *Genetics, Paleontology, and Evolution.* New York: Atheneum, 1963.

Jones, Greta. *Social Darwinism and English Thought: The Interaction Between Biological and Social Theory.* Brighton, Sussex: Harvester Press, 1980.

Jordanova, L. J. *Lamarck.* Oxford: Oxford University Press, 1984.

Kapterev, P. N. "Teleologiia neolamarkistov." *Bogoslovskii vestnik,* 1914, January, pp. 84–115; March, pp. 561–91; May, pp. 110–39; July–August, pp. 489–507.

Kareev, N. I. *Istoriko-filosofskie i sotsiologicheskie etiudy.* St. Petersburg, 1895.

———. "Novyi istoriko-filosofskii trud." *VFiP,* 1898, no. 45, pp. 388–415.

———. "P. L. Lavrov kak sotsiolog." In E. Radlov, ed. *P. L. Lavrov: stat'i,* pp. 193–248.

———. *Teoriia lichnosti P. L. Lavrova.* St. Petersburg, 1907.

———. *Vvedenie v izuchenie sotsiologii.* St. Petersburg, 1897.

Karpov, V. P. "Naturfilosofiia Aristotelia v nastoiashchee vremia." *VFiP,* 1911, no. 109, pp. 517–97; no. 110, pp. 725–814.

———. *Osnovnye cherty organicheskogo ponimaniia prirody.* Moscow, 1913.

———. "Vitalizm i zadachi nauchnoi biologii v voprose zhizni." *VFiP,* 1909, no. 3, pp. 341–92; no. 4, pp. 523–73.

Kaufman, I. M. *Russkie entsiklopedii.* Vol. 1. Moscow, 1960.

Kellogg, Vernon L. *Darwinism Today.* New York: Henry Holt, 1907.

Kennan, George. "A Visit to Count Tolstoi." *Century Illustrated Monthly Magazine,* 1887, vol. 34, pp. 252–65.

Kessler, K. F. "O zakone vzaimnoi pomoshchi." *TSPO,* 1880, vol. 11, no. 1, pp. 124–34.

Khakhina, L. N. "Eksperimental'nye istoki ucheniia o simbiogeneze (raboty A. S. Famintsyna)." *ITEU,* 1973, vol. 1, pp. 129–41.

———. "K istorii ucheniia o simbiogeneze." *Iz istorii biologii,* 1973, vol. 4, pp. 63–75.

————. "Kontseptsiia A. S. Famintsyna o znachenii simbioza v evoliutsii." In A.S. Kursanov et al., eds., *Andrei Sergeevich Famintsyn*, pp. 165–81.

————. "Problema simbiogeneza. Osnovnye etapy razrabotki problemy." In S. R. Mikulinskii and Iu. I. Polianskii, eds., *Razvitie evoliutsionnoi teorii*, pp. 421–35.

————. "Teoreticheskie vzgliady A. S. Famintsyna na rol' simbioza v evoliutsii organizmov." *ITEU*, 1973, vol. 1, pp. 142–49.

Kharakhorkin, L. R. "Iz istorii ideologicheskoi bor'by vokrug evoliutsionnogo ucheniia Darvina v Rossii." *VF*, 1959, no. 11, pp. 141–47.

Kholodkovskii, N. A. *Biologicheskie ocherki.* Moscow-Petrograd, 1923.

Khrushchov, G. K. "Il'ia Il'ich Mechnikov." In *LRN: biol.*, pp. 192–200.

Khvol'son, O. D. *Gegel', Gekkel', i dvenadtsataia zapovest.* St. Petersburg, 1911.

————. *Populiarnye stat'i i rechi.* Berlin, 1923.

————. "Pozitivnaia filosofiia i fizika." *MB*, 1898, no. 5, pp. 41–58.

Kirpotin, V. *Radikal'nyi raznochinets D. I. Pisarev.* Leningrad, 1929.

Kistiakovskii, B. A. *Sotsial'nye nauki i pravo.* Moscow, 1916.

————. "V zashchitu nauchno-filosofskogo idealizma." *VFiP*, 1907, no. 1, pp. 57–109.

Kiulpe, Osval'd. "Sovremennaia filosofiia v Germanii." *MB*, no. 9, sec. 3, pp. 17–22.

Klein, F. *Elementary Mathematics From An Advanced Point of View.* Translated by E. R. Hedrick and C. A. Noble. New York: Dover, n.d.

Kline, George L. "Darwinism in the Russian Orthodox Church." In Ernest Simmons, ed., *Continuity and Change in Russian and Soviet Thought*, pp. 307–28.

Kniazev, G. A. "Izbranie Ch. Darvina chlenom-korrespondentom Peterburgskoi Akademii nauk." *Pr.*, 1939, no. 11, pp. 117–20.

Kniazev, G. A., and B. E. Raikov, eds. *Pis'ma A. O. Kovalevskogo k I. I. Mechnikovu (1866–1900).* Moscow-Leningrad, 1955.

Knipovich, N. M. "Gekkel', Ernst." Brockhaus-Efron, 1892, vol. 13, p. 264.

————. "Lamarkizm." Brockhaus-Efron, 1895, vol. 17, pp. 290–91.

————. "Podbor." Brockhaus-Efron, 1898, vol. 24, pp. 48–50.

————. "Vzaimootnoshenie organizmov mezhdu soboiu i s okruzhaiushchim mirom." In M. M. Kovalevskii et al., eds., *Itogi nauki . . .*, vol. 6, pp. 385–469.

Kohn, David, ed. *The Darwinian Heritage.* Princeton, N.J.: Princeton University Press, 1985.

Kolosov, E. E. *Ocherki mirovozzreniia N. K. Mikhailovskogo.* St. Petersburg, 1902.

Kol'tsov, N. K. "Eksperimental'naia biologiia v SSSR." In A. F. Ioffe et al., eds., *Nauka i tekhnika . . .*, vol. 2, pp. 37–64.

————. "Ernst Gekkel'," *Pr.*, 1914, no. 2, pp. 201–8.

————. *K universitetskom voprosu.* Moscow, 1909.

————. "Organizatsiia kletki." *Pr.*, 1917, no. 2, pp. 191–212.

————. "Review of E. A. Bogdanov, *Mendelizm.*" *Pr.*, 1914, no. 11, pp. 1390–92.

———. "Review of Gans Drish, *Vitalizm.*" *Pr.*, 1914, no. 12, pp. 1527–31.

———. "Review of *Novye idei v biologii.*" *Pr.*, 1914, no. 1, pp. 121–24.

———. "Vzgliady Lotsi na evoliutsiiu organizmov." *Pr.*, 1915, no. 10, pp. 1253–64.

Kolubovskii, Ia. *Filosofskii ezhegodnik: 1893.* Moscow, 1895.

Komarov, V. L. "Review of V. I. Taliev, *Opyt issledovaniia vidoobrazovaniia.*" *Zhurnal Russkogo Botanicheskogo obshchestva,* 1916, vol. 1, no. 3–4, pp. 215–18.

———. "Vidoobrazovanie." In M. M. Kovalevskii et al., eds., *Itogi nauki . . .* , pp. 509–40.

———. "Zhizn' i tvorchestvo K. A. Timiriazeva." In K. A. Timiriazev, *Izbrannye sochineniia,* vol. 1, pp. 11–73.

Korf, N. "Teoriia Darvina i voprosy pedagogii." *VE*, 1873, no. 5, pp. 275–311.

Korzhinskii, S. I. *Chto takoe zhizn'?* Tomsk, 1888.

———. "Geterogenezis i evoliutsiia. K teorii proiskhozhdeniia vidov, I." *ZAN*, 1899, vol. 9, no. 2, Supplement, pp. 1–94.

———. "Geterogenezis i evoliutsiia. (Predvaritel'noe soobshchenie)." *IIAN*, 1899, vol. 10, no. 3, pp. 255–68.

"Korzhinskii, Sergei Ivanovich." *MBS-AN*, vol. 2, pp. 342–45.

Koshtoiants, Kh. S., ed. *Trudy soveshchaniia po istorii estestvoznaniia, 24–28 dekabria 1946 g.* Moscow, 1948.

Kotliarevskii, Nestor. "Ocherki iz istorii obshchestvennogo nastroeniia shestidesiatykh godov." *VE*, 1912, no. 11, pp. 268–94.

Kovalevskii, A. O. *Izbrannye raboty.* Moscow, 1951.

——— [Kowalevsky]. "Sitzungsberichte der zoologischen Abtheilung der III. Versammlung russischer Naturforscher in Kiev." *ZWZ*, 1872, vol. 22, pp. 283–304.

Kovalevskii, M. M. "Darvinizm v sotsiologii." In M. M. Kovalevskii et al., *Pamiati Darvina,* pp. 117–58.

———. *Ocherki proiskhozhdeniia i razvitiia sem'i i sobstvennosti.* Translated from the French by S. P. Moravskii. Moscow, 1939.

———. *Sotsiologiia.* 2 vols. St. Petersburg, 1910.

Kovalevskii, M. M., et al. *Pamiati Darvina.* Moscow, 1910.

Kovalevskii, M. M., N. N. Lange, N. Morozov, and V. M. Shimkevich, eds. *Itogi nauki v teorii i praktike.* Vols. 5–7. Moscow, 1912.

Kovalevskii, V. O. *Paleontologiia loshadei.* Moscow, 1948.

———. *Sobranie nauchnykh trudov.* 3 vols., Moscow, 1950–60.

[Kovalevskii, V. O.] *The Complete Works of Vladimir Kovalevskii.* Edited by Stephen Jay Gould. New York: ARNO Press, 1980.

Krasnosel'skii, A. I. *Mirovozzrenie gumanista nashego vremeni: osnovy ucheniia N. K. Mikhailovskogo.* St. Petersburg, 1910.

Krasovskii, I. "Noveishaia teoriia proiskhozhdeniia cheloveka." *Dukhovnyi vestnik,* 1865, vol. 4, September, pp. 1–38.

Kreps, M., et al., eds. *Neopublikovannye i maloizvestnye materialy I. P. Pavlova.* Leningrad, 1975.

Kropotkin, P. A. "The Direct Action of Environment and Evolution." *Annual Report Smithsonian Institution,* 1918, pp. 409–27.

————. *Ethics: Origin and Development.* New York: MacVeagh, 1924.

————. "Inherited Variation in Plants." *The Nineteenth Century and After,* 1914, October, pp. 816–36.

————. *Modern Science and Anarchism,* London: Freedom Press, 1912.

————. "The Morality of Nature." *The Nineteenth Century and After,* 1905, vol. 57, no. 337, pp. 407–26.

————. *Mutual Aid: A Factor of Evolution.* Boston: Extending Horizons, n.d.

————. *Revolutionary Pamphlets.* New York: Vanguard Press, 1927.

Kudriavtsev, P. *Absoliutizm ili reliativizm. Opyt istoricheskogo izucheniia chistogo empirizma noveishego vremeni v ego otnoshenii k nravstvennosti i religii.* Vol. 1. Kiev, 1908.

Kudriavtsev-Platonov, V. D. *Sochineniia.* 3 vols. Moscow, 1894.

Kulagin, N. M. "I. I. Mechnikov kak zoolog." *Pr.,* 1915, no. 5, pp. 703–6.

Külpe, O. *See* Kiulpe, O.

Kursanov, A. L., et al., eds. *Andrei Sergeevich Famintsyn: zhizn' i nauchnaia deiatel'nost'.* Leningrad, 1981.

Kuznetsov, F. *Publitsisty 1860kh godov.* Moscow, 1969.

K., V. "Review of V. I. Taliev, *Opyt issledovaniia protsessa vidoobrazovaniia.*" *Pr.,* 1915, no. 11, pp. 1450–51.

Lamarck, J. B. *Zoological Philosophy.* Translated by Hugh Elliot. Chicago: University of Chicago Press, 1984.

Lange, Friedrich Albert. *Geschichte des Materialismus und Kritik seiner Bedeutung in der Gegenwart.* 9th ed. Vol. 2. Leipzig, 1915.

Lankester, E. Ray. "Alexander Kowalevsky." *Nature,* 1902, vol. 66, no. 1712, pp. 394–95.

————. "Elias Metchnikoff." *Nature,* 1916, vol. 97, pp. 443–46.

————. "Heredity and the Direct Action of Environment." *The Nineteenth Century and After,* 1910, September, pp. 483–91.

Lavrenko, E. M. "Pamiati V. I. Talieva." *Botanicheskii zhurnal,* 1957, vol. 42, no. 9, pp. 1331–53.

Lavrov, P. L. *Filosofiia i sotsiologiia.* 2 vols. Moscow, 1965.

————. *Formula progressa N. K. Mikhailovskogo. Protivniki istorii. Nauchnye osnovy istorii tsivilizatsii.* St. Petersburg, 1906.

————. *Izbrannye sochineniia na sotsial'no-politicheskie temy.* 4 vols. Moscow, 1934–35.

————. *Opyt istorii mysli.* Vol. 1, parts 1 and 2. Geneva, 1894–96.

————. *Zadachi ponimaniia istorii.* 2nd ed. St. Petersburg, 1903.

————. *Zadachi pozitivizma i ikh resheniia. Teoretiki sorokovykh godov v nauke i verovaniiakh.* 2nd ed. St. Petersburg, 1906.

[Lavrov, P. L.] "Antropologi v Evrope i ikh sovremennoe znachenie." *OZ,* 1869, no. 3, sec. 2, pp. 1–57.

Lazarev, P. P. "Zakony fiziki i zakony biologii." *Pr.,* 1915, no. 6, pp. 771–84.

Lebedev, A. P. "Uchenie Darvina o proiskhozhdenii mira organicheskogo i cheloveka: filosofsko-kriticheskie etiudy." *RV,* 1873, no. 7, pp. 118–67; no. 8, pp. 431–509.

Lebedev, D. V. "Evoliutsionnye vzgliady V. I. Talieva." *Botanicheskii zhurnal,* 1957, vol. 42, no. 9, pp. 1338–53.

Lebedev, N. "P. A. Kropotkin kak teoretik bio-sotsiologicheskogo zakona vzaimnoi pomoshchi." *Byloe*, 1921, no. 17, pp. 95–99.

"Lektsiia Veismanna." *NO*, 1894, no. 21, pp. 661–64.

Lenin, V. I. *Materialism and Empirio-Criticism*. Moscow, 1970.

———. *Polnoe sobranie sochinenii*. 5th ed. 55 vols. Moscow, 1958–65.

Lenoir, Timothy. "Kant, von Baer und das kausal-historische Denken in der Biologie." *Berichte zur Wissenschaftsgeschichte*, 1985, vol. 8, pp. 99–114.

———. *The Strategy of Life*. Dordrecht, Holland: D. Riedel, 1982.

Leont'ev, K. *Vostok, Rossiia i slavianstvo*. 2 vols. Moscow, 1885.

Lesgaft, P. F. "Nasledstvennost'." *RB*, 1889, no. 9, sec. 2, pp. 33–80; no. 10, sec. 2, pp. 3–59.

———. "Pamiati Zhana Lamarka." In *Pamiati Petra Frantsevicha Lesgafta*, pp. 1–10.

Levin, M. G. *Ocherki po istorii antropologii v Rossii*. Moscow, 1960.

Lindeman, K. "Eshche somneniia v teorii Darvina." *OZ*, 1864, no. 10, pp. 938–48.

Liubimenko, V. N., and I. I. Liubimenko. "Ivan Parfen'evich Borodin. (Biograficheskii ocherk)." In A. A. Iachevskii, ed., *Iubileinyi sbornik* . . . , pp. 3–38.

Lopatin, L. M. "Filosofskoe mirovozzrenie N. V. Bugaeva." *VFP*, 1904, no. 72, pp. 172–95.

———. "K voprosu o matematicheskikh istinakh." *VFiP*, 1892, no. 14, sec. 2, pp. 55–70.

L-r, A. "Prazdnovanie stoletiia so dnia rozhdeniia Charlza Darvina v Khar'kove." *EiG*, 1909, no. 2, p. 87.

Luk'ianov, S. M. "O predelakh tsitologicheskogo issledovaniia pri normal'nykh i patologicheskikh usloviiakh." In R. K. Polenov, ed., *Dnevnik* . . . , pp. 662–78.

Lukina, T. A. "Ber i Darvin." *NiT*, 1977, vol. 9, pp. 111–13.

———. *Boris Evgen'evich Raikov*. Leningrad, 1970.

———, ed. *Kaspiiskaia ekspeditsiia K. M. Bera 1853–1857 gg.: dnevniki i materialy*. Leningrad, 1984.

Lunkevich, V. V. "Interesnaia kniga." *RB*, 1899, no. 4, sec. 2, pp. 30–54.

———. *Nereshennye problemy biologii*. St. Petersburg, 1904.

———. *N. K. Mikhailovskii*. Moscow, 1906.

———. *Osnovy zhizni*. 2 vols. 3rd ed. St. Petersburg, 1910.

———. "Staroe i novoe v evoliutsionnoi teorii." *RB*, 1910, no. 1, pp. 133–56; no. 6, pp. 100–23.

Lyell, Charles. "On the Occurrence of Works of Human Art in Post-Pliocene Deposits." *Report of the Twenty-Ninth Meeting of the British Association for the Advancement of Science, 1859*. London: Murray, 1860. Pp. 93–95.

McLean, Hugh. *Nikolai Leskov: The Man and His Art*. Cambridge: Harvard University Press, 1977.

MacMaster, Robert E. *Danilevsky, a Russian Totalitarian Philosopher*. Cambridge: Harvard University Press, 1967.

Maksimov, N. A. "Ocherk istorii fiziologii rastenii v Rossii." *TIIE*, 1947, vol. 1, pp. 21–79.

Maleshevskii, A. "Review of E. Baur, *Vvedenie v eksperimental'noe izuchenie naseledstvennosti.*" *Pr.*, 1914, no. 11, pp. 1389–90.

Malevanskii, G. "O dukhovnoi prirode cheloveka." *Pravoslavnoe obozrenie,* 1873, vol. 1, pp. 537–68, 721–69.

Manoilenko, K. V., and L. N. Khakhina. "Iz istorii razvitiia evoliutsionnoi teorii v Akademii nauk i vklad akademika A. S. Famintsyna." *ZhOB,* 1974, vol. 35, no. 2, pp. 308–14.

Matveev, B. S. "Aleksei Nikolaevich Severtsov." *LRN: biol.,* pp. 330–36.

———. "Mikhail Aleksandrovich Menzbir." *LRN: biol.,* pp. 268–73.

———. "Progress i regress v evoliutsii." *Pr.,* 1967, no. 5, pp. 71–78.

Matveev, B. S., and A. N. Druzhinin. "The Life and Work of Professor A. N. Severtsov." In I. I. Shmal'gauzen, ed., *Pamiati akademika A. N. Severtsova,* vol. 1, pp. 31–54.

Mayr, Ernst. "The Concept of Finality in Darwin and After Darwin." *Scientia,* 1983, vol. 118, pp. 97–117.

———. *The Growth of Biological Thought: Diversity, Evolution, and Inheritance.* Cambridge: Harvard University Press, 1982.

———. *Populations, Species, and Evolution.* Cambridge: Harvard University Press, 1970.

———. "The Triumph of Evolutionary Synthesis." *The Times Literary Supplement,* 1984, no. 4,257 (November, 2), pp. 1261–62.

Mayr, Ernst, and William P. Provine, eds. *The Evolutionary Synthesis: Perspectives on the Unification of Biology.* Cambridge: Harvard University Press, 1980.

Mazurmovich, B. N. "Materialy k biografii Nikolaia Vasil'evicha Bobretskogo." *TIIE-T,* 1960, vol. 31, pp. 304–25.

Mechnikov, I. I. *Akademicheskoe sobranie sochinenii.* 16 vols. Moscow, 1950–60.

———. "Antropologiia i darvinizm." *VE,* 1875, no. 1, pp. 159–95.

———. "Bor'ba za sushchestvovanie v obshirnom smysle." *VE,* 1878, no. 7, pp. 9–47.

———. "Darvinizm i meditsina." In M. M. Kovalevskii et al., *Pamiati Darvina,* pp. 112–16.

———. *Etiudy optimizma.* Moscow, 1964.

———. *Izbrannye biologicheskie proizvedeniia.* Moscow, 1950.

———. *Izbrannye proizvedeniia.* Moscow, 1956.

———. *Izbrannye raboty po darvinizmu.* Moscow, 1958.

———. *O darvinizme: sbornik statei.* Moscow-Leningrad, 1943.

———. *Sorok let iskaniia ratsional'nogo mirovozzreniia.* Moscow, 1913.

———. "Sovremennoe sostoianie nauki o razvitii zhivotnykh." *ZhMNP,* 1869 (March), vol. 142, pp. 158–86.

———. *Stranitsy vospominanii.* Moscow, 1946.

———. "Zadachi sovremennoi biologii." *VE,* 1871, no. 4, pp. 742–70.

———. "Zakon zhizni." *VE,* 1891, no. 9, pp. 228–60.

——— [Metchnikoff, Elie]. *Lectures on the Comparative Pathology of Inflammation.* Translated from the French by F. A. Starling and E. H. Starling. New York: Dover, 1968.

Menzbir, M. A. "Charlz Darvin i sovremennoe sostoianie evoliutsionnogo ucheniia." *RM*, 1882, no. 7, pp. 53–79.

———. "Estestvennyi i iskustvennyi podbor po otnosheniiu k cheloveku." In M. M. Kovalevskii et al., *Pamiati Darvina*, pp. 161–206.

———. "Evoliutsonnoe uchenie." Granat-*Ents.*, vol. 50, pp. 628–58.

———. "Glavneishie predstaviteli darvinizma v Zapadnoi Evrope: I. Al'fred Uolles." *RM*, 1900, no. 1, sec. 2, pp. 60–76.

———. "Glavneishie predstaviteli darvinizma v Zapadnoi Evrope: II. G. D. Romanes." *RM*, 1900, no. 2, pp. 45–57.

———. "Glavneishie predstaviteli darvinizma v Zapadnoi Evrope: III. Ernst Gekkel'." *RM*, 1900, no. 6, sec. 2, pp. 1–17.

———. "Glavneishie predstaviteli darvinizma v Zapadnoi Evrope: IV. Avgust Veismann." *RM*, 1900, no. 12, sec. 2, pp. 39–57.

———. "Istoricheskii ocherk vozzrenii na prirodu." *VFiP*, 1891, no. 10, sec. 1, pp. 151–74; no. 11, sec. 1, pp. 75–97.

———. "Korennoi vopros antropologii." *RM*, 1893, no. 7, sec. 2, pp. 112–42.

———. "Mnimyi krizis darvinizma." *RM*, 1902, no. 11, pp. 189–201.

———. "Ocherk uspekhov biologii v XIX stoletii." *RM*, 1901, no. 1, sec. 1, pp. 75–92.

———. "Opyt teorii nasledstvennosti." *RM*, 1893, no. 10, pp. 208–31.

———. "P. A. Kropotkin kak biolog." In A. Borovoi and N. Lebedev, eds., *Sbornik statei* . . . , pp. 99–107.

———. "Sovremennye zadachi biologii." *RM*, 1891, no. 9, pp. 172–87.

———. *Vvedenie v izuchenie zoologii i sravnitel'noi anatomii.* Moscow, 1897. (3rd ed., 1906.)

———. *Za Darvina.* Moscow-Leningrad, 1927.

Mikhailov, V. L. "Mekhanizm i vitalizm." *NO*, 1894, no. 52, pp. 1646–63.

Mikhailovskii, N. K. "Literatura i zhizn'." *RB*, 1895, no. 1, sec. 2, pp. 124–54; 1898, no. 2, sec. 2, pp. 132–62.

———. *Literaturnye vospominaniia i sovremennaia smuta.* Vol. 2. St. Petersburg, 1900.

———. *Polnoe sobranie sochinenii.* Vol. 1, 5th ed. St. Petersburg, 1911.

———. *Polnoe sobranie sochinenii.* Vols. 2, 4, and 5, 4th ed. St. Petersburg, 1907–9.

———. *Polnoe sobranie sochinenii.* Vol. 6, 2nd ed. St. Petersburg, 1909.

———. *Sochineniia.* 6 vols. St. Petersburg, 1896–98.

Mikulinskii, S. R. "Karl Frantsovich Rul'e." *LRN: biol.*, pp. 89–104.

———. *Razvitie obshchikh problem biologii v Rossii.* Moscow, 1961.

———. "Vzgliady K. M. Bera na evoliutsiiu v dodarvinskii period." *Annaly biologii*, 1959, vol. 1, pp. 287–362.

———, ed. *Istoriia biologii s drevneishikh vremen do nachala XX veka.* Moscow, 1972.

Mikulinskii, S. R., and A. P. Iushkevich, eds. *Razvitie estestvoznania v Rossii (XVII–nachalo XX veka).* Moscow, 1977.

Mikulinskii, S. R., and V. I. Nazarov. "Razvitie idei evoliutsii organicheskogo mira." In Bliakher, L. Ia., ed., *Istoriia biologii*, pp. 259–83.

Mikulinskii, S. R., and Iu. I. Polianskii, eds., *Razvitie evoliutsionnoi teorii v SSSR (1917–1970-e gody).* Leningrad, 1983.

Miliukov, P. "Universitety." Brockhaus-Efron, vol. 34, pp. 788–800.

Miliukov, P., Charles Seignobos, and L. Eisenmann. *History of Russia.* Vol. 3. New York: Funk & Wagnalls, 1969.

Miller, Martin A. *Kropotkin.* Chicago: University of Chicago Press, 1976.

Mirzoian, E. N. "Evoliutsionno-biokhimicheskie vzgliady A. S. Famintsyna v sviazi s ego filosofskimi i obshchebiologicheskimi vozzreniiami." In A. L. Kursanov et al., eds., *Andrei Sergeevich Famintsyn,* pp. 150–64.

———. "Istoriko-nauchnyi analiz i razvitie evoliutsionnogo napravleniia v morfologii." *Iz istorii biologii,* 1970, no. 2, pp. 123–43.

———. "K voprosu o formirovanii evoliutsionnykh vzgliadov A. O. Kovalevskogo." *TIIE-T,* 1962, vol. 40, pp. 283–91.

———. "Printsip istorizma i nekotorye voprosy razvitiia teorii ontogeneza." *Iz istorii biologii,* 1971, vol. 3, pp. 59–73.

———. *Razvitie sravnitel'no-evoliutsionnoi biokhimii v Rossii.* Moscow, 1984.

———. "Zakonomernosti evoliutsii ontogeneza i otnoshenii ontogeneza i filogeneza." In S.R. Mikulinskii and Iu. I. Polianskii, eds., *Razvitie evoliutsionnoi teorii,* pp. 380–92.

Mishchenko, P. "Review of N. Vavilov, *Genetika i ee otnoshenie k agronomii.*" *TBPB,* 1913, vol. 6, no. 1, p. 63.

Modestov, V. "Russkaia nauka v poslednie dvadtsat-piat let." *RM,* 1890, no. 5, sec. 2, pp. 73–91.

Modzalevskii, B. L., ed. *Perepiska L. N. Tolstogo s N. N. Strakhovym.* St. Petersburg, 1914.

Mokievskii, P. "Lavrov kak filosof." In E. Radlov, ed., *P. L. Lavrov: stat'i,* pp. 29–72.

Monod, Jacques. *Chance and Necessity.* New York: Vintage Books, 1972.

Morgan, Thomas Hunt. *Experimental Zoology.* New York: Macmillan, 1907.

Moser, A. Charles. *Antinihilism in the Russian Novel of the 1860s.* The Hague, Holland: Mouton, 1964.

"Moskovskoe Obshchestvo rasprostraneniia estestvennonauchnykh znanii." *EiG,* 1910, no. 7, p. 88.

Moskvin, A. "Charlz Robert Darvin." *Delo,* 1882, no. 2, sec. 2, pp. 1–21.

M-v, P. "Razvitie vyrazitel'nosti." *Znanie,* 1873, no. 1, sec. 3, pp. 1–26.

Naccache, Bernard. *Marx Critique de Darwin.* Paris, 1980.

Nasmyth, George. *Social Progress and the Darwinian Theory.* New York: Putnam, 1916.

Naturalist', "Strannoe napadenie na darvinizm." *RB,* 1889, no. 2, sec. 2, pp. 230–51.

Navashin, N. "Tsinger, Nikolai Vasil'evich." Granat-*Ents.,* vol. 45, pp. 432–33.

Nekrasov, A. D., and N. M. Artemov. "Aleksandr Onufrievich Kovalevskii." In A. O. Kovalevskii, *Izbrannye raboty,* pp. 536–621.

Nekrasov, P. A. *Moskovskaia filosofsko-matematicheskaia shkola i ee osnovateli.* Moscow, 1904.

Nemilov, A. V. "Analogii mezhdu iavleniiami zhivoi i mertvoi prirody i proiskhozhdenie zhizni na zemle." In M. M. Kovalevskii et al., eds., *Itogi nauki* . . . , vol. 5, pp. 1–48.

[Nikanor]. "Pouchenie na novyi god." *Strannik*, 1884, vol. 1, pp. 27–42.

Nikitin, S. N. "Darvinizm i vopros o vide v oblasti sovremennoi paleontologii." *Mysl'*, 1881, no. 7, pp. 144–70; no. 9, pp. 229–45.

Nikol'skii, B. V. *Nikolai Nikolaevich Strakhov: kritiko-biograficheskii ocherk.* St. Petersburg, 1896.

N. N. "O razbore Darvinovoi sistemy. (Pismo estestvoispytateleia)." In Mikhail Pogodin, *Prostaia rech'* . . . , pp. 436–47.

N. N. "Religiia i teoriia razvitiia." *Mysl'*, 1881, no. 4, pp. 218–26.

Noetzel, Hermann-Gerd. *Petr L. Lavrovs Vorstellungen vom Fortschritt für Russland aus den Jahren vor seiner Emigration.* Köln, 1968.

Nordenskiöld, Erik. *The History of Biology.* New York: Alfred Knopf, 1929.

"Notes." *Nature*, 1916, vol. 97, p. 386.

Novgorodtsev, P. I. "Nravstvennyi idealizm v filosofii prava." In P. I. Novgorodtsev, ed., *Problemy idealizma*, pp. 236–96.

———. *Ob obshchestvennom ideale.* 3rd ed. Vol. 1. Moscow, 1921.

———, ed. *Problemy idealizma.* Moscow, 1902.

Novicow, J. *La critique du Darwinisme social.* Paris, 1910.

Novikov, P. A. "Zoologiia." In L. Ia. Bliakher et al., eds., *Istoriia estestvoznaniia v Rossii*, pp. 272–346.

Nozhin, N. D. "Nasha nauka i uchenye." *Knizhnyi vestnik*, 1866, no. 1, pp. 18–21; no. 2, pp. 51–54; no. 3, pp. 73–79; no. 7, pp. 173–78.

Ognev, I. F. "Rech' E. diu-Bua-Reimona i ego nauchnoe mirovozzrenie." In A. I. Bachinskii et al., *Sbornik* . . . , pp. 42–70.

———. "Vitalizm v sovremennom estestvoznanii." *VFiP*, 1900, no. 4, pp. 686–730.

Ognev, S. I. "Mikhail Aleksandrovich Menzbir." *BMOIP: otdel biologii*, 1946, vol. 51, no. 1, pp. 5–15.

O., L. "Charlz Robert Darvin." *Mysl'*, 1881, no. 7, pp. 112–18.

Omelianskii, V. L. "Razvitie estestvoznaniia v Rossii v poslednuiu chetvert veka." In *Istoriia Rossii v XIX veke*, vol. 9. St. Petersburg: Granat, n.d. Pp. 116–44.

Oppenheimer, Jane. "An Embryological Enigma in the *Origin of Species*." In B. Glass et al., eds., *Forerunners of Darwin: 1745–1859*, pp. 292–322.

———. *Essays in the History of Embryology and Biology.* Cambridge: M.I.T. Press, 1967.

———. "Review of B. E. Raikov, *Karl von Baer*." *Bulletin of the History of Medicine*, 1971, vol. 45, pp. 85–88.

Osborn, H. F. *The Age of Mammals in Europe, Asia and North America.* New York: Macmillan, 1910.

Ostrianin, D. F. *Bor'ba za materializm i dialektiku v otechestvennom estestvoznanii.* Kiev, 1981.

Ostroumov, A. A. *Proiskhozhdenie vidov i estestvennyi podbor: k 50-letiiu darvinizma.* Kazan, 1909.

Ovsianiko-Kulikovskii, D. N. *Istoriia russkoi intelligentsii.* Vol. 2. Moscow, 1907.

Pamiati Petra Frantsevicha Lesgafta. St. Petersburg, 1912.

Pamiati Dmitriia Iosifovicha Ivanovskogo. Moscow, 1952.

Panteleev, L. F. *Iz vospominanii proshlogo.* 2 vols. St. Petersburg, 1905–1908.

Pauly, August. *Darwinismus und Lamarckismus.* Munich, 1905.

Pavlov, I. P. *Dvadtsatiletnii opyt ob'ektivnogo izucheniia vysshei nervnoi deiatel'nosti (povedeniia) zhivotnykh.* 7th ed. Moscow, 1951.

———. *Polnoe sobranie sochinenii.* Vol. 6. Moscow-Leningrad, 1952.

Pavlovskii, E. N. "Nikolai Aleksandrovich Kholodkovskii." *LRN: biol.,* pp. 313–18.

Peirce, Charles S. *Selected Writings (Value and Universe of Chance).* Edited by P. P. Wiener. New York: Dover, 1966.

"Pervoe prisuzhdenie premii tainogo sovetnika Bera." *ZAN,* 1867, vol. 11, pp. 147–92.

Petrovskii, A. V. *Voprosy istorii i teorii psikhologii: Izbrannye trudy.* Moscow, 1984.

Pisarev, D. I. *Izbrannye filosofskie i obshchestvenno-politicheskie stat'i.* Moscow, 1949.

———. *Selected Philosophical, Social and Political Essays.* Moscow, 1958.

———. *Sochineniia.* Vol. 3. St. Petersburg, 1894.

Plate, Ludwig. *Selektionsprinzip und Probleme der Artbildung: Ein Handbuch des Darwinismus.* 4th ed. Leipzig, 1913.

———, ed. *Ultramontane Weltanschauung und moderne Lebenskunde, Orthodoxie und Monismus.* Jena, 1907.

Plekhanov, G. V. *The Development of the Monist View of History.* Moscow, 1956.

———. *Essays in the History of Materialism.* Translated by Ralph Fox. New York: Fertig, 1967.

———. *Filosofsko-literaturnoe nasledie.* Vol. 3. Moscow, 1974.

———. *Fundamental Problems of Marxism.* New York: International Publishers, 1969.

———. *Fundamental Problems of Marxism.* Translated from the Russian by Julius Katzer. Moscow, 1974.

———. *Izbrannye filosofskie proizvedeniia.* 5 vols. Moscow, 1956–58.

———. *Selected Philosophical Works.* 5 vols. Moscow, 1974–81.

———. *Unaddressed Letters and Art and Social Life.* Moscow, 1957.

———. *Za dvadtsat' let.* 3rd ed. St. Petersburg, 1909.

Pogodin, Mikhail. *Prostaia rech' o mudrenykh veshchakh.* Moscow, 1873.

Polenov, B. K., ed. *Dnevnik XI S'ezda Russkikh estestvoispytatelei i vrachei,* no. 5. St. Petersburg, 1902.

Polevoi, V. V. "A. S. Famintsyn i fiziologiia rastenii v Peterburgskom-Leningradskom universitete." In A. L. Kursanov et al., eds., *Andrei Sergeevich Famintsyn,* pp. 56–85.

Poliakov, I. M. "Razrabotka osnovnykh problem darvinizma v trudakh I. I. Mechnikova." In I. I. Mechnikov, *Akademicheskoe sobranie sochinenii,* vol. 4, pp. 409–66.

Polianskii, Iu. I. "Osnovnye puti razvitiia darvinizma." In V. I. Polianskii and Ia. I. Polianskii, eds., *Istoriia . . . ,* pp. 286–318.

———. "Vydaiushchiisia russkii morfolog-darvinist." *Arkhiv anatomii, gistologii i embriologii,* 1963, no. 12, pp. 59–68.

Polianskii, Iu. I. et al., eds. *Pis'ma A. O. Kovalevskogo k I. I. Mechnikovu (1866–1900).* Moscow-Leningrad, 1955.

Polianskii, V. I., and Iu. I. Polianskii, eds. *Istoriia evoliutsionnykh uchenii v biologii.* Moscow-Leningrad, 1966.

Polynin, V. *Prorok v svoem otechestve.* Moscow, 1969.

Popov, L. "Charlz Robert Darvin." *Nabliudatel',* 1882, no. 7, pp. 68–102.

[Popov, L. K.] El'pe. "Posmertnyi trud." *Novoe vremia,* 1889, nos. 3525 and 3551.

"Pravitel'stvennye rasporiazheniia." *ZhMNP,* 1886, vol. 243 (Jan.), sec. 1, pp. 3–53.

Predtechenskii, A. V., and A. V. Kol'tsov. "Iz istorii Akademii nauk v period revoliutsii 1905–1907 godov." *VAN,* 1955, no. 3, pp. 82–89.

Priima, F. Ia., and N. I. Prutskov, eds. *Istoriia russkoi literatury.* Vol. 3. Leningrad, 1982.

"Protokoly." *ZAN,* 1867, vol. 11, pp. 76–79.

Protokoly VII S'ezda Russkikh estestvoispytatelei i vrachei v Odesse. Vol. 1. Odessa, 1883.

Puzanov, I. I. "M. A. Menzbir kak zoogeograf." *BMOIP: otdel biologii,* 1946, vol. 51, no. 1, pp. 16–31.

———. "Osnovopolozhniki russkoi zoogeografii (N. A. Severtsov, M. A. Menzbir, P. P. Sushkin)." In Kh. S. Koshtoiants, ed., *Trudy soveshchaniia . . . ,* pp. 286–309.

Quatrefages, A. de. *Les émules de Darwin.* 2 vols. Paris, 1894.

R. "Morgan, Lewis Henry." Brockhaus-Efron, vol. 19, 1896, pp. 833–35.

Rachinskii, S. "Tsvety i nasekomye." *RV,* 1863, no. 1, pp. 347–96.

Rádl, Emanuel. *The History of Biological Theories.* Translated from the German by E. J. Hatfield. London: Oxford University Press, 1930.

Radlov, E. "Review of Trudy S.-Peterburgskogo Filosofskogo obshchestva." *ZhMNP,* 1902, vol. 344, sec. 4, pp. 244–48.

———, ed. *P. L. Lavrov: stat'i, vospominaniia, materialy.* Petrograd, 1922.

Raikov, B. E. *Grigorii Efimovich Shchurovskii.* Moscow, 1965.

———. "Iz istorii darvinizma v Rossii." *TIIE-T,* part 1: 1957, vol. 16, pp. 3–33; part 2: 1960, vol. 31, pp. 17–81.

———. "Iz istorii zoologii v Leningradskom universitete." *Vestnik Leningradskogo gosudarstvennogo universiteta,* 1953, no. 4, pp. 73–86.

———. *Karl Ber.* Moscow-Leningrad, 1961.

———. *Russkie biologi-evoliutsionisty do Darvina.* 4 vols. Moscow-Leningrad, 1951–59.

———. "V. I. Shmankevich i ego raboty o vliianii sredy na organizm." *TIIE,* 1953, vol. 5, pp. 245–72.

Ranskii, S. P. *Sotsiologiia N. K. Mikhailovskogo.* St. Petersburg, 1901.

Regel', R. E. "K voprosu o vidoobrazovanii." *TBPB,* 1917, vol. 10, no. 1, pp. 157–81.

———. "Selektsiia s nauchnoi tochki zreniia." *TBPB,* 1912, no. 11, pp. 425–540.

Ren'iar, A. "Nauka i literatura v sovremennoi Anglii." *VE*, 1875, no. 1, pp. 330–70; no. 7, pp. 180–97.

Rensch, Bernhard. *Evolution Above the Species Level.* New York: Columbia University Press, 1959.

"Review of Charlz Darvin, *Proiskhozhdenie cheloveka.*" *Znanie*, 1871, no. 10, sec. 2, pp. 39–40.

"Review of Ch. Darvin, *Proiskhozhdenie vidov.*" *RB*, no. 2, sec. 2, pp. 36–37.

"Review of *Mendelizm* by R. K. Punnett." *Pr.*, 1913, no. 3, pp. 389–91.

"Review of N. Ia. Danilevskii, *Darvinizm.*" *ZhMNP*, 1886, vol. 243, January, sec. 2, pp. 186–87.

"Review of *Sovremennye znaniia o filogeneticheskom razvitii cheloveka.*" *RM*, 1900, no. 2, sec. 2, p. 61.

Rogers, James Allen. "Charles Darwin and Russian Scientists." *Russian Review*, 1960, vol. 19, pp. 371–83.

———. "Darwinism, Scientism, and Nihilism." *Russian Review*, 1960, vol. 19, pp. 10–23.

———. "Marxist and Russian Darwinism." *Jahrbücher für Geschichte Osteuropas*, 1965, n.s. vol. 13, no. 2, pp. 199–211.

———. "The Reception of Darwin's *Origin of Species* by Russian Scientists." *ISIS*, 1973, vol. 64, pp. 484–503.

———. "Russian Opposition to Darwinism in the Nineteenth Century," *ISIS*, 1974, vol. 65, pp. 487–505.

———. "Russia: Social Sciences." In T. I. Glick, ed., *The Comparative Reception of Darwinism*, pp. 256–68.

———. "The Russian Populists' Response to Darwin." *Slavic Review*, 1963, vol. 22, no. 3, pp. 456–68.

Romanes, George John. *Mental Evolution in Animals.* London: Kegan Paul, 1883.

Rosina, N. Ia. *Mikhail Aleksandrovich Menzbir.* Leningrad, 1985.

Royce, Josiah. *The Spirit of Modern Philosophy.* New York: Dover, 1983.

Rozanov, V. V. *O ponimanii. Opyt issledovaniia prirody, granits i vnutrennego stroeniia nauki kak tsel'nogo znaniia.* Moscow, 1886.

———. "Organicheskii protsess i mekhanicheskaia prichinost'." *ZhMNP*, 1889, vol. 236, no. 4, sec. 2, pp. 1–22.

———. *Priroda i istoriia.* St. Petersburg, 1900.

———. "Vopros o proiskhozhdenii organizmov: Review of N. Ia. Danilevskii, *Darvinizm.*" *RV*, 1889, no. 5, pp. 311–16.

Rudnitskaia, E. L. "Pis'ma A. O. Kovalevskogo k N. D. Nozhinu." *TIIE-T*, 1961, vol. 36, pp. 202–15.

Rul'e, K. F. *Izbrannye biologicheskie proizvedeniia.* Moscow, 1954.

Rumiantsev, N. "Darvinizm. (Kriticheskoe issledovanie)." *Vera i razum*, 1895, no. 1, part 2, pp. 1–31; no. 3, part 2, pp. 135–70; no. 6, part 2, pp. 291–302; no. 10, part 2, pp. 488–504; 1896, no. 1, part 2, pp. 25–45; no. 2, part 2, 90–110; no. 3, part 2, pp. 139–60; no. 5, part 2, pp. 240–68.

Safronov, B. G. *M. M. Kovalevskii kak sotsiolog.* Moscow, 1960.

Schönemann, J. "Teoriia Darvina," *Zagranichnyi vestnik*, 1864, no. 2, pp. 211–41.

Scudo, Francesco M., and Michele Acanfora. "Darwin and Russian Evolutionary Biology." In David Kohn, ed., *The Darwinian Heritage*, pp. 731–52.

Sechenov, I. M. *Selected Physiological and Psychological Works*. Translated by S. Belsky. Moscow, n.d.

Seidlitz, Georg. *Beiträge zur Descendenz-Theorie*. Leipzig, 1876.

———. *Die Darwin'sche Theorie*. Dorpat, 1871. 2nd ed., 1876.

Semenov, P. "O trude Nikolaia Iakovlevicha Danilevskogo *Darvinizm, kriticheskoe issledovanie*." *RV*, 1886, no. 12, pp. 783–88.

Serebrovskii, A. S. "Sovremennoe sostoianie teorii mutatsii." *Pr.*, 1915, no. 10, pp. 1239–54.

Sergeevich, V. I. "Issledovanie g. Zatyrkevicha v oblasti do-mongol'skogo perioda russkoi istorii." *ZhMNP*, 1876, vol. 183, no. 1, sec. 1, pp. 204–35.

Sergievskii, N. "O bibleiskoi istorii tvoreniia." *Pravoslavnoe obozrenie*, 1873, no. 2, pp. 187–205; no. 3, pp. 367–400.

Severtsov, A. N. "Evoliutsiia i embriologiia." In *Dnevnik XII-go S'ezda*, pp. 262–75.

———. *Evoliutsiia i psikhika*. Moscow, 1922.

———. *Morfologicheskie zakonomernosti evoliutsii*. Moscow-Leningrad, 1939.

———. *Sobranie sochinenii*. Vol. 3. Moscow-Leningrad, 1945.

[Severtsov, A. N.] Sewertzoff, A. N. *Morphologische Gesetzmässigkeiten der Evolution*. Jena, 1931.

Severtsov, N. A. *Puteshestviia po turkestanskomu kraiu*. 2nd ed. Moscow, 1947.

Severtsova, L. B. *Aleksei Nikolaevich Severtsov*. Moscow-Leningrad, 1946.

———. "Aleksei Nikolaevich Severtsov: biograficheskii ocherk." In A. N. Severtsov, *Morfologicheskie zakonomernosti evoliutsii*, pp. 11–52.

Seward, A. C., ed. *Darwin in Modern Science*. Cambridge University Press, 1909.

Shchapov, A. P. *Sochineniia*. Vol. 3. St. Petersburg, 1908.

———. *Sotsial'no-pedagogicheskie usloviia umstvennogo razvitiia russkogo naroda*. St. Petersburg, 1870.

Shchelkanovtsev, Ia. P. *Kratkii kurs zoologii*. 3rd ed. Moscow, 1917.

Shcherbakova, A. A. "A. N. Beketov, ego raboty i mirovozzrenie." *TIIE-T*, 1953, vol. 5, pp. 211–44.

Shcherbakova, A. A., et al. *Istoriia botaniki v Rossii (darvinskii period, 1861–1917)*. Novosibirsk, 1983.

Shchurovskii, G. "Ob obshchedostupnosti ili populiarizatsii estestvennykh nauk." In *Trudy I S'ezda . . .*, pp. 7–17.

Shelgunov, N. V. *Ocherki russkoi zhizni*. St. Petersburg, 1895.

Shimkevich, V. M. "A. O. Kovalevskii i ego trudy." *Dnevnik XI-go S'ezda*, pp. 655–62.

———. "A. O. Kovalevskii. (Nekrolog)." *Obr.*, 1901, no. 11, pp. 107–14.

———. *Biologicheskie osnovy zoologii*. 3rd ed. St. Petersburg-Moscow, 1907.

———. "Embrional'nye plasty i teoriia mutatsii." *Dnevnik XII-go S'ezda*, part 1, pp. 31–49.

———. "Evoliutsionnaia ideia v ee estestvenno-istoricheskom dvizhenii." *Severnyi vestnik*, 1895, no. 7, pp. 167–84.

———. "K teorii mutatsii." *TSPO*, 1906, vol. 35, no. 4, pp. 28–62.

———. "K voprosu o passivnykh strukturakh." *NO*, 1894, no. 18, pp. 553–55.

———. "Nasledstvennost'." In M. M. Kovalevskii et al., eds., *Itogi* . . . , vol. 5, pp. 338–76.

———. "Polemika po voprosu o nasledstvennosti." *Severnyi vestnik*, 1895, no. 3, pp. 94–105.

———. "Proiskhozhdenie pozvonochnykh." In M. M. Kovalevskii et al., eds., *Itogi* . . . , vol. 6, pp. 613–18.

———. "Spenser i Veismann k voprosakh o nasledstvennosti." *NO*, 1894, no. 15, pp. 459–65; no. 16, pp. 486–97.

———. "Teoriia mutatsii." *NS*, 1905, no. 5, pp. 16–28.

———. "Transformizm." Brockhaus-Efron, 1901, vol. 33, pp. 714–21.

———. "Zadachi sovremennoi biologii." *ZhMNP*, 1890, vol. 267, no. 1, sec. 2, pp. 72–94.

Shmal'gauzen, I. F. *O rastitel'nykh pomesiakh: nabliudeniia v peterburgskoi flory*. St. Petersburg, 1874.

Shmal'gauzen, I. I. "Nauchnaia deiatel'nost' A. N. Severtsova kak teoretika-evoliutsionista." In I. I. Shmal'gauzen, ed., *Pamiati akademika A. N. Severtsova*, vol. 1, pp. 55–61.

———. "Novoe v sovremennom darvinizme." *Pr.*, 1947, no. 12, pp. 31–44.

———. *Problemy darvinizma*. Moscow, 1946.

———. *Puti zakonomernosti evoliutsionnogo protsessa*. Moscow-Leningrad, 1940.

———, ed. *Pamiati akademika A. N. Severtsova*. 2 vols. Moscow-Leningrad, 1939.

[Shmankevich, V. M.] Schmankewitsch, W. "Über das Verhältniss der *Artemia salina* Miln. Edw. zur *Artemia Mühlhausenii* und dem Genus *Branchipus Schaeff*." *ZWZ*, 1875, vol. 25, pp. 103–16.

———. "Zur Kentniss des Einflusses der äusseren Lebensbedingungen auf die Organisation der Thiere." *ZWZ*, 1877, vol. 29, pp. 429–93.

Shmidt, G. "Review of A. N. Severtsov, *Morfologicheskie zakonomernosti evoliutsii*." *Sovetskaia nauka*, 1940, no. 6, pp. 145–52.

Shul'ts, E. A. "Irratsional'noe v biologii." *Pr.*, 1915, no. 6, pp. 783–92.

———. "Review of Lamark, *Filosofiia zoologii*." *Pr.*, 1912, no. 6, pp. 847–48.

Simmons, Ernest J., ed. *Continuity and Change in Russian and Soviet Thought*. Cambridge: Harvard University Press, 1955.

Simpson, George Gaylord. *Horses*. New York: Oxford University Press, 1951.

———. *The Meaning of Evolution*. New York: Bantam Press, 1967.

———. *Tempo and Mode in Evolution*. New York: Columbia University Press, 1944.

———. *This View of Life: The World of an Evolutionist*. New York: Harcourt, Brace & World, 1964.

Skabichevskii, A. M. *Istoriia russkoi literatury: 1848–1908 gg*. 7th ed. St. Petersburg, 1909.

———. *Sochineniia*. 2 vols. St. Petersburg, 1890.

Slonimskii, L. "Kniga o P. F. Lesgafte." *VE*, 1912, no. 6, pp. 350–57.

Smith, John Maynard. *The Theory of Evolution*. Baltimore: Penguin, 1958.

Sobol', S. L. "Bor'ba M. A. Menzbira za darvinizm." *BMOIP: otdel biologii*, 1946, vol. 51, no. 1, pp. 32–43.

———. "Evoliutsionnye vozzreniia D. I. Ivanovskogo." In *Pamiati Dmitriia Iosifovicha Ivanovskogo,* pp. 37–51.

———. "Iz istorii bor'by za darvinizm v Rossii." *TIIE-T,* 1957, vol. 14, pp. 195–226.

———. "Ob izdanii Charlza Darvina v SSSR." *VIE-T,* 1956, no. 1, p. 223.

———. "Pervye soobshcheniia o teorii Ch. Darvina v russkoi pechati." *BMOIP: otdel biologii,* 1945, vol. 50, no. 3–4, pp. 128–37.

———. "Voznikovenie i razvitie materialisticheskoi traditsii v russkoi evo- liutsionnoi mysli XVIII i pervoi poloviny XIX veka." *TIIE,* 1953, vol. 5, pp. 23–50.

Solov'ev, E. D. I. *Pisarev.* Berlin-Petrograd-Moscow, 1922.

[Solov'ev, E.] Andreevich, E., *Ocherki po istorii russkoi literatury XIX veka,* St. Petersburg, 1903.

———. *Opyt filosofii russkoi literatury.* St. Petersburg, 1905.

Solov'ev, N. M. *Nauchnyi ateizm.* Moscow, 1915.

Solov'ev, V. S. "Danilevskii (Nikolai Iakovlevich)." Brockhaus-Efron, vol. 10, pp. 77–80.

———. *Sobranie sochinenii.* Vol. 1. St. Petersburg, 1901.

Sorokin, Pitirim A. *Modern Historical and Social Philosophies.* New York: Dover, 1963.

Spasskii, I. Ia. "Razvitie idei Darvina v trudakh otechestvennykh paleontolo- gov." In D. L. Stepanov, ed., *Otechestvennaia paleontologiia,* pp. 32–36.

Stadlin, A., "Istoricheskaia teoriia Boklia," *RV,* 1876, no. 7, pp. 257–90.

Stepanov, D. L., ed. *Otechestvennaia paleontologiia za sto let (1870–1970 gg).* Leningrad, 1970.

Strakhov, N. N. *Bor'ba s Zapadom v nashei literature.* 3 vols. 2nd ed. Kiev. 1898.

———. "Durnye priznaki." *Vremia,* 1862, November, sec. 2, pp. 158–72.

———. *Filosofskie ocherki.* St. Petersburg, 1895.

———. *Iz istorii literaturnogo nigilizma, 1861–1865.* St. Petersburg, 1890.

———. *O metode estestvennykh nauk i znachenie ikh v obshchem obrazovanii.* St. Petersburg, 1865.

———. "Predislovie." In N. Ia. Danilevskii. *Darvinizm,* vol. 2, pp. 1–48.

———. "Review of Rozanov, *O ponimanii.*" *ZhMNP,* 1889, vol. 265, sec. 2, pp. 124–31.

[Strakhov, N. N.] "Poiavlenie cheloveka na zemle." *ZhMNP,* 1860, no. 1, sec. 7, pp. 1–9.

Strel'chenko, V. I. "V. A. Vagner kak zoopsikholog-darvinist." *ITEU,* 1975, vol. 3, pp. 101–9.

Stroganov, B. P. "Akademik A. S. Famintsyn—uchenyi i grazhdanin." In A. L. Kursanov et al., eds., *Andrei Sergeevich Famintsyn,* pp. 6–37.

Sushkin, P. P. "K 60-letnemu iubileiu M. A. Menzbira." *Pr.,* 1916, nos. 5–6, pp. 667–74.

Svatikov, S. G. "Nikolai Dmitrievich Nozhin." *Golos minuvshego,* 1914, no. 10, pp. 1–36.

Svetlov, P. Ia. "K voprosu o darvinizme." *Pravoslavnoe obozrenie,* 1887, vol. 1, pp. 498–530, 641–76.

Taliev, V. I. "Biologicheskie idei vtoroi poloviny 19-go veka." In M. M. Filippov, ed., *Deviatnadtsatyi vek*, pp. 270–78.

———. *Ch. Darvin: K 100-letnemu iubileiu so dnia rozhdeniia*. Kharkov,
   1910.

———. "Darvinizm, lamarkizm i teoriia mutatsii." In *Dnevnik XII-go S'ezda*,
   pp. 174–75.

———. *Opyt issledovaniia vidoobrazovaniia v zhivoi prirode*. Kharkov, 1916.

———. "O tselesoobraznosti v prirode." *NO*, 1902, no. 1, pp. 107–14;
   no. 2, pp. 129–39.

Tarkhanov, I. "Vitalizm." Brockhaus-Efron, vol. 6, pp. 549–51.

Tax, Sol, ed. *Evolution After Darwin*. Vol. 2. Chicago: University of Chicago
   Press, 1960.

"Teoriia Darvina i iazykoznanie. (Pis'mo Avgusta Shleikhera k d-ru Ernstu Gekkeliu)." *Zagranichnyi vestnik*, 1864, vol. 2, pp. 241–63.

Terray, Emmanuel. *Marxism and "Primitive" Societies*. Translated by Mary
   Klopper. New York: Monthly Review Press, 1972.

Tikhomirov, A. A. "Nashe universitetskoe delo." *Moskovskie vedomosti*, 1906,
   nos. 306 and 307.

———. *Osnovnoi vopros evoliutsionizma v biologii*. St. Petersburg, 1911.

———, *Polozhenie cheloveka v prirode*. Moscow, 1906.

———. *Samoobman v nauke i iskusstve*. Moscow, 1911.

———. *Sud'ba darvinizma*. St. Petersburg, 1907.

Tikhomirov, P. "Matematicheskii proekt reformy sotsiologii na nachalakh filosofskogo idealizma." *Bogoslovskii vestnik*, 1903, no. 2, pp. 341–402.

Timiriazev, K. A. "Charlz Darvin, 12 fevralia (n.s.) 1809 g.-12 fevralia 1909 g."
   VE, 1909, no. 2, pp. 781–804.

———. "Charlz Darvin." In M. M. Kovalevskii et al., *Pamiati Darvina*,
   pp. 8–38.

———. "Faktory organicheskoi evoliutsii." *VIII S'ezd* . . . , pp. 62–69.

———. "God itogov i pominok. (Iz nauchnoi letopisi 1909 goda)." *VE*, 1910,
   no. 11, pp. 315–31; no. 12, pp. 261–83.

———. "Istoricheskii metod v biologii." *RM*, 1892, no. 8, pp. 83–99; no. 10,
   pp. 142–63; 1893, no. 8, pp. 38–58; 1894, no. 6, pp. 81–95; no. 7,
   pp. 90–101; 1895, no. 8, pp. 72–89.

———. *Izbrannye sochineniia*. 4 vols. Moscow, 1948–49.

———. "Kembridzh i Darvin: iz vospominanii o prazdnestvakh 22–24 iiuna."
   VE, 1909, no. 11, pp. 238–63; no. 12, pp. 682–706.

———. *Kratkii ocherk teorii Darvina*. Moscow, 1865.

———. *Nauka i demokratiia: Sbornik statei, 1904–1919 gg*. Moscow, 1963.

———. "Otboi mendel'iantsev." *VE*, 1913, no. 5, pp. 267–83.

———. "Otbor estestvennyi." Granat-*Ents.*, 7th ed., vol. 30, pp. 721–41.

———. "Piatidesiatiletnii iubilei darvinizma." *Russkie vedomosti*, June 18,
   1908.

———. "Sezon nauchnykh s'ezdov." *VE*, 1911, no. 11, pp. 353–68.

———. *Sochineniia*. 10 vols. Moscow, 1937–40.

———. *Zhizn' rastenii*. 11th ed. Moscow, n.d.

[Timiriazev, K. A.] K. T. "Kniga Darvina, eia kritiki i kommentatory." *OZ*,
   1864, no. 8, pp. 880–912; no. 12, pp. 859–82.

Tkachev, P. N. *Izbrannye sochineniia.* 6 vols. Moscow, 1936–37.

Todes, Daniel P. "V. O. Kovalevskii: The Genesis, Content, and Reception of Paleontological Work." *Studies in the History of Biology,* 1878, no. 2, pp. 99–165.

Tolstoy, L. N. "Protivorechiia empiricheskoi nravstennosti." *Severnyi vestnik,* 1985, no. 1, sec. 1, pp. 279–96.

[Tolstoy, L. N.] *L. N. Tolstoy v vospominaniiakh sovremennikov.* 2 vols. Moscow, 1978.

———. *L. N. Tolstoy o literature.* Moscow, 1955.

Trudy I-go S'ezda Russkikh estestvoispytatelei i vrachei. St. Petersburg, 1868.

Trudy III-go S'ezda Russkikh estestvoispytatelei v Kieve. Kiev, 1873.

Tsetlin, L. S. "Akademik M. A. Menzbir." *Izvestiia Akademii nauk SSSR: seriia istorii i filosofii,* 1947, no. 2, pp. 163–76.

Tsibenko, V. A. *Mirovozzrenie D. I. Pisareva.* Moscow, 1969.

Tsinger, N. V. "O zasoriaiushchikh posevy l'na vidakh *Camelina i Spregula* i ikh proiskhozhdenii." *Trudy Botanicheskogo muzeia Imperatorskoi Akademii nauk,* 1909, vol. 6, 303 pp.

Tsion, I. "Proiskhozhdenie cheloveka po Gekkeliu." *RV,* 1878, no. 1, pp. 5–71.

Ul'iankina, T. I. "I. I. Mechnikov i zarozhdenie immunologii." *VIE-T,* 1985, no. 3, pp. 11–22.

Umov, N. A. "Evoliutsiia mirovozzrenii v sviazi s ucheniem Darvina." *BMOIP,* 1908, n.s. vol. 22, sec. 2, pp. 35–38.

———. "Fiziko-mekhanicheskaia model' zhivoi materii." In B. K. Polenov, ed., *Dnevnik XI S'ezda . . . ,* pp. 678–90.

———. "Po povodu sbornika." In M. M. Kovalevskii et al., *Pamiati Darvina,* pp. 1–7.

Urmantsev, Iu. A. "Ob opytno-filosofskom mirovozzrenii akademika A. S. Famintsyna." In A. L. Kursanov et al., eds., *Andrei Sergeevich Famintsyn,* pp. 36–55.

Uschmann, Georg. "Haeckel's Biological Materialism." *History and Philosophy of the Life Sciences,* 1979, vol. 1, no. 1, pp. 101–18.

Uspenskaia, N. V. "'Priroda' do 'Prirody'." *Pr.,* 1981, no. 11, pp. 64–75.

Utkina, N. F. *Pozitivizm, antropologicheskii materializm i nauka v Rossii.* Moscow, 1975.

Vagner, N. P. "Chuvstva i ikh vyrazheniia." *VE,* 1873, no. 3, pp. 316–33.

———. "Zhorzh Kiuv'e i E. Zh. Sent-Iler." *Zapiski Kazanskogo universiteta,* 1860.

Vagner, V. A. *Biologicheskie osnovaniia sravnitel'noi psikhologii: Biopsikhologiia.* 2 vols. Moscow, 1910–13.

———. "Gorodskaia lastochka *Chelidon urbica,* ee industriia i zhizn' kak material sravnitel'noi psikhologii." *ZAN:F-MO,* series VIII, 1900, vol. 10, no. 6, pp. 1–125.

———. "Granitsy i oblast' biologii." *Severnyi vestnik,* 1898, no. 6–7, pp. 79–101; no. 8–9, pp. 80–89.

———. "Izuchenie dushevnoi zhizni zhivotnykh." *NO,* 1894, no. 17, pp. 517–32.

———. "Lamark i Darvin kak tipy uchenykh." *VE,* 1910, no. 7, pp. 122–33; no. 8, pp. 3–18.

———. "Novye i starye puti v biologii." *Novoe slovo*, 1896, no. 11, pp. 90–117.

———. "Vliianie Darvina na prepodavanie sravnitel'noi psikhologii v shkole." *EiG*, 1909, no. 5, pp. 1–17.

———. *Voprosy zoopsikhologii*. St. Petersburg, 1896.

———. "Voprosy zoopsikhologii: genezis instinkta." *VFiP*, 1892, no. 14, pp. 29–38; no. 15, pp. 39–54.

———, ed. *Novye idei v biologii: Nasledstvennost'*. Vol. 4. St. Petersburg, 1914.

Varsanof'eva, V. A. *Aleksei Petrovich Pavlov*. 2nd ed. Moscow, 1947.

Vasetskii, G. S., and S. R. Mikulinskii, eds. *Izbrannye proizvedeniia russkikh estestvoispytatelei pervoi poloviny XIX veka*. Moscow, 1959.

Vatson, E. K. *Etiudy i ocherki po obshchestvennym voprosam*. St. Petersburg, 1892.

Vavilov, N. I. "Immunity to Fungous Diseases as a Physiological Test in Genetics and Systematics, Exemplified in Cereals." *Journal of Genetics*, 1914–15, vol. 4, no. 1, pp. 49–68.

Venadskii, V. I. *Izbrannye trudy po istorii nauki*. Moscow, 1981.

———. "O nauchnom mirovozzrenii." In A. I. Bachinskii et al., *Sbornik po filosofii estestvoznaniia*, pp. 104–57.

Verner, A. "Iubilei Darvina v Kembridzhe." *EiG*, 1909, no. 7, pp. 91–96.

Verworn, Max. *See* Fervorn, M.

Veselov, E. A. *A. N. Severtsov: zhizn', deiatel'nost' i nauchnye trudy*. Moscow, 1975.

Veselovskii, K. S. "Otchet po Fiziko-matematicheskomu i Istoriko-filologicheskomu otdeleniiam za 1861 god." *ZAN*, 1862, vol. 1, pp. 1–45.

———. "Otchet po Fiziko-matematicheskomu i Istoriko-filologicheskomu otdeleniiam za 1869 god." *ZAN*, 1870, vol. 17, pp. 1–24.

———. "Otchet po Fiziko-matematicheskomu i Istoriko-filologicheskomu otdeleniiam za 1878 god." *ZAN*, 1879, vol. 34, part 1, pp. 1–24.

Vilenskaia, E. S. *N. K. Mikhailovskii i ego rol' v narodnicheskom dvizhenii 70kh-nachala 80-kh godov XIX veka*. Moscow, 1979.

Viner, E. N. *Bibliograficheskii zhurnal 'Knizhnyi vestnik' (1860–1867)*. Leningrad, 1950.

Vinogradov, N. "Biologicheskii mekhanizm i materializm." *VFiP*, 1897, no. 38, pp. 413–42.

———. "Review of A. S. Famintsyn, *Sovremennoe estestvoznanie*." *VFiP*, 1898, no. 5, pp. 416–23.

Vinogradov, P. G. "O progresse." *VFiP*, 1898, no. 42, sec. 1, pp. 254–313.

Vladislavlev, M. "Review of Charles Darwin, *The Expression of the Emotions in Man and Animals*." *ZhMNP*, 1873, vol. 166, April, sec. 2, pp. 163–200.

Vol'fson, V. D. "Sovremennaia biologiia i Ernst Gekkel'." *Znanie*, 1875, no. 3, sec. 1, pp. 47–65; no. 6, sec. 2, pp. 1–18; 1876, no. 3, sec. 2, pp. 60–80; no. 5, sec. 2, pp. 22–62.

Vucinich, A. "Russia: Biological Sciences." In T. Glick, ed., *The Comparative Reception of Darwinism*, pp. 217–55.

————. *Science in Russian Culture: 1861–1917*. Stanford: Stanford University Press, 1970.

————. *Social Thought in Tsarist Russia: The Quest for a General Science of Society*. Chicago: University of Chicago Press, 1976.

Vvedenskii, Aleksandr. *Filosofskie ocherki*. Prague, 1924.

Wagner, Adolph. *Geschichte des Lamarckismus*. Stuttgart, 1909.

Walicki, A. *The Controversy Over Capitalism: Studies in Social Philosophy of the Russian Populists*. Oxford: Clarendon Press, 1969.

————. *A History of Russian Thought From the Enlightenment to Marxism*. Stanford: Stanford University Press, 1979.

————. *Russian Social Thought: An Introduction to the Intellectual History of Nineteenth-Century Russia*. Stanford: *Russian Review*, 1977.

Wallace, Alfred Russel. *Darwinism*. London: Macmillan, 1891.

Wasmann, Erich. *Die moderne Biologie und die Entwicklungstheorie*. 2nd ed. Freiburg, 1904.

Watson, D. M. S. "The Evidence Afforded by Fossil Vertebrates on the Nature of Evolution." In G. L. Jepsen et al., eds., *Genetics, Paleontology, and Evolution*, pp. 45–63.

Weiner, Douglas R. "The Roots of 'Michurinism': Transformist Biology and Acclimatization as Currents in the Russian Life Sciences." *Annals of Science*, 1985, vol. 42, pp. 243–60.

Weismann, August. *The Evolution Theory*. Translated by J. A. Thomson and M. A. Thomson. 2 vols. London: E. Arnold, 1904.

Zalenskii, V. V. "Il'ia Il'ich Mechnikov (Nekrolog)." *IIAN*, 1916, vol. 10, no. 18, pp. 1713–30.

————. "O nauchnoi deiatel'nosti A. O. Kovalevskogo." In *Dnevnik XI-go S'ezda*, pp. 646–54.

————. *Osnovnye nachala obshchei zoologii*. Novorossisk, 1896.

Zalkind, S. *Ilya Mechnikov: His Life and Work*. Moscow, 1959.

Zavadovskii, M. M. "Osnovnye etapy v istorii eksperimental'noi biologii (zoologii) v Rossii." *Uchenye zapiski Moskovskogo universiteta*, 1946, no. 103, pp. 47–60.

Zavadskii, K. M. *Razvitie evoliutsionnoi teorii posle Darvina, 1859–1920-e gody*. Leningrad, 1973.

————. *Vid i vidoobrazovanie*. Lenigrad, 1968.

Zavadskii, K. M., and M. T. Ermolenko, "Evoliutsionnaia teoriia v vtoroi polovine XIX veka." In L. Ia. Bliakher, ed., *Istoriia biologii*, pp. 492–509.

Zavadskii, K. M., and E. I. Kolchinskii, *Evoliutsiia evoliutsii*. Leningrad, 1977.

Zelinskii, N. D. "Nauchnyi podvig M. A. Menzbira." *BMOIP: otdel biologii*, 1946, vol. 51, no. 1, pp. 3–4.

Zograf, N. Iu. "Noveishiia techeniia v zoologii, namechaiushchiiasia v novom, XX veke." *Dnevnik XII-go S'ezda*, sec. 1, pp. 175–95.

Zolotnitskaia, R. L. "Nikolai Alekseevich Severtsov." In N. A. Severtsov, *Puteshestviia . . .* , pp. 8–64.

VIII S'ezd russkikh estestvoispytatelei i vrachei v S. Peterburge ot 28 dekabria 1889 g. do 7 ianvaria 1890 g. St. Petersburg, 1890.

Index

Academic autonomy, 78, 135, 146, 178, 218, 262, 326
Academic liberals, 135, 148
Academy of Military Medicine, 274
Adams, Mark, 229
Adaptation, 17, 112–13, 160, 199, 223, 234, 236, 264, 318, 366, 368
Aesthetics, 2, 8, 10, 29, 153
Agassiz, Louis, 18, 57, 63, 73, 80, 124, 142, 198
Alekseev, V. G., 266
Ammonites, 67
Anarchism, 330, 346, 353
Angell, Norman, 355
Anikin, V. F., 159
Annelids, 38, 56–57, 61
Anthropoid origin of man, 69, 90–91, 106, 188, 241–42, 324–25, 381
Anthropological museums, 68
Anthropology, 2, 8, 28, 48, 50, 52, 55, 68–70, 91, 148, 155, 208, 311
Antonovich, M. A., 10, 19, 31, 174, 204, 362
Anuchin, D. N., 15, 55, 67, 69–70, 310–11, 324–25
"Archeological" orientation, 101
Archeology, 23, 69
Aristotelian entelechies, 173
Aristotelian logic, 254
Aristotle, 267, 270
Arithmology, 265–67
Arnol'di, M. V., 307
Aromorphosis, 223

Artemia Mühlhausenii, 84–85, 158–59, 163
Artemia salina, 84–85, 158, 163
Artificial selection, 53, 206, 232, 286
Ascidians, 36–38, 56–59, 61, 73, 194
Atheism, 93, 95, 188, 266
Atomism, 169
Autocracy, 90, 99, 101, 104, 109, 149–50, 266–67, 373, 484
Autogenesis, 276, 300–310
Avenarius, R., 380

Bachinskii, A. I., 250–51, 265–66, 270, 379
Bacon, F., 26, 29, 102, 191, 254, 338, 384
Baer, Karl von, 32, 36, 99, 104–5, 108, 111, 124, 133, 142, 242, 383; precursor of Darwin, 1, 13–15, 47, 94–95; A. O. Kovalevskii and, 38, 57–58, 93; on Mechnikov and A. O. Kovalevskii, 42; against Darwin, 47–49, 92–98; A. O. Kovalevskii on, 58; on V. O. Kovalevskii's work, 66–67; and N. Ia. Danilevskii, 119–20, 371; A. N. Severtsov on, 226; Mikhailovskii's comment on, 335; alternative theory of evolution proposed by, 371
Balfour, F. M., 231
Baranetskii, O. V., 292–93
Barrande, J., 124
Bateson, William, 5, 84, 158, 229, 294, 303–4, 306–10, 315, 322, 371

455

466

Index

Semenov, P. P., 11
Serebrovskii, A. S., 294
Seventh Congress of Russian Naturalists
and Physicians, 59, 87
Severtsov, A. N., 158, 163, 176, 197,
217–32, 234, 312, 326, 373; on
Haeckel, 194, 221–22; on Darwin
and Darwinism, 217; theoretical con-
cerns of, 219–20; I. I. Shmal'gauzen
on, 223; on genetics, 229; and evolu-
tionary morphology, 370; on struggle
for existence, 387
Severtsov, Nikolai Alekseevich, 217, 353
Sexual selection, 209
Shaniavskii University, 327, 378
Shatskii, E. A., 155
Shchapov, A. P., 9–10, 25, 101
Shchelkanovtsev, Ia. P., 323
Shchurovskii, G. I., 12, 33
Shelgunov, N. V., 142
Shiller, N. N., 249
Shimkevich, V. M., 58, 153, 167, 194,
273, 282–85, 289–90, 295, 311,
322, 373, 386
Shmal'gauzen, I. F., 82
Shmal'gauzen, I. I., 223–230
Shmankevich, V. I., 84–85, 158–59,
277, 371
Shul'ts, Evgenii, 290–91
Simpson, George Gaylord, 170, 370
Slavophiles, 121
Smith, Adam, 103, 122
Social Darwinism, 20, 30, 90, 114,
205–6, 336, 339, 352, 355–56, 363,
384–85
Social democracy, 4, 258
Social evolution, 26, 44, 77, 205, 257,
340–45, 349, 356, 361, 366
Social progress, 26, 112, 119, 257, 331,
334–35, 341, 343, 345, 354, 385
Social sciences, 31, 50, 112–13, 148, 237,
252, 254, 331, 364–65
Society of the Friends of Natural Science,
Anthropology, and Ethnography, 33
Sociology, 2, 20, 27, 31, 50, 52, 60,
75–76, 148, 155, 181, 236, 252–53,
260, 265, 332, 339, 368; biological
reductionism in, 334; Lavrov's definition of, 341; Darwinian, 345, 347,
360; Ia. A. Novikov's view of, 355;
Marxist, 356, bourgeois, 358; evolutionary, 360
Sokolov, N. V., 30
Solov'ev, V. S., 108, 112, 135, 141, 253,
259, 382
Sorokin, Pitirim, 121
Spencer, Herbert, 9, 29–31, 70–71, 88,

154, 157, 237, 287, 344, 384–85;
Kropotkin's opposition to, 349–50;
Mikhailovskii's criticism of, 334–35
Spencer-Weismann dispute, 151, 157,
166–68, 283
Spontaneous generation, 22, 247, 321
Standfuss, Max, 158, 210, 277, 290
Stern, William, 254
Strakhov, N. N., 11, 96, 129, 142, 149,
174, 189, 215, 384; on antitransformist views, 17; opposed to C. Royer,
19–20; against Darwin and in defense of N. Ia. Danilevskii, 129–36
Struggle for existence, 20–21, 26, 30, 50,
55, 70, 72, 77, 79, 83, 87, 90, 134,
153, 184, 203, 212–13, 368, 386;
Nozhin's rejection of, 44, 332; Anuchin on, 70; Mechnikov on, 75–77,
279–80, 282; Flerovskii's view of,
112; N. Ia. Danilevskii on, 122, 125;
Timiriazev on, 132, 205, 386; Chrny-
shevskii's critique of, 146–47; Famintsyn on, 181; Filippov's comments
on, 204, 239; Tsinger's empirical support of, 231–32; Nietzsche's view of,
251; rejected by Chicherin, 258;
Korzhinskii against, 264; Shimkevich
in defense of, 284; Taliev on, 289;
Regel' on, 296; Knipovich on, 324,
Mikhailovskii's view of, 334, 336;
Lavrov's view of, 341; in Kropotkin's
view of, 347–49, 372; M. M. Kovalevskii's view of, 354; Marx on, 367;
Russian biologists on, 385–87
Sushkin, P. P., 209–24
Svetlov, P. Ia., 127–28
Symbiogenesis, 178–79, 182
Synthesis of evolutionary theories, 167,
230, 285, 297, 304–5, 318, 325, 329,
351–52, 371, 373
Systematics, 207, 282, 340

Taliev, V. I., 166, 273, 285–89, 294,
311–12, 316, 320
Tarde, G., 237
Tarkhanov, Ivan, 180
Teleology, 7, 32, 46, 94, 97, 111, 115–16,
127–43, 169–70, 173–74, 177, 197,
210, 245–46, 248, 259; Aristotelian,
48; von Baer's, 97; according to
Kholodkovskii, 278; criticized by Taliev, 287; metaphysical, 299; "internal," 364
Terray, E., 362
Theological Herald (*Bogoslovskii vestnik*), 149, 155, 315
Theological scholars, 45–46, 71, 104–

Compositor: G & S Typesetters, Inc.
Text: 10/13 Sabon
Display: Sabon
Printer: Braun-Brumfield, Inc.
Binder: Braun-Brumfield, Inc.